# Advanced Structural Concrete Materials in Bridges

# Advanced Structural Concrete Materials in Bridges

Editor

**Eva O.L. Lantsoght**

MDPI • Basel • Beijing • Wuhan • Barcelona • Belgrade • Manchester • Tokyo • Cluj • Tianjin

*Editor*
Eva O.L. Lantsoght
Politecnico
Universidad San Francisco de Quito
Quito
Ecuador
Delft University of
Technology
Delft
The Netherlands

*Editorial Office*
MDPI
St. Alban-Anlage 66
4052 Basel, Switzerland

This is a reprint of articles from the Special Issue published online in the open access journal *Materials* (ISSN 1996-1944) (available at: www.mdpi.com/journal/materials/special_issues/Structural_Concrete_Material_Bridge).

For citation purposes, cite each article independently as indicated on the article page online and as indicated below:

LastName, A.A.; LastName, B.B.; LastName, C.C. Article Title. *Journal Name* **Year**, *Volume Number*, Page Range.

**ISBN 978-3-0365-6058-8 (Hbk)**
**ISBN 978-3-0365-6057-1 (PDF)**

# Contents

# About the Editor

**Eva O.L. Lantsoght**

Eva Lantsoght is a Professor of Structural Engineering at Universidad San Francisco de Quito, Quito, Ecuador, and a part-time tenured Assistant Professor at Delft University of Technology. Her field of research includes the design and analysis of concrete bridges, topics related to shear, punching, and torsion in concrete members, plasticity-based methods, and the load testing of bridges.

MDPI

*Editorial*

# Advanced Structural Concrete Materials in Bridges

**Eva Olivia Leontien Lantsoght** [1,2]

[1] Concrete Structures, Department of Engineering Structures, Civil Engineering and Geosciences, Delft University of Technology, 2628 CN Delft, The Netherlands; e.o.l.lantsoght@tudelft.nl

[2] Colegio de Ciencias e Ingenierías—Politécnico, Universidad San Francisco de Quito, Quito EC170901, Ecuador

**Citation:** Lantsoght, E.O.L. Advanced Structural Concrete Materials in Bridges. *Materials* **2022**, *15*, 8346. https://doi.org/10.3390/ma15238346

Received: 21 November 2022
Accepted: 22 November 2022
Published: 24 November 2022



Many existing and newly constructed bridges are made of reinforced and prestressed concrete. Advanced concrete materials play an increasingly important role in concrete bridges, facilitating the strengthening and repair of existing bridges, fast replacement solutions for parts of existing bridges, and for the design of novel challenging bridge projects. The development of advanced concrete materials and their structural applications is, thus, an important topic in the built environment.

The articles presented in this Special Issue bring together research insights, practical applications, and discussions on how to develop better structural concrete materials for bridge engineering practices. Since articles were written both by materials scientists and bridge engineers, a wide view on the topic is presented.

One example of an advanced structural concrete material is ultra-high-performance fiber-reinforced concrete (UHPFRC). With research on this topic having reached a high level of maturity, practical applications of this material are becoming increasingly more common. A key practical issue is related to the resulting fiber content, fiber orientation, and efficiency factor of the structural elements, particularly relevant for thin elements. This property determines the post-cracking tensile strength of the UHPFRC. In [1], the authors propose a nondestructive evaluation (NDE) method for the quality control of UHPFRC in industrial settings and provide recommendations for the efficient implementation of this method. In [2], the authors study the interface shear strength of UHPFRC by testing Z-shaped specimens. They found that the steel fiber type had little effect on the shear strength and ductility, while increasing the length of the steel fibers improved the ductility and slightly reduced the shear strength. Ultimately, the authors proposed an interaction formula for the shear and compressive strength to predict the shear capacity of cast-in-place UHPFRC structures.

Advances in steel-fiber-reinforced concrete include the development of steel-fiber-reinforced expanded-shale lightweight concrete (SFRELC) [3]. In particular, the authors explored the fatigue strength of composite SFRELC-RC beams through experiments. The results showed that with the increase in SFRELC depth and the volume fraction of steel fibers, the fatigue life of the test beams was prolonged. The outcome of the research was a method for predicting the stress level, the stress amplitude of the longitudinal tensile rebar, and the degenerated flexural stiffness of SFRELC superposed beams with a fatigue life. These insights are important for the development of composite SFRELC-RC bridge elements, making optimal use of both materials.

Another example of an advanced structural concrete material is concrete with an MgO expansive agent and steel fibers added to the mix. This improved concrete mix [4] was applied in the construction of the Xiaoqing River Bridge to provide protection against shrinkage cracking. Measurements on the deck after casting showed that the MgO expansive agent could effectively prevent shrinkage cracks. The reinforcement ensured that the MgO expansive agent did not further expand in time, and adding steel fibers resulted in a three-dimensional restraint of further MgO expansion. The proposed improved mix not only counteracted shrinkage cracking, but also improved the mechanical properties of the bridge deck.

For the prefabrication industry, a better understanding of the structural behavior can result in the optimization of the resulting prefabricated elements. A topic of particular interest is the interface bond strength and anchorage performance of the steel reinforcement (ribbed and plain bars) in the prefabricated concrete. The experimental study of [5] resulted in a proposal for the anchorage length in prefabricated elements, which could be applied to prefabricated bridge elements. An application of precast, prestressed concrete deck slabs with high-performance concrete (HPC) is given in [6]. The developed deck slabs had a smaller cross-section, and, thus, reduced weight, as well as better durability than traditional solutions. However, the joint between the slab panels was considered a weak spot in the deck. Therefore, the authors experimentally evaluated slabs with joints. The experiments showed that the deck slabs fulfilled all the static and fatigue performance requirements, as well as serviceability requirements. This solution could, thus, be applied in bridges.

For existing bridges, the first challenge is often the proper assessment of the structure; the information is important for the evaluation of various possible solutions, such as for the strengthening of the structure. In [7], the authors determine the durability and strength of a 95-year-old concrete built-in arch bridge based on the mechanical, physical, and chemical properties of the concrete. The insights of these experiments, as well as the literature review carried out by the authors, provide guidance to bridge engineers who are faced with the problem of assessing existing bridges using historical concrete mixes. Then, the authors used this information in [8] to perform a structural analysis of the bridge. Using insights from the material investigation resulted in the conclusion that the bridge fulfilled the code requirements and that it could remain in service. Such evaluations are of the utmost importance before determining whether a replacement or the application of a strengthening technique to existing bridges would be needed.

On the topic of strengthening existing bridges, various advanced structural concrete materials are proposed. Experiments on reinforced concrete beams strengthened with textile-reinforced concrete (TRC) and fiber–textile-reinforced concrete (F/TRC) showed the beneficial effects of this strengthening technique at the ultimate limit state, as well as in the serviceability limit state [9]. A second paper [10] also addresses the use of TRC for strengthening, and, in particular, for bridge deck slabs. The proposed strengthening solution, SMART-DECK, consists of a carbon-fiber-reinforced polymer reinforcement together with a high-performance mortar. The experiments addressed both the flexural and shear strength of the TRC-strengthened slab segments, and the authors found a high increase in the bending and shear capacity when using the proposed strengthening approach.

UHPFRC is also proposed for the strengthening of existing bridges [11]. Experiments have shown the applicability of UHPFRC for strengthening, but a key issue that remains is the estimation of the bond strength to the existing concrete, especially for numerical applications. In this paper, the authors experimentally and numerically investigated this bond strength as a function of the roughening of the existing concrete surface. From this research, the authors proposed to use a zero-thickness volume model in nonlinear finite element analyses to properly model the bond properties.

Polyvinyl alcohol fiber-reinforced engineering cementitious composites (PVA-ECCs) are proposed in another paper [12], where the authors address the challenge of closing and repairing a portion of a bridge while the other portion is left open to traffic. Consequently, newly placed PVA-ECC bridge repairs (NP-ECC-BRs) are exposed to continuous traffic vibrations (TRVs), even during the setting periods. Therefore, they address the TRVs and their influence on the flexural properties of the PVA-ECC repairs. The authors found that, indeed, the flexural capacity decreased when the PVA-ECC repairs were subjected to TRVs, and that the flexural deformation capacity was not affected.

Another novel method for strengthening was proposed in [13], in which carbon-fiber-reinforced polymer (CFRP) was applied as a near-surface-mounted reinforcement (NSMR). The resulting system had a high ductility and, thus, an advantage over typical adhesively-bonded CFRP systems that can result in a brittle concrete delamination failure, reduced warning, and the consequent inefficient use of the CFRP. Thanks to the high ductility of the

CFRP NSMR system, a high utilization of the CFRP can be reached as well. The proposed strengthening technique was studied experimentally, showing promising results.

Ultimately, [14] proposes a conceptual approach for the replacement of existing bridges. As high numbers of bridges are reaching the end of their originally devised service life, smart concepts for this replacement task are needed. The authors propose a fast and hindrance-free method for their replacement. A combination of recent innovations in construction technology, such as advanced cementitious materials (ACMs), structural health monitoring (SHM) techniques, advanced design methods (ADMs), and accelerated bridge construction (ABC), is being used. Thus, the authors show how all recent research insights can be combined to tackle this major societal challenge at hand in various countries around the world.

These presented papers provide insight into the current state-of-the-art of advanced structural concrete materials in bridge applications, and show the way towards the practical application of these materials. I appreciate all author contributions and sincerely value the time and effort spent in preparing these articles. I would also like to thank all reviewers who contributed to this Special Issue for their time, effort, and valuable remarks to the articles. Finally, I would like to thank the academic editors of *Materials*, who helped handle the manuscripts of this Special Issue, as well as the journal staff, who provided expert assistance to the development of this Special Issue at every step of the way.

**Conflicts of Interest:** The authors declare no conflict of interest.

## References


1. Nunes, S.; Pimentel, M.; Sine, A.; Mokhberdoran, P. Key Factors for Implementing Magnetic NDT Method on Thin UHPFRC Bridge Elements. *Materials* **2021**, *14*, 4353. [CrossRef] [PubMed]
2. Li, C.; Feng, Z.; Ke, L.; Pan, R.; Nie, J. Experimental Study on Shear Performance of Cast-In-Place Ultra-High Performance Concrete Structures. *Materials* **2019**, *12*, 3254. [CrossRef] [PubMed]
3. Qu, F.; Li, C.; Peng, C.; Ding, X.; Hu, X.; Pan, L. Flexural Fatigue Performance of Steel Fiber Reinforced Expanded-Shales Lightweight Concrete Superposed Beams with Initial Static-Load Cracks. *Materials* **2019**, *12*, 3261. [CrossRef] [PubMed]
4. Jiang, F.; Deng, M.; Mo, L.; Wu, W. Effects of MgO Expansive Agent and Steel Fiber on Crack Resistance of a Bridge Deck. *Materials* **2020**, *13*, 3074. [CrossRef] [PubMed]
5. Hu, Z.; Shah, Y.I.; Yao, P. Experimental and Numerical Study on Interface Bond Strength and Anchorage Performance of Steel Bars within Prefabricated Concrete. *Materials* **2021**, *14*, 3713. [CrossRef] [PubMed]
6. Bae, J.-H.; Hwang, H.-H.; Park, S.-Y. Structural Safety Evaluation of Precast, Prestressed Concrete Deck Slabs Cast Using 120-MPa High-Performance Concrete with a Reinforced Joint. *Materials* **2019**, *12*, 3040. [CrossRef] [PubMed]
7. Ambroziak, A.; Haustein, E.; Niedostatkiewicz, M. Chemical, Physical, and Mechanical Properties of 95-Year-Old Concrete Built-In Arch Bridge. *Materials* **2021**, *14*, 20. [CrossRef] [PubMed]
8. Ambroziak, A.; Malinowski, M. A 95-Year-Old Concrete Arch Bridge: From Materials Characterization to Structural Analysis. *Materials* **2021**, *14*, 1744. [CrossRef] [PubMed]
9. Rossi, E.; Randl, N.; Mészöly, T.; Harsányi, P. Effect of TRC and F/TRC Strengthening on the Cracking Behaviour of RC Beams in Bending. *Materials* **2021**, *14*, 4863. [CrossRef] [PubMed]
10. Adam, V.; Bielak, J.; Dommes, C.; Will, N.; Hegger, J. Flexural and Shear Tests on Reinforced Concrete Bridge Deck Slab Segments with a Textile-Reinforced Concrete Strengthening Layer. *Materials* **2020**, *13*, 4210. [CrossRef] [PubMed]
11. Valikhani, A.; Jahromi, A.J.; Mantawy, I.M.; Azizinamini, A. Numerical Modelling of concrete-to-UHPC Bond Strength. *Materials* **2020**, *13*, 1379. [CrossRef] [PubMed]
12. Zhang, X.; Liu, S.; Yan, C.; Wang, X.; Wang, H. Effects of Traffic Vibrations on the Flexural Properties of Newly Placed PVA-ECC Bridge Repairs. *Materials* **2019**, *12*, 3337. [CrossRef] [PubMed]
13. Schmidt, J.W.; Christensen, C.O.; Goltermann, P.; Sena-Cruz, J. Activated Ductile CFRP NSMR Strengthening. *Materials* **2021**, *14*, 2821. [CrossRef] [PubMed]
14. Reitsema, A.D.; Luković, M.; Grünewald, S.; Hordijk, D.A. Future Infrastructural Replacement Through the Smart Bridge Concept. *Materials* **2020**, *13*, 405. [CrossRef] [PubMed]

*Article*

# Experimental Study on Shear Performance of Cast-In-Place Ultra-High Performance Concrete Structures

**Chuanxi Li [1,2], Zheng Feng [1,2,*], Lu Ke [1,2], Rensheng Pan [1,2,*] and Jie Nie [1,2]**

1 Key Laboratary of Bridge Engineering Safety Control by Department of Education, Changsha University of Science and Technology, Changsha 410114, China; lichuanxi2@163.com (C.L.); clkelu@foxmail.com (L.K.); niejie23@126.com (J.N.)

2 School of Civil Engineering, Changsha University of Science and Technology, Changsha 410114, China

* Correspondence: fzllufr@hotmail.com (Z.F.); panrshc@hnu.edu.cn (R.P.); Tel.: +86-731-85256006 (Z.F. & R.P.)

Received: 23 August 2019; Accepted: 3 October 2019; Published: 5 October 2019

**Abstract:** In order to study the direct shear properties of ultra-high performance concrete (UHPC) structures, 15 Z-shaped monolithic placement specimens (MPSs) and 12 Z-shaped waterjet treated specimens (WJTSs) were tested to study the shear behavior and failure modes. The effects of steel fiber shape, steel fiber volume fraction and interface treatment on the direct shear properties of UHPC were investigated. The test results demonstrate that the MPSs were reinforced with steel fibers and underwent ductile failure. The ultimate load of the MPS is about 166.9% of the initial cracking load. However, the WJTSs failed in a typical brittle mode. Increasing the fiber volume fraction significantly improves the shear strength, which can reach 24.72 MPa. The steel fiber type has little effect on the shear strength and ductility, while increasing the length of steel fibers improves its ductility and slightly reduces the shear strength. The direct shear strength of the WJTSs made from 16 mm hooked-type steel fibers can reach 9.15 MPa, which is 2.47 times the direct shear strength of the specimens without fibers. Finally, an interaction formula for the shear and compressive strength was proposed on the basis of the experimental results, to predict the shear load-carrying capacity of the cast-in-place UHPC structures.

**Keywords:** ultra-high performance concrete (UHPC); shear property; shear strength; cast-in-place; steel fiber

---

## 1. Introduction

Shear failure may occur near the geometric discontinuity or joint interfaces of concrete structures, where the cracks are usually perpendicular to the axis of the member, without a bending moment. This shear behavior is known as "direct shear" [1]. Direct shear failure is a sudden and catastrophic failure mode in traditional concrete structures [2]. Although this behavior has been studied on ordinary concrete structures for more than 40 years, it was still not clear if the empirical models can accurately predict the actual shear behaviors [1]. Ultra-high performance concrete (UHPC) is a novel fiber reinforced concrete (FRC) with high strength, excellent service durability and low permeability [3–6]. It has been extensively used in buildings, bridges and other structural projects with a thin-walled structure [7–9]. Thin-walled reinforced concrete structures subjected to a distributed load of short duration (such as explosive loading and seismic load) may not behave plastically at the mid-span and fail there. Some of the beams might fail at positions very close to the support owing to direct shear failure [10]. Hence, it is very important to study the direct shear properties of UHPC structures. Through the numerous experimental studies of FRC [11–13], it had turned out that steel fibers have a great effect on the improvement of the shear properties of concrete. Due to the poor interface properties between the coarse aggregates and cements and less fiber volume fraction used in FRC, the shear failures mostly show brittle modes. However, unlike ordinary concrete, FRC and high-strength concrete, UHPC has no coarse aggregates and possesses high compactness, as well as a high fiber volume fraction. Thus, the shear behavior of UHPC may be different from those of traditional concrete, and its shear failure process is worth studying.

Owing to the limitation of the mixing of UHPC, transportation and maintenance ability, structures will inevitably show joint connection problems [14–16]. Even in precast UHPC structures, there still remains some components or joints of segments to be cast in situ [17]. Thus, the study of shear properties at joint interfaces is very important for all composite structures [17–20]. At present, the ultimate limit and serviceability limit state that calculation for traditional concrete structures, because of its low tensile strength, do not need to take into account concrete tensile strength [21,22]. However, the tensile strength of UHPC is very well owing to the bridging effect of continuous steel fibers, and the utilization of the tensile capacity of UHPC has a significant impact on its economic rationality. For segmental precast UHPC structures, current practice is to use multiple key joints that are generally unreinforced and may be dry or epoxied [23–25]. For this reason, it could not utilize the beneficial effect of the continuity of steel fiber distribution at the interface and the natural occlusion between the UHPC aggregate matrix well. Hyun-O Jang [26] studied the shear properties of UHPC specimens with Z-shaped specimens. Their results show that the shear strength of the waterjet treated specimens (WJTSs) can reach 32.2% of the monolithic placement specimens (MPSs). However, it is insufficient that each group of specimens contain only one specimen without considering different types of steel fibers. Direct shear performance with different interface treatments for segmental cast-in-place UHPC is worthy of further research.

The objective of this paper is to obtain the failure modes, shear strength and shear slip properties of UHPC in situ wet joints. The variables include fiber types (13 mm straight-type steel fibers (13SSF), 13 mm hooked-type steel fibers (13HSF) and 16 mm hooked-type steel fibers (16HSF)), fiber volume fractions (2.0%, 2.5% and 3.0%) and interface treatment. The direct shear tests were performed on 15 MPSs and 12 WJTSs (flat joints). Besides, in order to evaluate the load-carrying capacity of cast-in-place UHPC structures, an interaction formula with respect to shear and compressive strength and the relative reduction of the shear strength ratio for the WJTSs are offered on the basis of the experimental results.

## 2. Experimental Program

### 2.1. Experimental Specimens

Herein, to make the shear transfer of the specimens more consistent with that of the segmental concrete structure, the direct shear test of Z-shaped specimens was used. Dimensions of each specimen are given by 200 mm × 400 mm × 100 mm, in which the shear plane size is 100 mm × 200 mm. In order to avoid the failure of other parts prior to the shear plane, the reinforced bar with a diameter of 8 mm was arranged in these specimens for strengthening. The dimensions for the specimens are illustrated in Figure 1.

The UHPC mixture used in the tests is composed of cementitious material (mix of Portland cement, silica fume and mineral powder), quartz sand and solid polycarboxylate superplasticizer (water reducing efficiency of 30%), and the mix proportion of UHPC is shown in Table 1.

To study the effect of steel fiber types on the shear properties of UHPC structures, three kinds of steel fibers were selected (see Figure 2), namely 13 mm straight-type steel fibers (13SSF), 13 mm hooked-type steel fibers (13HSF) and 16 mm hooked-type steel fibers (16HSF), respectively. The physical and mechanical properties of the steel fibers are shown in Table 2. Besides, three kinds of fiber volume fraction were selected to study the effect of steel fiber volume fraction on the UHPC shear properties, which is 2.0%, 2.5% and 3.0%, respectively (Table 3). In this way, 15 MPSs (Figure 3a) and 12 WJTSs (Figure 3b) were fabricated.

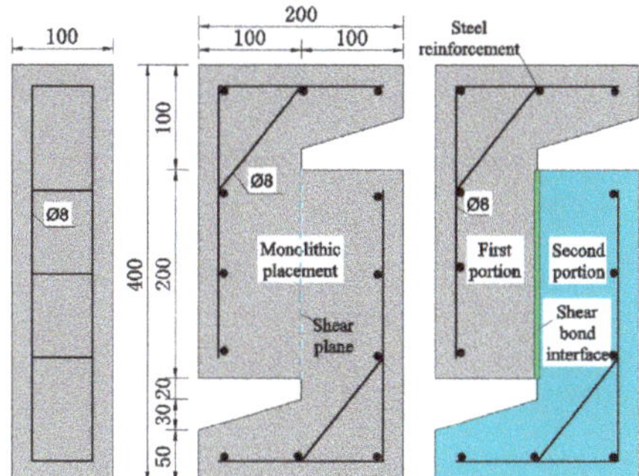


**Figure 1.** Direct shear test specimen.

**Table 1.** Mix proportion of the ultra-high performance concrete (UHPC; mass).

| Cement | Quartz Sand | Cementitious Material | Water | Superplasticizer |
|---|---|---|---|---|
| 1.00 | 1.00 | 1.690 | 0.304 | 0.041 |

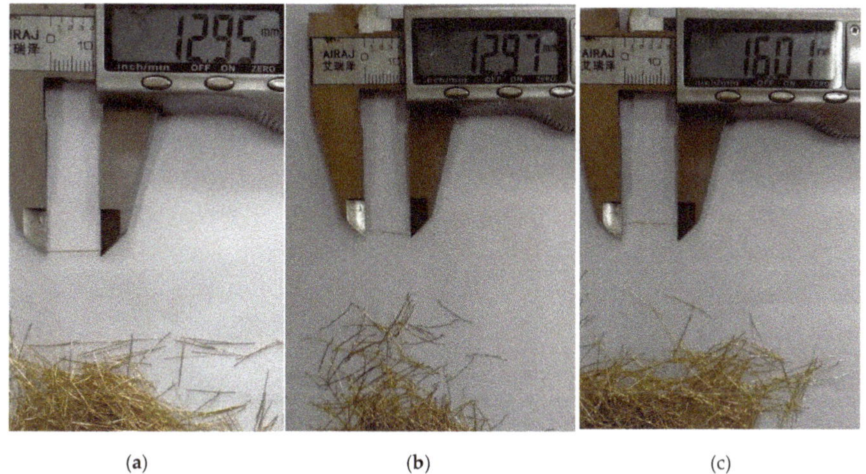


(**a**) (**b**) (**c**)

**Figure 2.** Steel fiber types: (**a**) 13SSF, (**b**) 13HSF and (**c**) 16HSF.

**Table 2.** Physical and mechanical properties of the steel fibers.

| Types | Density ($g \cdot cm^{-3}$) | Length (mm) | Diameter (mm) | Ratios of Length–Diameter | Modulus of Elasticity (GPa) | Tensile Strength (MPa) | External Features |
|---|---|---|---|---|---|---|---|
| 13SSF | 7.8 | 13 | 0.2 | 65 | 205 | 2850 | straight-type |
| 13HSF | 7.8 | 13 | 0.2 | 65 | 205 | 2850 | hooked-type |
| 16HSF | 7.8 | 16 | 0.2 | 80 | 205 | 2850 | hooked-type |

**Table 3.** Shear specimen parameters.

| Specimen Type | Specimen Number | Steel Fiber Volume Fraction | Types of Steel Fibers |
|---|---|---|---|
| MPSs | MP25S13 | 2.5% | 13SSF |
| | MP25H13 | 2.5% | 13HSF |
| | MP20H16 | 2.0% | 16HSF |
| | MP25H16 | 2.5% | 16HSF |
| | MP30H16 | 3.0% | 16HSF |
| WJTSs | WJ25S13 | 2.5% | 13SSF |
| | WJ25H13 | 2.5% | 13HSF |
| | WJ25H16 | 2.5% | 16HSF |
| | WJ-NN | 0 | / |

Number Description: "MP" stands for the condition of monolithic placement, "WJ" stands for the condition of waterjet treatment; "20", "25" and "30" stands for the volume fraction of steel fibers; "H" stands for the hooked-type fibers, "S" stands for the straight-type fibers; "13" and "16" stands for the length of the fibers; and "NN" stands for non-doped fibers.

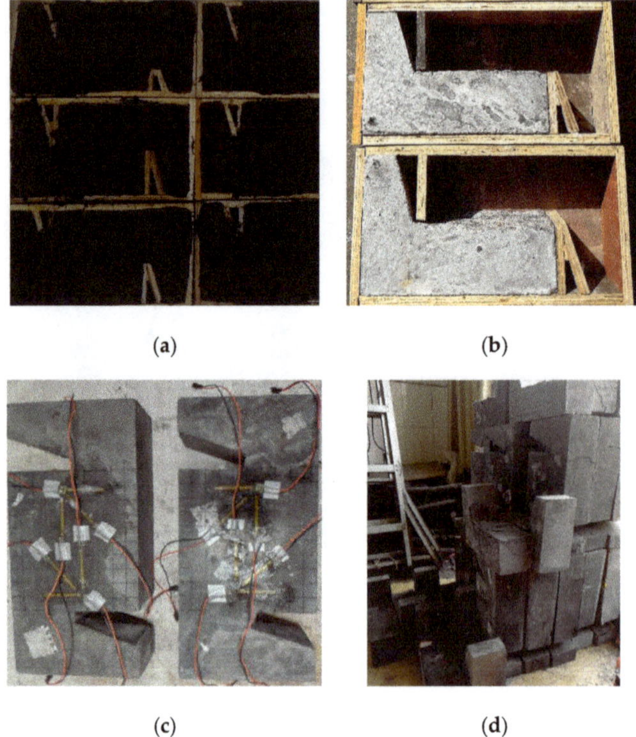

**Figure 3.** Manufacturing of Z-shaped specimens: (**a**) monolithic placement specimens (MPSs), (**b**) waterjet treated specimens (WJTSs), (**c**) finished specimens after maintenance and (**d**) all completed specimens.

The following steps were conducted to mix the UHPC ingredients:

(1) In a dry mixer (pre-wetting), dry components (cement, silica fume and mineral powder) were added and mixed for 2 min.

(2) Then the mixer was suspended, and fine quartz sand added and stirred for 1 min.

(3) The required solid superplasticizer was poured into the total water outside of the mixer and the solution was added to the mix gradually and stirred for 4 min.

(4) Finally, steel fibers were added manually by slowly sprinkling them into the mixer, to avoid balling and to produce a concrete with uniform material consistency and good workability. Stirring occurred until the steel fibers were well encapsulated and evenly distributed in the slurry. The stirring process lasted about 3 min.

The MPSs were poured at one time and cured by high-temperature steam above 95 ± 3 °C for 48 h after conventional standard curing (temperature 20 ± 2 °C, relative humidity 95%) for two days. In contrast, the preparation of WJTSs appears more complex and follows the procedures below. First, the first portion was poured and cured in the standard environment for two days. Secondly, the shear bond interface of the first portion was treated by the high-pressure waterjet, and they were cured in high-temperature steam condition for 36 h. Then, the second portion was poured and cured in the standard environment for 48 h. Finally, whole specimens, including the first and second portions, were cured in high-temperature steam conditions for 48 h.

In order to utilize the beneficial effect of the steel fibers continuity distribution and the natural occlusion between the UHPC aggregate matrix at the shear plane, the treatment of the shear bond interface needs to ensure the retention of steel fibers and the length of fiber exposure as far as possible. As shown in the Figure 4, the roughing on the concrete surface with a high-pressure waterjet is converging water flow at a point through a high-pressure device and the energy will be greatly weakened once the water flow impacts the specimen surface. Therefore, the process will not cause damage to the inside of the specimen. It is theoretically possible to retain a certain amount of steel fibers at the UHPC shear plane, which has been widely accepted by engineers [26,27].

(a) (b)

**Figure 4.** Interface treatment with a high-pressure waterjet. (**a**) Waterjet; (**b**) surface treatment.

Figure 5 illustrates the variation of cubic compressive strength of UHPC with curing days under normal temperature. UHPC strength develops rapidly after initial solidification. After 4 days of maintenance, the cubic compressive strength of UHPC is close to 90 MPa. Therefore, in order to ensure the shear bond interface possesses excellent chiseling effect, the specimens should be controlled to chisel after 1.5~4 days of maintenance in practical engineering.

Owing to the existence of a large number of steel fibers at the interface, it is difficult to accurately measure the interface roughness by conventional measurement methods. Thus, the effect of the roughness of the interface is not considered in this experiment. Considering the complexity of steel fiber dispersion, the distribution quantity of steel fiber types is also not considered.

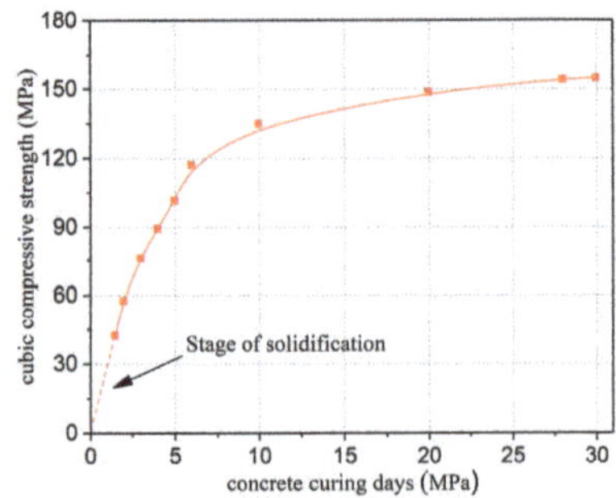


**Figure 5.** The change in cubic compressive strength of UHPC with curing days under normal temperature.

### 2.2. Material Properties

Three cubic compressive specimens (100 mm × 100 mm × 100 mm) and three flexural specimens (100 mm × 100 mm × 400 mm) were prepared to obtain the UHPC material properties. All specimens were cured under the same environment to determine the actual strength of UHPC materials during the test. As illustrated in Figure 6, some material properties experiments of UHPC have been carried out. Table 4 summarizes the material properties of UHPC, including the cubic compressive strength ($f_{cu}$) and flexural strength ($f_{cf}$).

(a)

(b)

**Figure 6.** Performance test of UHPC material: (**a**) cubic compression test and (**b**) four-point bending test.

**Table 4.** Test results of the UHPC materials (unit: MPa).

| Specimens | Cubic Compressive Strength ($f_{cu}$) | Average Cubic Compressive Strength ($f_{cu}$) | Coefficient of Variation | Flexural Strength ($f_{cf}$) | Average Flexural Strength ($f_{cf}$) | Coefficient of Variation |
|---|---|---|---|---|---|---|
| MN25S13 | 163.90<br>164.72<br>167.89 | 165.50 | 0.010 | 36.92<br>39.09<br>40.25 | 38.75 | 0.036 |
| MN25H13 | 164.10<br>160.77<br>159.42 | 161.43 | 0.012 | 35.50<br>35.68<br>38.11 | 36.43 | 0.033 |
| MN20H16 | 152.21<br>150.38<br>148.15 | 150.25 | 0.011 | 31.78<br>32.24<br>34.42 | 32.81 | 0.035 |

**Table 4.** *Cont.*

| Specimens | Cubic Compressive Strength ($f_{cu}$) | Average Cubic Compressive Strength ($f_{cu}$) | Coefficient of Variation | Flexural Strength ($f_{cf}$) | Average Flexural Strength ($f_{cf}$) | Coefficient of Variation |
|---|---|---|---|---|---|---|
| MN25H16 | 158.24<br>155.60<br>164.21 | 159.35 | 0.023 | 44.52<br>38.45<br>38.51 | 40.49 | 0.070 |
| MN30H16 | 184.38<br>173.12<br>178.01 | 178.50 | 0.026 | 42.17<br>46.89<br>38.92 | 42.66 | 0.077 |

### 2.3. Loading Process and Measuring Arrangement

The shear test was conducted on a 2000 kN universal testing machine. In order to examine deformation characteristics, a set of two linear variable differential transducers (LVDTs) was installed on the vertical direction of the specimen to measure the relative deformation under direct shear load at the construction joint. Besides, a set of two LVDTs was arranged at the center of the horizontal shear plane of the specimen for the purpose of measuring the variation of crack width along with the increase of load. To ensure that the LVDTs and strain gauges were fixed firmly as well as the test device was connected reliably, preloading was carried out before formal loading. Besides, when the load-carrying capacity of the specimen drops sharply, the test machine will automatically stop loading. In the process of formal loading, the condition of the crack initiation and extension was observed directly by a high-power magnifier (zoom in 30 times). The test set-up is shown in Figure 7.

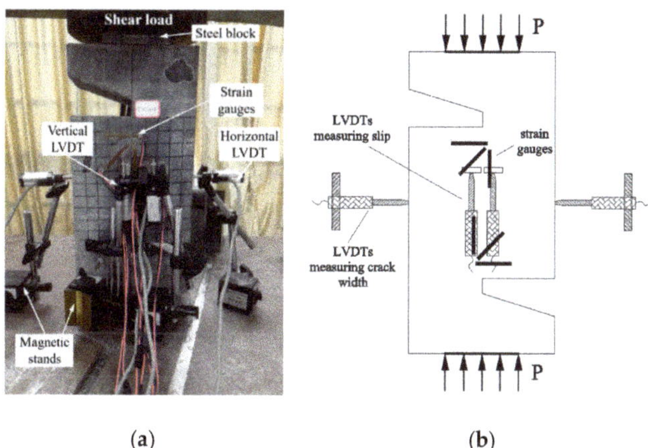


**Figure 7.** Test set-up. (**a**) Photo of test set-up and (**b**) arrangement of the transducers.

After loading, the shear strength of the UHPC specimens under ultimate load can be obtained from Equation (1):

$$\tau = \frac{F_{cr}}{A} \tag{1}$$

where $\tau$ represents the shear strength (MPa), $F_{cr}$ represents the ultimate load (kN) and $A$ represents the shear plane area of specimens (mm$^2$).

## 3. Experimental Results and Discussion

### 3.1. Test Results and Analysis of the MPSs

#### 3.1.1. Load-Carrying Capacity and Failure Modes

The failure modes of the MPSs are similar. The development of cracks was analyzed based on the MP25H16 specimen as an example. Firstly, there were no changes in the surface of these specimens at the initial loading stage. As the loading continued, the fragments at the shear plane began to exfoliate. After that, the initial cracks appeared on the shear plane and several small cracks appeared instantaneously. Herein, it should be noted that the crack width of 0.05 mm is adopted as the criterion of visible initial cracking [28]. With the increase in loading, fine cracks further spread, connected and penetrated to form a crack zone along the shear failure surface, and the fibers between the crack zones were gradually pulled out or pulled off. Finally, with the further increase in the load, along with a huge sound, specimens were sheared and damaged. The condition of the crack development is shown in Figure 8. In order to verify whether the UHPC specimens still possess the bearing capacity after the main crack occurs, the test machine was restarted to continue loading. It turns out the load could still reach 1/2~2/3 of the ultimate load. On the basis of the testing results, it can be seen that there are two main crack modes in the failure modes for these specimens, namely the single main crack (Figure 9a) and multiple diagonal cracks (Figure 9b), respectively.

Since the failure characteristics of the specimens are similar, we took the MP25H13 specimen as an example to describe the shear load–slip relationship of the MPSs. The shear load–slip curve of MP25H13 specimen is shown in Figure 10. The results show that the direct shear cracking failure characteristics of UHPC can be divided into three domains. The first domain is a linear development stage in which the relationship between load and slip is basically linear (ideal status). The second domain is a crack growth stage. With the initial stiffness degenerating, the slope of the load–slip curve begins to decline. After the micro-crack occurs, the stiffness gradually decreases, and then the load enters a linear deviation domain. As a whole, because the bridging action of the steel fibers suppresses the crack expansion, the load increases in the alternation as the slip increases. The third domain is the shear failure stage. At this stage, large amounts of steel fibers are broken or pulled out from the matrix, and the specimens are rapidly destroyed.

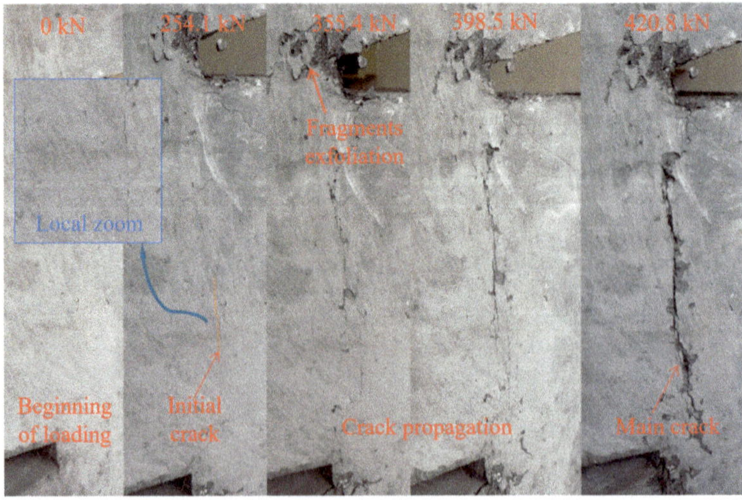


**Figure 8.** Failure modes and the crack propagation process.

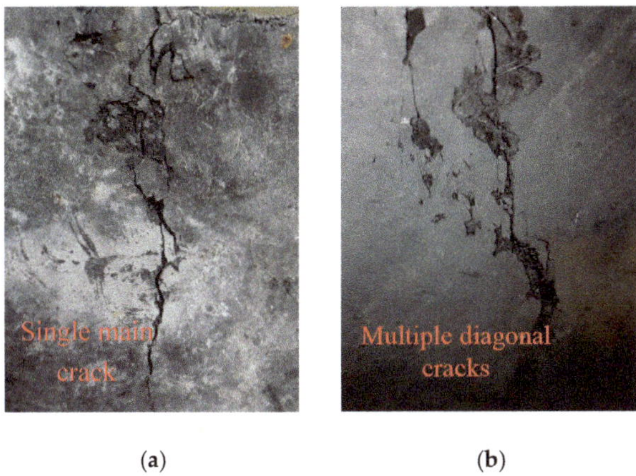


**Figure 9.** Two typical crack forms: (**a**) single main crack and (**b**) multiple diagonal cracks.

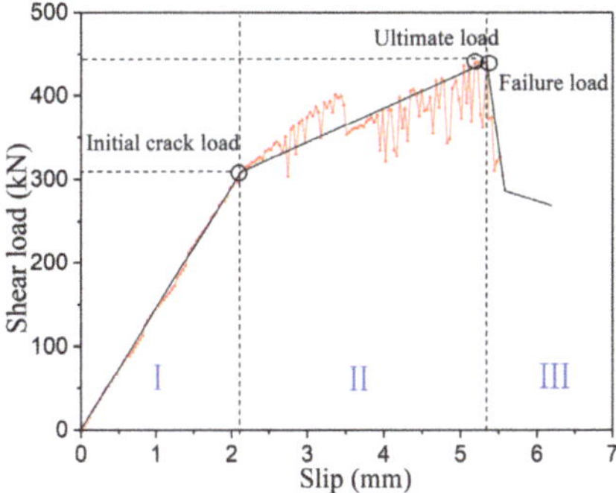


**Figure 10.** UHPC shear load–slip curve.

### 3.1.2. Shear Strength Results

For the MPSs, the baroclinic bar and the steel fibers passing through the shear plane act together to form a truss to resist the shear force along the shear plane. The effect of steel fibers on the shear properties of UHPC was analyzed, and the test results of fifteen MPSs are listed in Table 5.

**Table 5.** Direct shear test results of the MPSs.

| Specimen Number | Initial Crack Load $F_{ci}$/kN | Ultimate Load $F_{cr}$/kN | Ultimate Load $F_{cr}$/kN (Average) | Shear Strength $\tau$/MPa | Shear Strength $\tau$/MPa (Average) | Coefficient of Variation | $F_{cr}/F_{ci}$ | $F_{cr}/F_{ci}$ (Average) |
|---|---|---|---|---|---|---|---|---|
| MP20H16 | 262.2 | 340.7 | 360.1 | 17.04 | 18.01 | 0.055 | 1.299 | 1.296 |
| | 294.3 | 387.5 | | 19.38 | | | 1.317 | |
| | 276.6 | 352.0 | | 17.60 | | | 1.273 | |
| MP25H16 | 278.6 | 405.6 | 430.6 | 20.28 | 21.53 | 0.059 | 1.456 | 1.604 |
| | 254.1 | 420.8 | | 21.04 | | | 1.656 | |
| | 273.8 | 465.5 | | 23.28 | | | 1.700 | |
| MP30H16 | 286.4 | 515.6 | 494.5 | 25.78 | 24.72 | 0.035 | 1.800 | 1.669 |
| | 296.4 | 494.8 | | 24.74 | | | 1.669 | |
| | 307.3 | 473.0 | | 23.65 | | | 1.539 | |

**Table 5.** *Cont.*

| Specimen Number | Initial Crack Load $F_{ci}$/kN | Ultimate Load $F_{cr}$/kN | Ultimate Load $F_{cr}$/kN (Average) | Shear Strength τ/MPa | Shear Strength τ/MPa (Average) | Coefficient of Variation | $F_{cr}/F_{ci}$ | $F_{cr}/F_{ci}$ (Average) |
|---|---|---|---|---|---|---|---|---|
| | 291.0 | 469.8 | | 23.49 | | | 1.614 | |
| MP25S13 | 307.2 | 473.6 | 462.7 | 23.68 | 23.13 | 0.028 | 1.542 | 1.560 |
| | 291.5 | 444.6 | | 22.23 | | | 1.525 | |
| | 284.2 | 454.3 | | 22.72 | | | 1.599 | |
| MP25H13 | 261.3 | 481.5 | 459.8 | 24.08 | 22.98 | 0.035 | 1.843 | 1.630 |
| | 306.2 | 443.5 | | 22.18 | | | 1.447 | |

According to Table 5, the maximum shear strength of UHPC can reach 24.72 MPa (MP30H16), which is 1.37 times the shear strength of MP20H16. It can be seen that the direct shear strength of UHPC increases with the increase of the fiber volume fractions. In addition, the shear strength of MP25H16 is 19.6% higher than that of MP20H16, while the shear strength of MP30H16 is only 14.8% higher than that of MP25H16. This shows that the shear strength of UHPC increases more obvious at the initial phase of the fiber volume fraction increment. According to the ratio of ultimate load to initial crack load of the specimens and shear slip relationship along the shear plane under different fiber volume fraction (Figure 11), when the fiber volume fraction is fewer, cracking failure characteristics of UHPC are close to brittle failure. Based on the above analysis, it can be considered that the ductile characteristic and direct shear strength of the monolithic placement specimens with appropriate steel fibers can be improved in direct shear load.

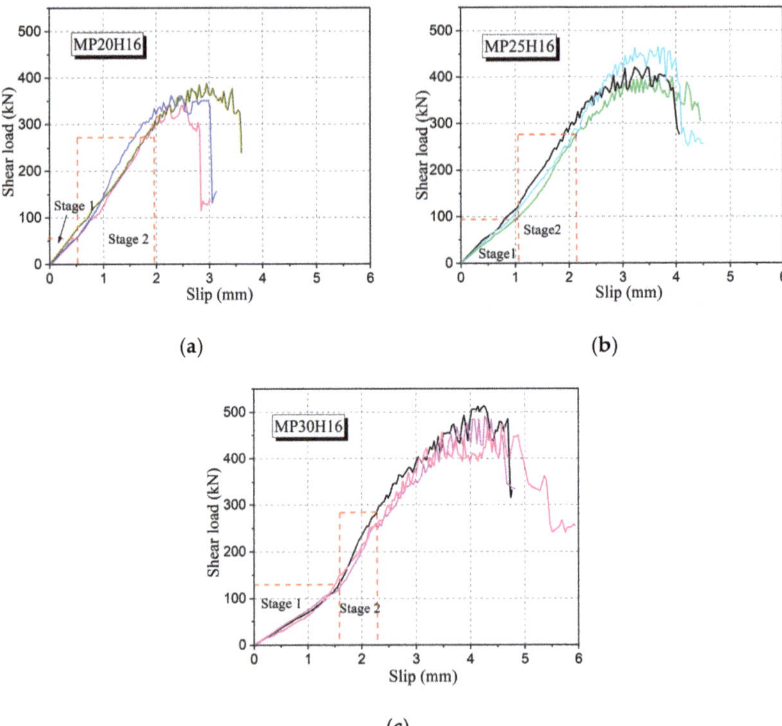


**Figure 11.** Shear load–slip curves of UHPC specimens with a different fiber volume fraction: (**a**) MP20H16, (**b**) MP25H16 and (**c**) MP30H16.

As shown in Figure 11, the initial stiffness of the specimens (Stage 1) is reduced with the increase in steel fibers. This can be explained as follows. On the one hand, the uniformity of fiber mixing will be affected with the increase of steel fiber content, and the probability of internal defects (voids) of the UHPC will also be increased. On the other hand, the shear stiffness at the early stage of loading is predominantly dependent on the overall modulus of the matrix. Only after micro-cracks occur, the shear force will be balanced by the steel fibers and the matrix. Thus, with the increase in steel fiber content, the overall modulus of the matrix itself is reduced, leading to a reduced shear stiffness at Stage 1. With the further increase in load, micro-cracks occur, and the influence of the internal defects is gradually eliminated. The subsequent stiffness (Stage 2) mainly depends on the comprehensive modulus of the steel fiber and matrix. It should be noted that these phenomena are only reflected when the fiber content is more than 2.5%.

Table 5 also shows the effect of different types of fibers on the UHPC shear strength. Under the same volume fraction of fibers (2.5%), the shear strengths of MP25S13, MP25H13 and MP25H16 are almost at the same level. It shows that the shear strength is not significantly influenced by the shape of the steel fibers. In the case of a certain fiber length, whether or not the hooked-type steel fiber is used has little effect on the shear strength of UHPC. However, when using the same shape fibers, increasing the fiber length will slightly reduce the shear strength of UHPC. This is because the UHPC specimens with short fibers possess more fibers at the same volume fraction. From the previous analysis, increasing the fiber numbers is an important way to enhance the compressive strength of UHPC. In addition, according to the ratio of ultimate load to the initial crack load of the specimens and shear load–slip curves under different fiber types (Figure 12), the use of hooked-type fibers has little effect on the ductile failure characteristics of UHPC specimens.

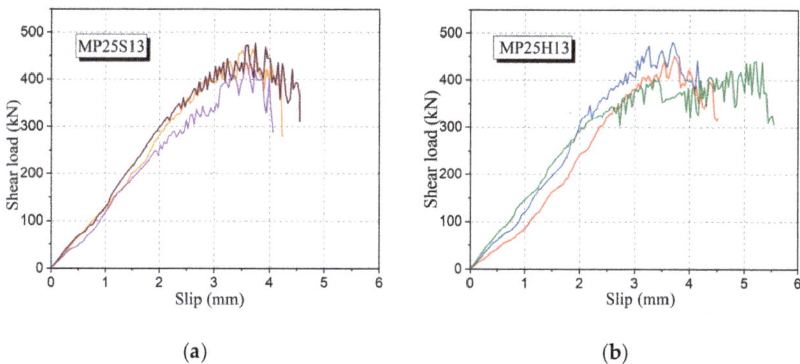


**Figure 12.** Shear load–slip curves of UHPC specimens with different fiber types: (**a**) MP25S13 and (**b**) MP25H13.

### *3.2. Test Results and Analysis of the WJTSs*

#### 3.2.1. Load-Carrying Capacity and Failure Modes

The shear failure characteristic of the WJTSs is a typical brittle failure, which occurs abruptly and without obvious symptoms. Similarly, the shear failure of the specimens is accompanied by a loud noise, especially for the WJ25H16, while the sound of the specimens without steel fibers will not be harsh. From the failure of the bond interface of the specimens, the failure surface of the specimens without fibers is relatively smooth (Figure 13a). But parts of the poured UHPC are embedded together successively, and the peeling phenomenon between the UHPC blocks interfaces is not obvious. The bonding effect of steel fiber reinforced specimens (especially for the WJ25H16) is excellent due to the bridging effect of steel fibers (Figure 13b), which is similar to the MPSs. When the specimens are damaged, obvious fiber pullout marks and UHPC fragments can be seen. Thus, steel fibers can play an important role in enhancing the shear plane strength for the WJTSs.

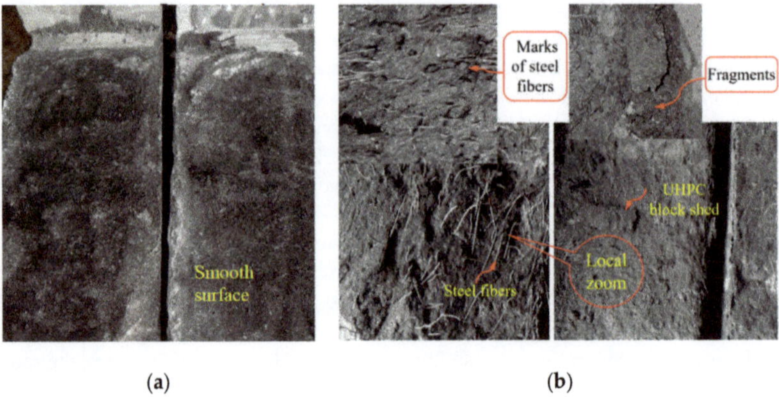


**Figure 13.** Direct shear failure of cast-in-place specimens: (**a**) without steel fibers and (**b**) doped with steel fibers.

### 3.2.2. Shear Strength Results

Because of the uneven surface of the UHPC flat joint interface, the interface subjected to shear force can provide resistance through the friction between the poured aggregates at the interface. When the applied load increases, some UHPC aggregates are sheared off, which results in a decrease in shear stiffness and rapid deformation of the interface. Therefore, the shear failure of the specimen without fibers occurs directly once the aggregates are crushed. The test results of 12 WJTSs are given in Table 6.

**Table 6.** Direct shear test results of the WJTSs.

| Specimen Number | Initial Crack Load $F_{ci}$/kN | Ultimate Load $F_{cr}$/kN | Ultimate Load $F_{cr}$/kN (Average) | Shear Strength $\tau$/MPa | Shear Strength $\tau$/MPa (Average) | Coefficient of Variation | $F_{cr}/F_{ci}$ | $F_{cr}/F_{ci}$ (Average) |
|---|---|---|---|---|---|---|---|---|
| WJ25H16 | 159.7 | 205.60 | 182.9 | 10.28 | 9.15 | 0.139 | 1.287 | 1.188 |
| | 160.5 | 195.40 | | 9.77 | | | 1.217 | |
| | 139.5 | 147.75 | | 7.39 | | | 1.059 | |
| WJ25S13 | 115.9 | 128.4 | 141.8 | 6.42 | 7.09 | 0.110 | 1.108 | 1.042 |
| | 154.9 | 163.7 | | 8.18 | | | 1.057 | |
| | 129.3 | 133.2 | | 6.66 | | | 1.030 | |
| WJ25H13 | 141.2 | 146.9 | 157.1 | 7.35 | 7.86 | 0.046 | 1.040 | 1.153 |
| | 140.1 | 161.3 | | 8.06 | | | 1.151 | |
| | 128.8 | 163.2 | | 8.16 | | | 1.267 | |
| WJ-NN | 67.2 | 70.2 | 74.1 | 3.51 | 3.71 | 0.039 | 1.045 | 1.040 |
| | 70.0 | 75.2 | | 3.76 | | | 1.074 | |
| | 77.0 | 77.0 | | 3.85 | | | 1.000 | |

From Table 6 as well as the shear load–slip curves of UHPC specimens with different fiber types (Figure 14), the cracking load of the specimens without fibers is very close to the failure load, and it is difficult to accurately distinguish the cracking load from the failure load. The difference is that the ratio of initial crack load to ultimate load of WJ25H16 is significantly greater than that of other conditions. Besides, the direct shear strength of WJ25H16 can reach 9.15 MPa, which is 2.47 times that of WJ-NN and 1.29 times that of WJ25S13, respectively. Thus, increasing the length of the fibers and using profiled fibers can significantly improve the interfacial bonding force. From the above analysis, it can be seen that the interface shear strength of UHPC can be significantly increased by using 16HSF.

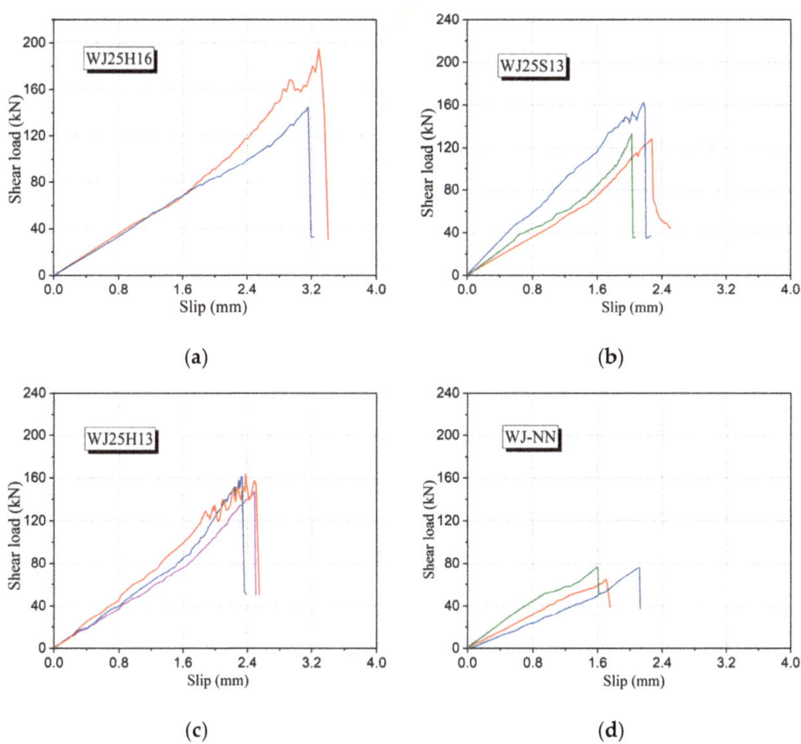


**Figure 14.** Shear load–slip curves of UHPC specimens with different fiber types: (**a**) WJ25H16, (**b**) WJ25S13, (**c**) WJ25H13 and (**d**) WJ-NN.

## 4. Analytical Study

### *4.1. Interaction of Compressive and Shear Strength of the MPSs*

The experimental results demonstrate that the influence of concrete strength should be considered when calculating the ultimate shear strength of the MPSs without pre-cracking. The experimental data are shown in the Table 7. Considering that the shear strength and shear failure characteristics of UHPC are closely related to the compressive strength of concrete and the bridging effect of steel fibers [29], the relationship between the compressive strength and shear strength of concrete is proposed based on the test results, as shown in Equation (2).

**Table 7.** Tests of UHPC shear specimens.

| Specimen Number | Ratios of Length-Diameter $l_f/d_f$ | Volume Fraction of Fibers $\rho_f$ | Characteristic Coefficient of Steel Fibers $\lambda_f$ | Shear Strength $\tau_{exp}$/MPa | Cubic Compressive Strength $f_{cu}$/MPa | Shear Strength $\tau_{cal}$/MPa | $\tau_{exp}/\tau_{cal}$ |
|---|---|---|---|---|---|---|---|
| MP20H16 | 80 | 2.0% | 1.6 | 18.01 | 150.25 | 20.11 | 0.90 |
| MP25H16 | 80 | 2.5% | 2.0 | 21.53 | 159.35 | 22.58 | 0.95 |
| MP30H16 | 80 | 3.0% | 2.4 | 24.72 | 178.50 | 24.69 | 1.00 |
| MP25H13 | 65 | 2.5% | 1.625 | 22.98 | 161.43 | 22.02 | 1.04 |
| MP25S13 | 65 | 2.5% | 1.625 | 23.13 | 165.50 | 22.29 | 1.04 |

Owning to the orientation of the steel fibers inside the concrete matrix, it is affected by a number of parameters, which are essentially the geometry of the fibers and their interaction effects (fibers–aggregates–formwork), the flowability of the concrete, the means of pouring and compacting of the concrete [12]. In addition, the distribution and orientation of fibers is, in turn, the parameter which most influences the ductility of UHPC. Hence, Equation (2) draws into the influence coefficient of steel fiber dispersion $\beta_{cr}$ on direct shear bearing capacity to weaken the above effects.

$$\tau_{mn} = (1 + \beta_{cr}\lambda_f)\sqrt{f_{cu}/0.45}, \tag{2}$$

$$\lambda_f = \frac{\rho_f l_f}{d_f}, \tag{3}$$

where $\tau_{mn}$ is the shear strength of the MPSs, $\rho_f$ is the volume fraction of steel fibers, $l_f/d_f$ stands for the ratio of length-diameter of steel fibers, $\lambda_f$ is the characteristic coefficient of steel fibers, and $\beta_{cr}$ is the influence coefficient of steel fiber dispersion; the experimental value of 0.1 was chosen in this paper.

As can be seen from Table 7, the variations between the calculated shear strength and the experimental shear strength are insignificant, in which the mean value of $\tau_{exp}/\tau_{cal}$ is 0.986 and the difference is only 0.014. For this reason, Equation (2) can provide a reference for the design of cast-in-place UHPC structures. It is worth mentioning that the configuration of shear reinforcement is not considered in this paper. Therefore, Equation (2) is only applicable to the condition of no shear reinforcement, while other situations, including the vertical shear stress direction of steel bar arrangement, need to be further studied.

### *4.2. Relative Reduction of Shear Strength Ratio of the WJTSs*

Figure 15 shows the relative reduction of the shear strength ratio with different fiber types. Compared with the MPSs, the shear strength of WJ25H16 can reach 42.5% of MP25H16, that of WJ25S13 can reach 30.6% of MP25S13, and that of WJ25H13 can reach 34.2% of MP25H13. Therefore, the maximum shear strength of the UHPC shear bond interface treated by a waterjet can reach 42.5% of the monolithic placement. It can be concluded that the shear strength at the interface of the UHPC specimens reinforced with steel fibers (because of the contribution of the aggregate biting force, interface friction force and steel fiber drawing force) can be effectively utilized under the appropriate interface treatment. What is insufficient is that the number of specimens in this study is still fewer, and the study of direct shear strength under a different fiber volume fraction has not been carried out, so the relative reduction formula of the shear strength ratio under various conditions cannot be obtained accurately.

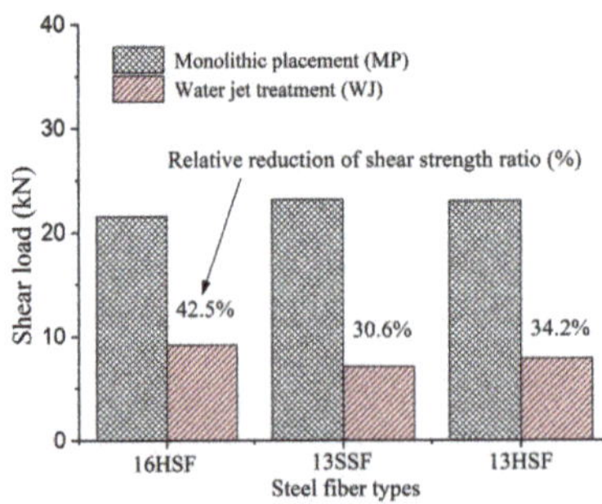


**Figure 15.** Relative reduction of the shear strength ratio with different fiber types.

## 5. Conclusions

This study investigates the direct shear strength and failure mechanism of Z-shaped specimens through the push-off test. A total of 27 specimens with the test parameters of steel fiber shape, steel fiber volume fraction and interface treatment, were designed to test their shear strength, load-carrying capacity and failure modes. Based on the testing results, the following conclusions can be drawn:

(1) Ductile characteristics of the monolithic placement specimens with appropriate steel fibers can be improved in direct shear load, and the ultimate load can reach 166.9% of the initial cracking load.

(2) Increasing the steel fiber volume fraction can significantly improve the shear strength of UHPC structure. Direct shear strength of UHPC specimens with 3.0% volume fraction can reach 24.72 MPa. In addition, the steel fiber shape has little effect on the shear strength and ductility, while increasing the length of steel fibers improves its ductility and slightly reduces the shear strength.

(3) The waterjet treatment for the interface of the adjacent segment is an effective way to improve the direct shear performance of cast-in-place segmental UHPC structures. The direct shear strength of the specimens with the waterjet treatment can reach 42.5% of the monolithic placement specimens.

(4) The formula proposed for predicting the direct shear strength of the cast-in-place UHPC structures shows good agreement with the test results.

Although the feasibility of the segmental cast-in-place UHPC structure has been validated and the influence of steel fiber on shear performance has been obtained through the experimental studies in this paper, further experimental studies are still needed, including the configuration of shear reinforcement at the interface, the type of key joints at the interface (shape, structure size), the number of key joints, the level of normal stress and the roughness of the interface.

**Author Contributions:** C.L. and R.P. planned and managed the project. Z.F. conducted the experiments, analyzed the data, and compiled the final manuscript. L.K. and J.N. contributed to the literature review for this study. Moreover, all authors reviewed the data and the final manuscript.

**Funding:** This research was funded by the National Natural Science Foundation of China (NO. 51778069, NO. 51808055), Excellent Youth Project of Hunan Education Department (NO. 18B131), the Hunan Provincial Innovation Foundation for Postgraduate, China (NO. CX2018B522), the Open Fund of the Key Laboratory of Provincial and Ministerial Level of Bridge Engineering, China (NO.18KE04, No. 18KB03).

**Acknowledgments:** The authors gratefully acknowledge the financial support of the National Natural Science Foundation of China and the Hunan Provincial Innovation Foundation for Postgraduate.

**Conflicts of Interest:** The authors declare no conflict of interest.

## References

1. French, R.; Maher, E.; Smith, M.; Stone, M.; Kim, J.; Krauthammer, T. Direct shear behavior in concrete materials. *Int. J. Impact Eng.* **2017**, *108*, 89–100. [CrossRef]
2. Zhou, X.; Mickleborough, N.; Li, Z. Shear strength of joints in precast concrete segmental bridges. *ACI Struct. J.* **2005**, *102*, 3–11.
3. Park, J.-S.; Kim, Y.-J.; Cho, J.-R.; Jeon, S.-J. Early-age strength of ultra-uigh performance concrete in various curing conditions. *Materials* **2015**, *8*, 5537–5553. [CrossRef] [PubMed]
4. Yang, J.; Peng, G.-F.; Shui, G.-S.; Zhang, G. Mechanical Properties and Anti-Spalling Behavior of Ultra-High Performance Concrete with Recycled and Industrial Steel Fibers. *Materials* **2019**, *12*, 783. [CrossRef]
5. French Association of Civil Engineering-French Authorities of Civil Engineering Structure Design, and Control (AFGC-Sétra). *Ultra High Performance Fibre-Reinforced Concretes*; French Association of Civil Engineering: Paris, France, 2013.
6. Tam, C.M.; Tam, V.W.; Ng, K. Assessing drying shrinkage and water permeability of reactive powder concrete produced in Hong Kong. *Constr. Build. Mater.* **2012**, *26*, 79–89. [CrossRef]
7. Zhou, M.; Lu, W.; Song, J.; Lee, G.C. Application of Ultra-High Performance Concrete in bridge engineering. *Constr. Build. Mater.* **2018**, *186*, 1256–1267. [CrossRef]
8. Blais, P.Y.; Couture, M. Precast, Prestressed Pedestrian Bridge World's First Reactive Powder Concrete Structure. *PCI J.* **1999**, *44*, 60–71. [CrossRef]
9. Almansour, H.; Lounis, Z. Innovative design approach of precast–prestressed girder bridges using ultra high performance concrete. *Can. J. Civ. Eng.* **2010**, *37*, 511–521. [CrossRef]
10. Hsin, Y.L.; Hong, H. Reliability analysis of direct shear and flexural failure modes of RC slabs under explosive loading. *Eng. Struct.* **2002**, *24*, 189–198. [CrossRef]
11. Yang, M.; Huang, C.K.; Liu, Y. Experimental research on shear behavior of high-strength SFRC. *J. Dalian Univ. Tech.* **2005**, *45*, 842–846.

12. Boulekbache, B.; Hamrat, M.; Chemrouk, M.; Amziane, S. Influence of yield stress and compressive strength on direct shear behaviour of steel fibre-reinforced concrete. *Constr. Build. Mater.* **2012**, *27*, 6–14. [CrossRef]
13. Ronald, F.Z. Fiber-reinforced concrete: An overview after 30 years of development. *Cement Concrete Comp.* **1997**, *19*, 107–122.
14. Kim, Y.-J.; Chin, W.-J.; Jeon, S.-J. Interface Shear Strength at Joints of Ultra-High Performance Concrete Structures. *Int. J. Concr. Struct. Mater.* **2018**, *12*, 2–14. [CrossRef]
15. Lee, C.H.; Kim, Y.J.; Chin, W.J.; Choi, E.S. Shear Strength of Ultra High Performance Fiber Reinforced Concrete (UHPFRC) Precast Bridge Joint. In *High Performance Fiber Reinforced Cement Composites*; Parra-Montesinos, G.J., Reinhardt, H.W., Naaman, A.E., Eds.; Springer: Dordrecht, The Netherlands, 2012; pp. 413–420.
16. Shao, X.; Pan, R.; Zhan, H.; Fan, W.; Yang, Z.; Lei, W. Experimental Verification of the Feasibility of a Novel Prestressed Reactive Powder Concrete Box-Girder Bridge Structure. *J. Bridg. Eng.* **2017**, *22*, 4017015. [CrossRef]
17. Shamass, R.; Zhou, X.; Alfano, G. Finite-Element Analysis of Shear-Off Failure of Keyed Dry Joints in Precast Concrete Segmental Bridges. *J. Bridg. Eng.* **2015**, *20*, 4014084. [CrossRef]
18. Ke, L.; Li, C.X.; Luo, N.H.; He, J.; Jiao, Y.; Liu, Y. Enhanced comprehensive performance of bonding interface between CFRP and steel by a novel film adhesive. *Comp. Struct.* **2019**, *229*, 111393. [CrossRef]
19. Li, C.; Ke, L.; He, J.; Chen, Z.; Jiao, Y. Effects of mechanical properties of adhesive and CFRP on the bond behavior in CFRP-strengthened steel structures. *Compos. Struct.* **2018**, *211*, 163–174. [CrossRef]
20. Lee, H.-S.; Jang, H.-O.; Cho, K.-H. Evaluation of Bonding Shear Performance of Ultra-High-Performance Concrete with Increase in Delay in Formation of Cold Joints. *Materials* **2016**, *9*, 362. [CrossRef]
21. AASHTO Highway Subcommittee on Bridges and Structures, AASHTO LRFD bridge design specifications. *American Association of State Highway and Transportation Officials*, 7th ed.; AASHTO: Washington, DC, USA, 2015.
22. *JTG 3362-201, Specifications for Design of Highway Reinforced Concrete and Prestressed Concrete Bridges and Culverts*; Ministry of Transport of the People's Republic of China: Beijing, China, 2018; (In Chinese).
23. Kim, H.S.; Chin, W.J.; Cho, J.R.; Kim, Y.J.; Yoon, H. An Experimental Study on the Behavior of Shear Keys According to the Curing Time of UHPC. *Engineering* **2015**, *7*, 212–218. [CrossRef]
24. Hussein, H.H.; Sargand, S.M.; Al Rikabi, F.T.; Steinberg, E.P. Laboratory evaluation of ultrahigh performance concrete shear key for prestressed adjacent precast concrete box girder bridges. *J. Bridge Eng.* **2016**, *22*, 04016113. [CrossRef]
25. Sargand, S.M.; Walsh, K.K.; Hussein, H.H.; Al Rikabi, F.T.; Steinberg, E.P. Modeling the shear connection in adjacent box-beam bridges with ultrahigh performance concrete joints. II: Load transfer mechanism. *J. Bridge Eng.* **2017**, *22*, 04017044. [CrossRef]
26. Jang, H.-O.; Lee, H.-S.; Cho, K.; Kim, J. Experimental study on shear performance of plain construction joints integrated with ultra-high performance concrete (UHPC). *Constr. Build. Mater.* **2017**, *152*, 16–23. [CrossRef]
27. Islam, M.A.; Farhat, Z.; Bonnell, J. High Pressure Water-Jet Technology for the Surface Treatment of Al-Si Alloys and Repercussion on Tribological Properties. *J. Surf. Eng. Mater. Adv. Technol.* **2011**, *1*, 112–120. [CrossRef]
28. Zhang, Z.; Shao, X.D.; Li, W.G. Axial tensile behavior test of ultra high performance concrete. *China J. Highw. Transp.* **2015**, *8*, 50. (in Chinese).
29. Mansur, M.A.; Vinayagam, T.; Tan, K.-H. Shear Transfer across a Crack in Reinforced High-Strength Concrete. *J. Mater. Civ. Eng.* **2008**, *20*, 294–302. [CrossRef]

MDPI

*Article*

# Key Factors for Implementing Magnetic NDT Method on Thin UHPFRC Bridge Elements

**Sandra Nunes *, Mário Pimentel, Aurélio Sine and Paria Mokhberdoran**

CONSTRUCT-LABEST, Faculty of Engineering (FEUP), University of Porto, 4200-465 Porto, Portugal; mjsp@fe.up.pt (M.P.); aurelio.sine@fe.up.pt (A.S.); pmokhber@fe.up.pt (P.M.)

* Correspondence: snunes@fe.up.pt

**Abstract:** This paper provides an overview of the use of the magnetic NDT method for estimating the fibre content, and fibre orientation and efficiency factors in thin UHPFRC elements/layers, along any two orthogonal directions. These parameters are of utmost importance for predicting the post-cracking tensile strength in the directions of interest. After establishing meaningful correlations at the lab-specimen scale, this NDT method can be effectively implemented into quality control protocols at the industrial production scale. The current study critically addresses the influence of key factors associated with using this NDT method in practice and provides recommendations for its efficient implementation.

**Keywords:** ultra-high performance fibre reinforced cementitious composite (UHPFRC); non-destructive method (NDT); magnetic inductance; quality control

**Citation:** Nunes, S.; Pimentel, M.; Sine, A.; Mokhberdoran, P. Key Factors for Implementing Magnetic NDT Method on Thin UHPFRC Bridge Elements. *Materials* **2021**, *14*, 4353. https://doi.org/10.3390/ma14164353

Academic Editor: Karim Benzarti

Received: 7 July 2021
Accepted: 30 July 2021
Published: 4 August 2021



## 1. Introduction

Ultra-high performance fibre reinforced composites (UHPFRC) are used increasingly for a wide range of structural applications, including innovative solutions for new and existing bridges [1–5]. The economic feasibility and interest of UHPFRC have been demonstrated mainly for new (foot)bridges (structures of extended spans and reduced sections) [1] and in the rehabilitation/strengthening of existing reinforced concrete bridge structures [5]. However, using these advanced materials raises new questions, such as how to control/predict the final fibres distribution in a given structural element, which motivated the development of rational tools to optimise casting processes and fibre distribution control [6,7].

UHPFRC elements can be unreinforced in many circumstances, especially in thin UHPFRC elements such as footbridge deck slabs [3,4,8], watertight jackets or layers over existing concrete elements [5,9], among others. In these cases, the tensile strength of UHPFRC is critical and depends primarily on the properties of the matrix and fibres—such as fibre content, geometry, fibre tensile strength and fibre-to-matrix interface bond—but can also vary significantly according to fibre orientation [10,11]. While the former can be tailored during the mix-design stage, the latter depends on the mould geometry, on the casting procedure, the vibration mode (when applied) and the rheological properties of the material in the fresh state, for example, whether the concrete is self-compacting or not [7]. This means that it is very complicated to define an intrinsic tensile response for UHPFRC, and it is necessary to characterise the representative fibre orientation in the actual structure [11,12]. Non-destructive (NDT) evaluation methods are particularly suited for this purpose.

The most comprehensive method for characterising the distribution and orientation of fibres in a specimen is X-ray tomography [13,14]. However, due to its high cost and limitations in terms of the size of the specimen analysed (which depends on the capacity of the tomograph) it is not currently applicable on the industrial scale. Conversely, the methods based on magnetic induction [15–18] and electrical resistivity [19,20] are more effective because of their low cost, ease of use, portability and non-destructive nature. The magnetic method consists of generating a local magnetic field and measuring the influence of the fibres on this magnetic field by measuring the inductance. The electrical resistivity method has similarities in its implementation, although it involves measuring a voltage difference between two electrodes. This method is however very sensitive to environmental conditions, mixture proportions and age of testing [20]; for this reason, the magnetic method is preferred to the detriment of the resistivity method.

The authors recently developed a procedure for estimating the post-cracking tensile strength of a thin UHPFRC layer in the directions of interest, eliminating the need for extracting cores or samples from the structure [21]. This procedure is based on magnetic induction [18] and relies on the ferromagnetic properties of the steel fibres for estimating the parameters of the underlying physical model, namely, the fibre content ($V_f$), the fibre orientation and efficiency factors [21]. The significance of this NDT method for the quality control of thin UHPFRC elements/layers was demonstrated, and this NDT method is now being used by other research groups [12,22,23].

The performance of the magnetic NDT method on field conditions was assessed during the strengthening of the deck slab of the Chillon Viaducts, in Switzerland, with UHPFRC [9,24]. The main objective of the present work is to critically address the influence of key factors associated with the use of this NDT method in practice and find ways of mitigating their effects on inspection performance. In order to accomplish this objective, a new experimental campaign was carried out aiming to understand the impact of the following key factors: element geometry (thickness and area around the probe), surface condition (roughness of the surface), and fibres segregation (in-depth). Recommendations for the proper implementation of the magnetic NDT method in practice are provided.

To the best of our knowledge, this is the first time that a systematic study has been undertaken on the influence of these key factors for an efficient implementation of the magnetic method, contributing to the advancement of knowledge and promoting its wider use in field applications.

## 2. Non-Destructive Method

### *2.1. Principles*

UHPFRC material is constituted by two phases: cementitious matrix and steel fibres. The relative magnetic permeability (Magnetic permeability represents the ease with which a magnetic field is created in an object. The relative magnetic permeability is the ratio between the magnetic permeability of the material and that of the vacuum) of the cementitious matrix equals 1.0 (similar to air) [25], while that of the fibres material (steel) is significantly greater than 1. Therefore, when incorporating short steel fibres in the cementitious matrix, the magnetic permeability of the resulting UHPFRC becomes higher than 1.0.

A magnetic probe was developed for performing inductance measurements [18]. It consists of a U-shaped ferrite core (Siemens ferrite N47) comprising a single copper wire coil around both core legs (see Figure 1). Table 1 summarises the main characteristics of the magnetic probe used in the current study. A high-precision Agilent E4980A LCR Meter (see Figure 1) with a test frequency of 20 Hz and signal level of 2 V (AC) was used to measure both $L_{UHPFRC}$ and $L_{air}$ using a magnetic probe. As shown in Figure 1, it is possible to attach an acetate sheet to the probe legs, with reference lines drawn in different directions, making it easier to align the probe with the direction of interest.

**Table 1.** Characteristics of the magnetic probe.

| U-Shape Ferrite Core | | | | Copper Wire Coil | |
|---|---|---|---|---|---|
| **Reference** | **Relative Magnetic Permeability** | **Length** | **Cross-Section** | **Wire Diameter** | **Number of Turns** |
| Siemens ferrite N47 | ~2000 | 189 mm | $28 \times 30$ mm$^2$ | 0.5 mm | 1454 |

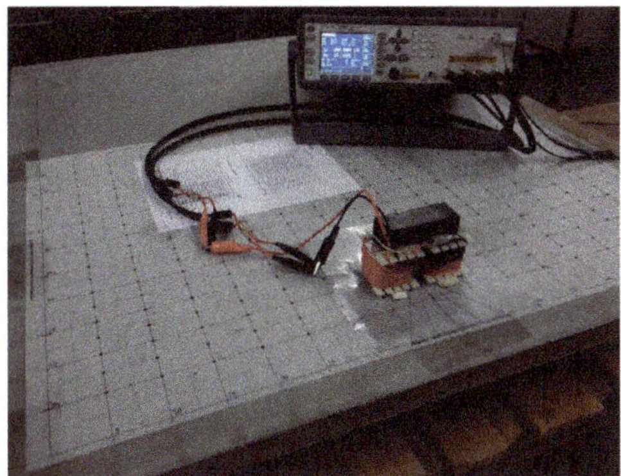

**Figure 1.** Measurements performed on a UHPFRC large plate using the magnetic probe and an LCR meter.

Nunes et al. [18] showed that, for the usual fibre contents, the relative magnetic permeability of UHPFRC ($\mu_r$) could be readily computed using the following approximation:

$$\mu_r \approx \frac{L_{UHPFRC}}{L_{air}} \quad (1)$$

where $L_{UHPFRC}$ is the inductance measured through the UHPFRC material and $L_{air}$ is the inductance measured when the sample is removed (in the air).

$L_{UHPFRC}$, and so the relative permeability of UHPFRC, is governed by the properties of the fibre steel, fibre content ($V_f$) and fibre orientation. $\mu_r$ increases with $V_f$ and changes with the measuring direction ($\theta_i$) when there is a preferential orientation of the fibres in the UHPFRC layer [18]. When the axis of the magnetic field (direction of probe legs) and the main direction of the fibres are aligned, the measured inductance (and thus $\mu_{r,\,i}$) is at its maximum, and vice-versa. As the relative permeability of the cementitious matrix is equal to 1.0, irrespective of its composition and age, $\mu_r$ does not change with cementitious matrix composition or age.

### 2.2. Determination of Fibre Content and Fibre Orientation

Nunes et al. [18] proposed the fibre content ($\mu_{r,mean}$) and the fibre orientation $(\rho_X - \rho_Y)$ indicators for UHPFRC, computed from the inductance measurements in any two orthogonal directions (herein designated as X and Y).

The simplified physical model proposed in [18] was recently generalised, allowing to predict the relative magnetic permeability of a thin UHPFRC layer in any direction and for different types of in-plane fibre distribution functions (herein, we will assume fibres are distributed in XY plane) [26].

#### 2.2.1. Fibre Content

Based on the generalised simplified model [26], it was confirmed that $\mu_{r,mean}$ computed from Equation (2) is practically independent of the fibre orientation (Figure 2a) and increases linearly with the fibre content with a proportionality constant $k_V$ as depicted in Figure 2b. For $V_f = 0\%$, $\mu_{r,mean}$ should equal 1.0 (the relative magnetic permeability of the cementitious matrix); therefore, the regression line should intersect the origin. For a given fibre type (or fibres mixture), after calibrating $k_V$ by means of representative tests in the laboratory, $V_f$ can be estimated using Equation (3).

$$\mu_{r,mean} = \frac{1}{2}(\mu_{r,X} + \mu_{r,Y}) \quad (2)$$

$$V_f = (\mu_{r,mean} - 1)/k_V \quad (3)$$

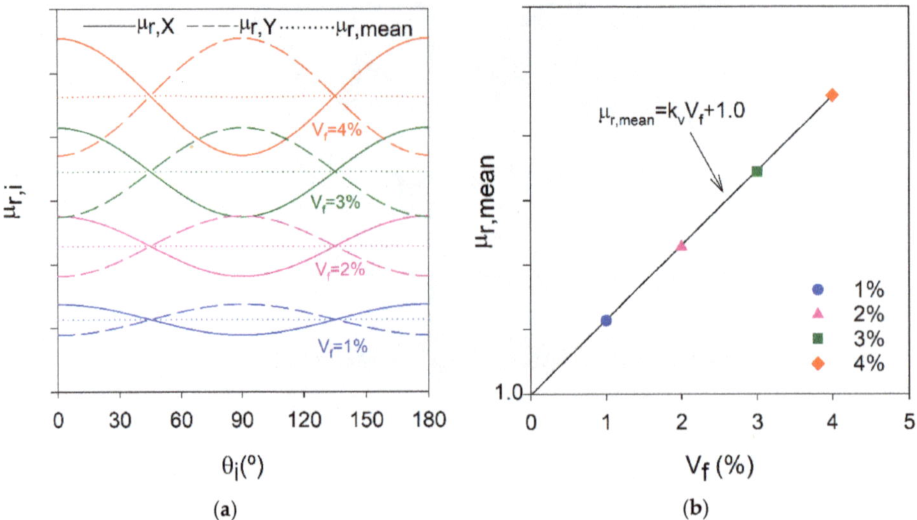


**Figure 2.** (**a**) Variation of $\mu_{r,i}$ and $\mu_{r,mean}$ with $V_f$ and the measuring direction $\theta_i$ (preferential fibre orientation coincides with $\theta_i = 0°$) (**b**) $\mu_{r,mean}$ linear increase with $V_f$.

2.2.2. Fibre Orientation

From the same $\mu_{r,X}$ and $\mu_{r,Y}$ values, the orientation indicator $(\rho_X - \rho_Y)$ can be computed from Equation (4), which was shown to be practically independent of the fibre content [18,26]. This indicator allows to identify the direction of preferential orientation of the fibres. Positive or negative $(\rho_X - \rho_Y)$ values indicate a preferential orientation along the X and Y directions, respectively. Values of $(\rho_X - \rho_Y)$ close to zero indicate a uniform fibre distribution (The only exception is when all fibres are perfectly aligned along a given direction and X/Y measuring directions form a 45° with the preferential orientation direction of the fibres, but this situation rarely occurs in practice, in real applications. This situation can be easily detected by adding a third measuring direction at each measuring point, for example, along the direction forming 45° with X). $(\rho_X - \rho_Y)$ value also provides a scalar measure of the fibre anisometry with respect to the X- and Y- directions, as described in the following paragraphs.

$$(\rho_X - \rho_Y) = 0.5\frac{\mu_{r,X} - \mu_{r,Y}}{\mu_{r,mean} - 1} \tag{4}$$

The fibre orientation factor ($\alpha_0$) has been used by several researchers and proved to be an effective tool to characterise the orientation of steel fibres in cracked sections. It was proposed by Krenchel [27] and can be computed as follows:

$$\alpha_0 = N_f \frac{A_f}{V_f A} \tag{5}$$

where $N_f$ is the number of fibres counted in the cross-section, A the cross-section area analysed, $V_f$ is the fibre content and $A_f$ the cross-section area of one fibre. Based on the generalised simplified model [26], for the most common fibre distribution functions in UHPFRC layers, the relation between the fibre orientation indicator $(\rho_X - \rho_Y)$ and the fibre orientation factor along X- and Y- directions ($\alpha_{0,X}$ and $\alpha_{0,Y}$) can be linearly approximated as follows (see Figure 3a).

$$\alpha_{0,X} = k_\alpha\,(\rho_X - \rho_Y) + \alpha_{0,iso}\,;\ \alpha_{0,Y} = -k_\alpha\,(\rho_X - \rho_Y) + \alpha_{0,iso} \tag{6}$$

where $k_\alpha$ is a constant that needs to be calibrated experimentally for a given fibre type (or fibres mixture), and $\alpha_{0,iso}$ is the fibre orientation factor corresponding to fibre isometry with respect to X- and Y- directions ($\alpha_{0,X}= \alpha_{0,Y}$) [26].

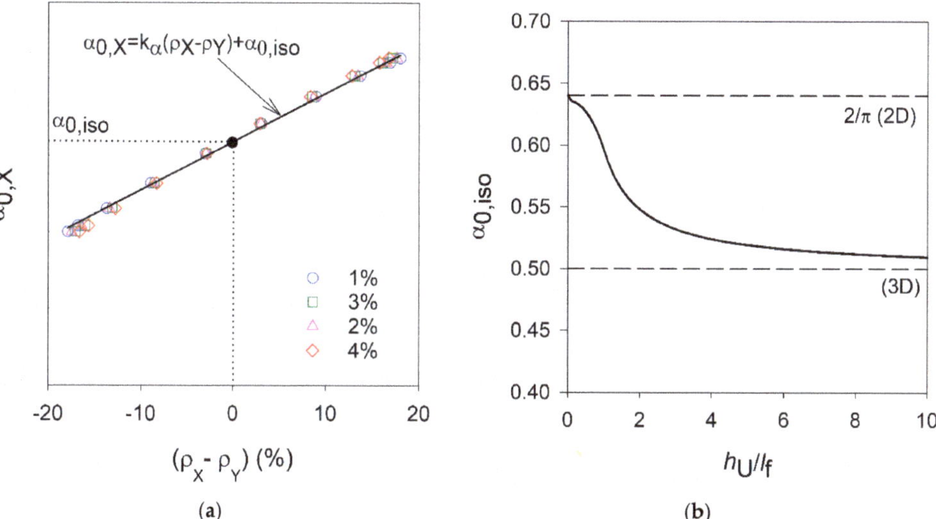


**Figure 3.** (**a**) Linear increase of $\alpha_{0,X}$ with $(\rho_X - \rho_Y)$; (**b**) Variation of $\alpha_{0,iso}$ with $h_U/l_f$ ratio [26].

$k_\alpha$ can be calibrated based on experimental data obtained, for example, after image analysis of cross-sections normal to X- or Y- directions, using specimens with different fibre contents and a wide range of fibre orientation profiles [21]. $\alpha_{0,iso}$ can be determined from Figure 3b as a function of the ratio between the UHPFRC element thickness and the fibre length ratio ($h_U/l_f$). When all fibres are lying in the same plane ($h_U/l_f \rightarrow 0$), that is, fibres have a perfect 2D distribution $\alpha_{0,iso} \approx 0.64$. Otherwise, when fibres have a 3D distribution ($h_U/l_f \rightarrow \infty$), then $\alpha_{0,iso} \rightarrow 0.5$, as shown in Figure 3b.

### *2.3. Estimation of the (Post-Cracking) Tensile Strength*

The post-cracking tensile strength of UHPFRC on a given i-direction, $f_{Utu,i}$, is often estimated as

$$f_{Utu,i} = \tau_f\, \alpha_{0,i}\, \alpha_{1,i} V_f \frac{l_f}{d_f} = \tau_f . \lambda_i \quad (7)$$

where $\tau_f$ is the equivalent (rigid-plastic) fibre-to-matrix bond strength and $\lambda_i$ is the so-called 'fibre structure parameter' [11,21,28], which encompasses fibre parameters related to orientation ($\alpha_{0,i}$), efficiency ($\alpha_{1,i}$), content ($V_f$) and geometry ($l_f/d_f$). $\lambda_i$ can be obtained from the analysis of an image of the cross-section under consideration. This requires cutting several samples from larger specimens and in different directions, carrying a surface treatment (polishing) and, finally, identifying the fibres, which is a time-consuming process and may not be feasible in actual structures. $\lambda_i$ can also be determined by X-ray tomography, which allows the fibres orientation and distribution to be determined very precisely. However, it is a much more expensive process and not currently applicable on an industrial scale. Alternatively, Nunes et al. [21] proposed to estimate $\lambda_i$ from the NDT measurements. $V_f$ and $\alpha_{0,i}$ are estimated using the regression lines presented in Section 2.2, and the fibre efficiency factor ($\alpha_{1,i}$) can be estimated from $\alpha_{0,i}$ using the relation proposed in [11,21].

Previous work by the authors showed that $f_{Utu,i}$ estimates based on $\lambda_i$ results obtained from image analysis or the NDT measurements have equivalent accuracy, thereby demonstrating the significance of the proposed NDT method for the quality control of thin UHPFRC elements/layers [21,26].

## 3. Experimental Programme

### *3.1. Materials and Mix-Proportions*

The non-proprietary UHPFRC mixes employed in this study are presented in Table 2. The binder phase consisted of ordinary Portland cement (Type I and class 42.5 R), limestone filler (98% $CaCO_3$) and silica fume (in suspension with 50% solids content, $SiO_2 > 90\%$,) with a specific gravity of 3.16, 2.68 and 1.38, respectively. As aggregate, a fine siliceous sand (maximum size of 1 mm) was used with a specific gravity of 2.63 and 0.3% water absorption. Besides the water included in the silica fume, the water amount indicated in Table 2 and a polycarboxylate-based superplasticiser (specific gravity of 1.08 and 40% solid content) were added to the mixture. Concerning the fibres phase, a hybrid mixture of two types of straight steel fibres was used, both having high tensile strength (2100 MPa) and circular cross-section ($d_f = 0.175$ mm), but different slenderness ($l_f/d_f = 51$ and 69). The performance of this hybrid mixture of fibres was evaluated in a previous experimental campaign that involved performing uniaxial tensile tests on specimens with 1.5% and 3% fibre content, and covering a wide range of orientation distributions [11].

**Table 2.** Mix-proportions of UHPFRC employed in this study (kg/m$^3$).

| UHPFRC | Constituent Materials | $V_f = 1\%$ | $V_f = 2\%$ | $V_f = 3\%$ | $V_f = 4\%$ |
|---|---|---|---|---|---|
| Cementitious matrix | Cement | 794.9 | | | |
| | Silica fume | 79.49 | | | |
| | Limestone filler | 311.43 | | | |
| | Water | 145.36 | | | |
| | Superplasticizer | 30 | | | |
| | Sp/c * | 1.51% | | | |
| | Sand | 993.56 | 967.26 | 940.96 | 914.66 |
| Steel fibres (straight) | $l_f = 9$ mm/$d_f = 0.175$ mm | 39.25 | 78.5 | 117.5 | 157 |
| | $l_f = 12$ mm/$d_f = 0.175$ mm | 39.25 | 78.5 | 117.5 | 157 |

(*) percentage of superplasticizer (solid content) by weight of cement.

### *3.2. Mixing, Workability and Specimen Preparation*

The mixing sequence consisted of the following steps: (1) mixing the binder materials, aggregate and 80% of the mixing water for 2.5 min; (2) stopping to scrape material adhering to the mixing bowl; (3) mixing for another 2.5 min; (4) adding the rest of the water with 75% of the superplasticiser, mixing for 2.5 min and repeating step 2); (5) adding the rest of the superplasticiser; mixing for another 1.5 min; (6) adding the fibres and finally mixing during a further 2 min. In the mixtures with higher fibre contents (3% and 4%), fibres were incorporated in two steps, half the amount at a time.

Immediately after mixing, the mortar flow test was carried out (using a mini-cone according to EFNARC recommendations [29]) to assess the mixture's flowability by calculating the flow diameter (Dflow) as the mean of two diameters of the spread area.

In order to understand the effect of specimens geometry (thickness and area around the probe), surface condition (roughness of the surface) and fibres segregation (in-depth) on the inductance measurements, and thus relative magnetic permeability, three test series were developed according to Table 3. The geometry, position of the moulds during casting and number of specimens prepared for each test series is indicated in Table 3. In general, special care was taken during the casting of specimens to promote a random orientation of fibres within the specimens. The discharge point was moved randomly along the mould not to let the material flow, preventing a preferential orientation of the fibres. In some specimens of test series C, the orientation of fibres was forced by pouring UHPFRC with the mould placed inside an orientation set-up, where an electromagnetic field was produced, promoting preferential orientation of the fibres, as described in [21]. After casting, the specimens were covered with a plastic sheet and demoulded after one day. Then, the samples were maintained in a chamber under controlled environmental conditions (temperature = 20 ± 2 °C and relative humidity > 95%).

**Table 3.** NDT testing programme.

| **Test Series:** | **A** | **B** | **C** | |
|---|---|---|---|---|
| **Effect Being Evaluated:** | **Thickness** | **Area** | **Surface Roughness** | **Fibres Segregation (In-Depth)** |
| specimen's geometry | cylinder<br>h = 100 to 20 mm<br>ϕ = 150 mm | square plate<br>l = 350 to 100 mm<br>h = 30 mm | square plate<br>l = 200 mm<br>h = 30 mm | |
| total fibre content | 1%<br>2%<br>3%<br>4% | 3% | 1%<br>2%<br>3%<br>4% | |
| number of specimens | 4 | 3 | 12 | |
| fibres orientation | random | random | random<br>oriented * | |
| mould position | horizontal | horizontal<br>vertical | horizontal<br>vertical | |
| condition of the test surface | moulded | moulded | moulded surface<br>casting surface<br>polished surface | |

(*) fibres orientation was achieved by means of an external electromagnetic field.

As the relative permeability of the cementitious matrix is equal to 1.0, irrespective of its composition and age, the inductance measurements could be carried out immediately after demolding.

### 3.3. NDT Testing

#### 3.3.1. Test Series A

In order to assess the effect of the specimen's thickness on the relative magnetic permeability evaluated using the NDT method, inductance measurements were performed on cylindrical specimens (150 mm diameter) with decreasing thickness. The probe was placed in the centre of the moulded surface of the cylinder and inductance measurements were taken every 15°. The choice of a cylindrical specimen for this series of tests was dictated by the greater ease of handling during the cutting process. The specimens had an original thickness of 100 mm (see Figure 4), from which layers were successively cut (using a diamond blade saw) and removed until the height was reduced to 20 mm. The first five layers to be removed had 10 mm thickness, while the remaining had just about 5 mm (see Figure 5). Inductance measurements, with the probe placed in the centre of the specimen, were repeated at each stage, always using the same surface (moulded face).

#### 3.3.2. Test Series B

In order to assess the effect of specimen area, three large plates of 350 × 350 × 30 $mm^3$ were produced. Two of these plates were cast horizontally (here referenced as H1 and H2), while the other was cast in the vertical position (here referenced as V1). The plate size was reduced from 300 mm to 100 mm by cutting it with a diamond blade saw, 10–25 mm at a time (Figure 6a). Inductance measurements were repeated at each stage, with the probe placed in the centre of the plate and always using the same surface (moulded face). Again inductance measurements were taken every 15° (see Figure 6b).

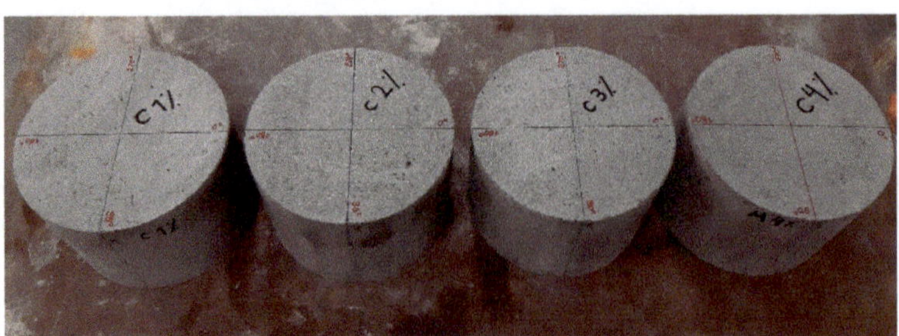

**Figure 4.** Specimens used in test series A (with original thickness).

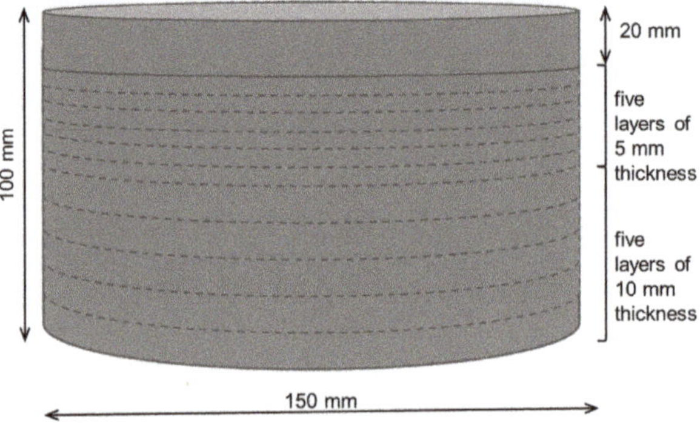


**Figure 5.** Successive removal of layers of the original cylinder.

**Figure 6.** (**a**) Successive reduction of plate size; (**b**) Inductance measurements in the centre of the plate (original size).

3.3.3. Test Series C

Within test series C, 12 plate specimens of 200 × 200 × 30 mm$^3$ were produced. For each fibre content, two specimens were cast horizontally (one with random and the other with preferential fibre orientation), and one specimen was cast vertically, as shown in

Figure 7. Inductance measurements were taken on both surfaces of each specimen: one casting surface and one moulded surface, in the case of horizontal specimens, and two moulded surfaces in vertical specimens. Additionally, a polishing machine was used in the casting surface of horizontal specimens to reduce their roughness (smoothed surface), and measurements were repeated on the casting surfaces after polishing (see Figure 8). The testing conditions and specimen referencing are detailed in Table 4.

**Figure 7.** Casting of UHPFRC plates with moulds in (**a**) horizontal position; (**b**) vertical position.

**Figure 8.** (**a**) Process of polishing the casting surface; (**b**) Surface after polishing.

**Table 4.** Referencing of specimens from test series C.

| Mould Position | Fibres Orientation | Condition of the Test Surface | Specimens Reference * | |
|---|---|---|---|---|
| Horizontal | Random | Moulded surface<br>Casting surface<br>Smoothed surface | HRM_i<br>HRC_i<br>HRS_i | HR_i |
| Horizontal | Oriented | Moulded surface<br>Casting surface<br>Smoothed surface | HOM_i<br>HOC_i<br>HOS_i | HO_i |
| Vertical | Random | Moulded surface A<br>Moulded surface B | VRMA_i<br>VRMB_i | VR_i |

(*) In the specimen reference i refers to the fibre content.

For each test condition, the magnetic probe was centred in the plate and rotated by 15° to take the inductance measurements in different directions ($\theta_i = 0°$ up to 180°), similar to the test series A and B.

## 4. Results and Discussion

### *4.1. Workability*

Table 5 presents the slump-flow diameter results of different mixes tested at different dates. All tested mixtures are self-compacting, exhibiting good flowability. An increase in fibre content leads to a decrease in spread diameter. However, the decrease is not drastic; for instance, tripling the fibres content from 1% to 3% leads to a decrease in the average spread diameter from 290.0 to 287.8 mm or about −0.8%. A higher loss of workability was observed for the fibre content of 4%. Figure 9a,b show pictures of the spread areas for two mixes with a fibre content of 1.0% and 4.0%, respectively. It can be observed that the mixture with the highest fibre content shows some signs of fibre agglomeration within the spread area, while the other mixture had no such effect. In addition, the shape of the spread area assumes a more irregular shape.

**Table 5.** Slump-flow diameter results (mm).

| Test Series | Casting Date | Reference | $V_f = 1\%$ | $V_f = 2\%$ | $V_f = 3\%$ | $V_f = 4\%$ |
|---|---|---|---|---|---|---|
| A | 22 May 2019 | – | 294.5 | 295.0 | 291.5 | 263.0 |
| C | 12 June 2019 or 14 June 2019 | HR_i | 295.5 | 289.0 | 287.5 | 281.0 |
| | | HO_i | 286.0 | 287.5 | 284.0 | 280.5 |
| | | VR_i | 288.5 | 287.0 | 288.0 | 274.0 |
| Average | | | 290.0 | 289.6 | 287.8 | 274.6 |
| Standard deviation | | | 4.9 | 3.7 | 3.1 | 8.4 |
| Coefficient of variation | | | 1.7% | 1.3% | 1.1% | 3.1% |

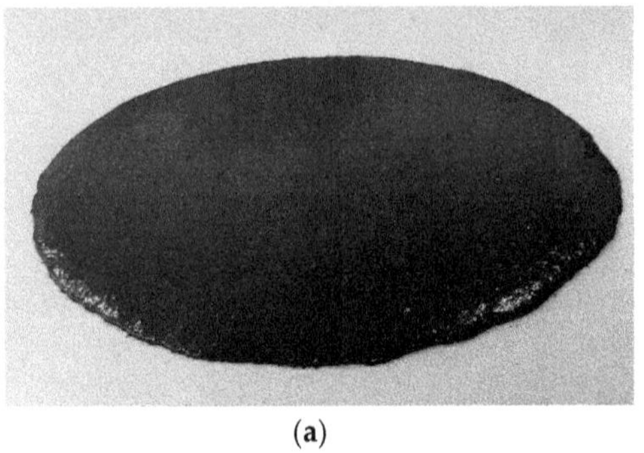

(**a**)

(**b**)

**Figure 9.** Final spread area of UHPFRC mixes with a fibre content of (**a**) 1% and (**b**) 4%.

It should be stressed that UHPFRC mixes with improved flowability, such as those used in this study, facilitate the casting process but have an increased risk of fibres segregation [30]. This subject will be addressed again in Section 4.4. Based on the authors' own experience, the spread diameter should not exceed 265 mm to minimise the risk of fibres segregation.

### *4.2. Effect of Specimen Thickness*

Figure 10a shows the variation of the relative magnetic permeability when the probe was aligned in different directions ($\theta_i$), with $\theta_i$, ranging from 0° to 180° by steps of 15°. The graphs in Figure 10a were obtained considering $\mu_{r,\theta i} = \mu_{r,(180°\text{-}\theta i)}$. Figure 10a clearly shows the increase of $\mu_{r,\theta i}$ with the fibre content. In most specimens, $\mu_r$ does not change significantly with the measuring direction (data points draw approximately a circle, Figure 10a), confirming a random distribution of fibres. The only exception is the specimen with $V_f = 4\%$, where some preferential fibre orientation is observed along $\theta_i = 30°$ direction (data points draw approximately an ellipse with its major axis aligned with $\theta_i = 30°$ direction, Figure 10a). This is also clearly indicated by the variation of the orientation indicator in Figure 10b.

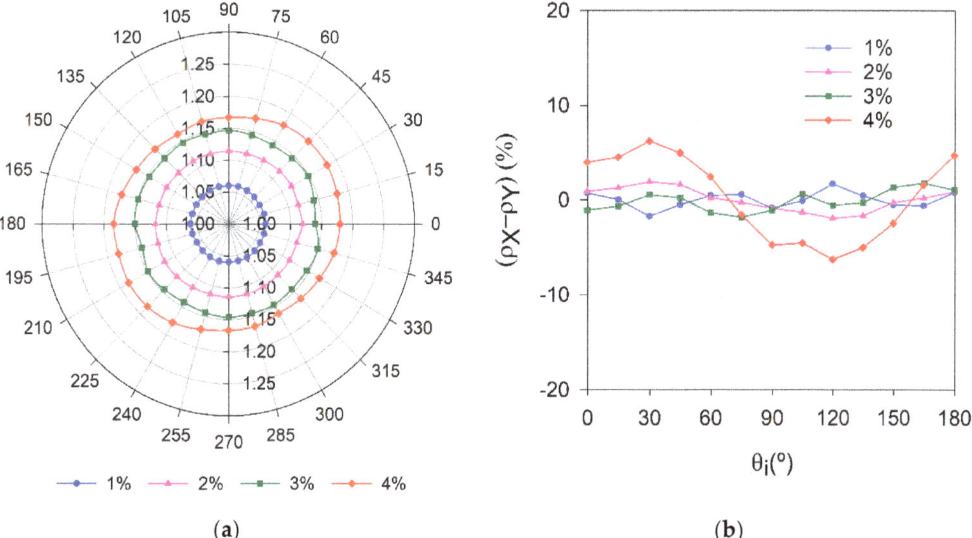


**Figure 10.** Variation of (**a**) $\mu_r$ and (**b**) $(\rho_X - \rho_Y)$ with the measuring direction (specimens with $h_U$ = 100 mm).

Figure 11a shows the variation of the mean relative magnetic permeability for all four specimens as a function of the specimen's thickness. $\mu_{r,mean}$ decreases with the thickness of the specimen, for $h_U < 70$ mm, thus the calibrating constant $k_V$ in Equation (3) will also change with $h_U$. From Equation (3), for each specimen thickness, $k_V$ can be estimated from

$$k_V = \frac{V_f}{(\mu_{r,mean} - 1)} \tag{8}$$

and then, by normalising the results dividing them by the constant obtained for $h_U$ = 100 mm ($k_{V,100}$). Figure 11b was obtained. It can be observed that all data points can be fitted by a Chapman-Richards growth function of the type

$$y(x) = y_{max}\left[1 - e^{-kx}\right]^p \tag{9}$$

where $y_{max}$ = 1.0, and k and p are the fitting parameters. Therefore, when the thickness of the element under analysis does not match the thickness of the specimens used to fit the line in Figure 2b, $k_V$ should be corrected using a function similar to the one shown in Figure 11b.

### 4.3. *Effect of the Specimen Area*

Figure 12a shows the variation of the relative magnetic permeability with the measuring direction of specimens obtained from plate H1. These results reveal a preferential orientation of the fibres in the direction corresponding to $\theta_i = 15°$. Figure 12b shows the change in $\mu_{r,mean}$ with the reduction of the plates' size. A similar effect of the specimen's size was found on the three plates. Figure 11b shows that $\mu_{r,mean}$ remains unchanged for specimens with sizes up to 200 mm, decreasing significantly for smaller sizes. Comparing $\mu_{r,mean}$ results obtained in specimens of 200 mm size with those of 150 mm (the size of the notched DEWST specimens used in previous campaigns [18,21]), the difference in $\mu_{r,mean}$ is relatively small (0.27% in plate H1).

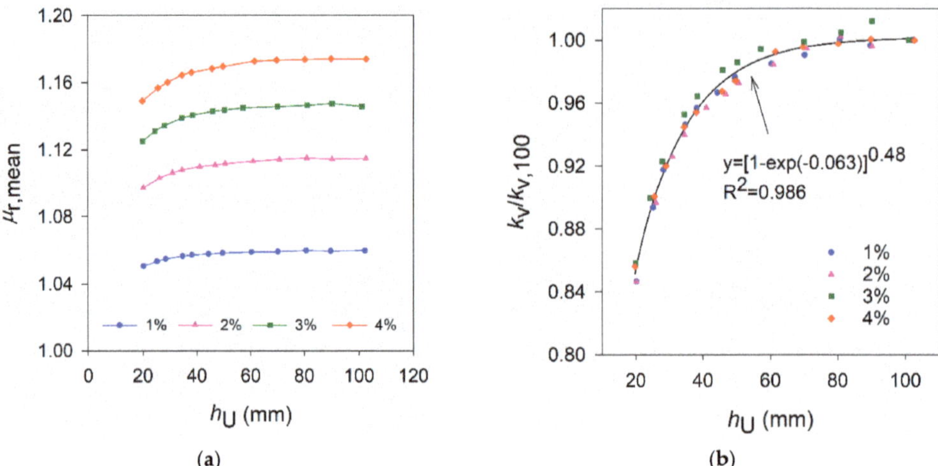


**Figure 11.** (**a**) Variation of $\mu_{r,mean}$ with $h_U$; (**b**) Variation of $k_V$ normalised to $k_{V,100}$ with $h_U$.

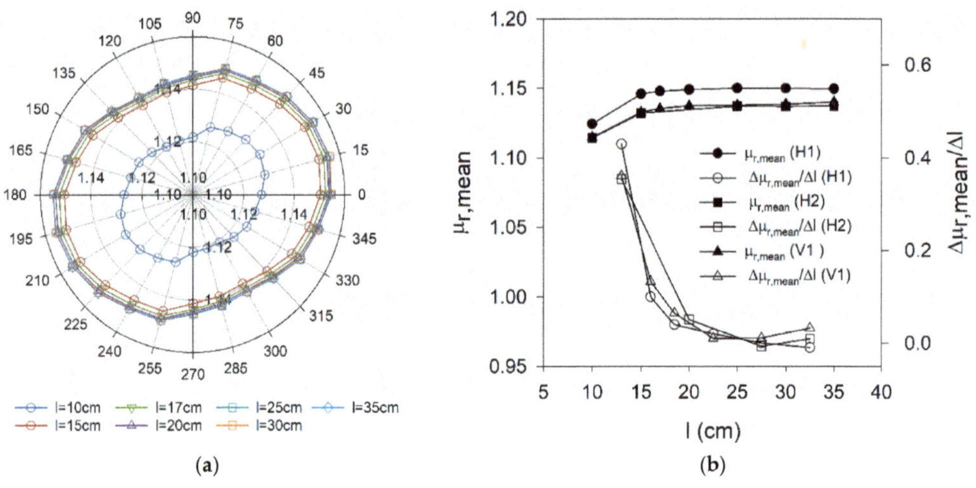


**Figure 12.** (**a**) Variation of $\mu_r$ with the measuring direction ($V_f$ = 3%) for square specimens with varying size (plate H1); (**b**) Variation of $\mu_{r,mean}$ with the specimen size l (plates H1, H2 and V1).

To perform measurements with a probe similar to the one used in this study (Section 2.1), the measuring points over a UHPFRC layer/specimen should be selected so that a minimum area of 200 × 200 $mm^2$ surrounding the measuring point exists (the probe is centred on the measuring point). Consequently, NDT measuring points should be at a minimum distance of 100 mm from the sides of the test specimens/elements.

Dog-bone or dumbbell shape specimens are often used in direct tensile tests to characterise the tensile behaviour of UHPFRC [11,12,28]. This type of specimen usually has reduced dimensions in its central part, making it impossible to comply with the recommended minimum distance between the measuring point and the sides of the specimen. Thus, to characterise the fibre content and orientation in the central part of dog-bone or dumbbell shape specimens employing this NDT method, it is recommended to cast a large slab and perform the inductance measurements in the zones of interest (corresponding to the central part of the specimens), and only then cutting the dog-bone/dumbbell shape specimens from the large slab. This procedure was followed in the work of Shen and Brühwiler [12] for scaling the representative tensile response of UHPFRC.

### 4.4. Effect of Surface Roughness

The effectiveness of the orientation set-up [21] used to promote a preferential orientation of the fibres in test series C can be confirmed by comparing the polar plots in Figure 13a,b or the evolution and magnitude of the orientation indicator in Figure 14a,b, corresponding to HRM_i and HOM_i specimens, respectively. A more pronounced orientation of fibres can be found in specimens HOM_i along $\theta_i = 0°$ (or X direction). Some of the HRM_i specimens also exhibit preferential fibre orientation, originated by the casting conditions, but much less pronounced than in HOM_i specimens. HRM_4% is a good example of a uniform fibre distribution.

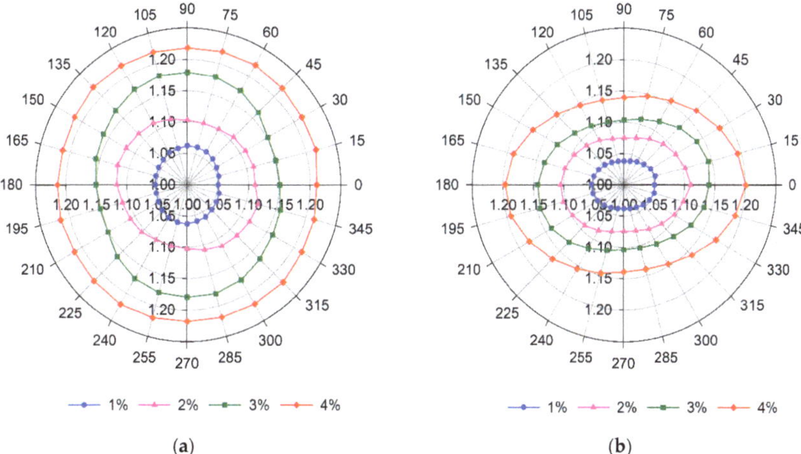


**Figure 13.** Variation of $\mu_r$ with the measuring direction for (**a**) HRM_i and (**b**) HOM_i specimens (moulded surface).

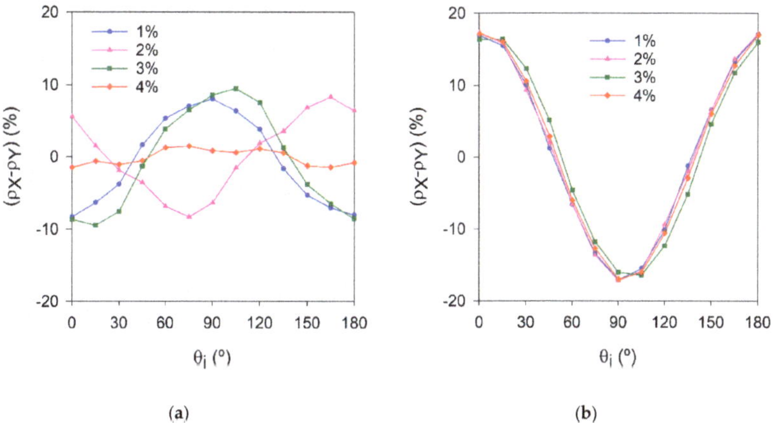


**Figure 14.** Variation of $(\rho_X - \rho_Y)$ with the measuring direction for (**a**) HRM_i and (**b**) HOM_i specimens (moulded surface).

Figures 15 and 16 show the variation of $\mu_{r,mean}$ computed according to Equation (2) as a function of $V_f$. These figures include results obtained on surfaces with different types of roughness (moulded, casting, smoothed) and specimens cast both horizontally and vertically. Table 6 summarises the calibrating constant $k_V$ and the squared correlation coefficient ($R^2$) of the fitting lines presented in Figures 15 and 16.

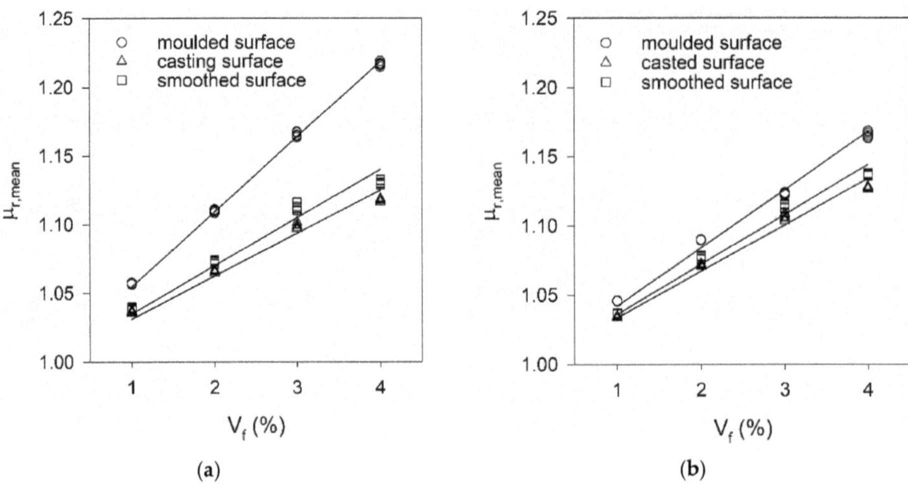


**Figure 15.** Increase of $\mu_{r,mean}$ as a function of $V_f$ and corresponding fitting lines for (**a**) HR_i and (**b**) HO_i specimens.

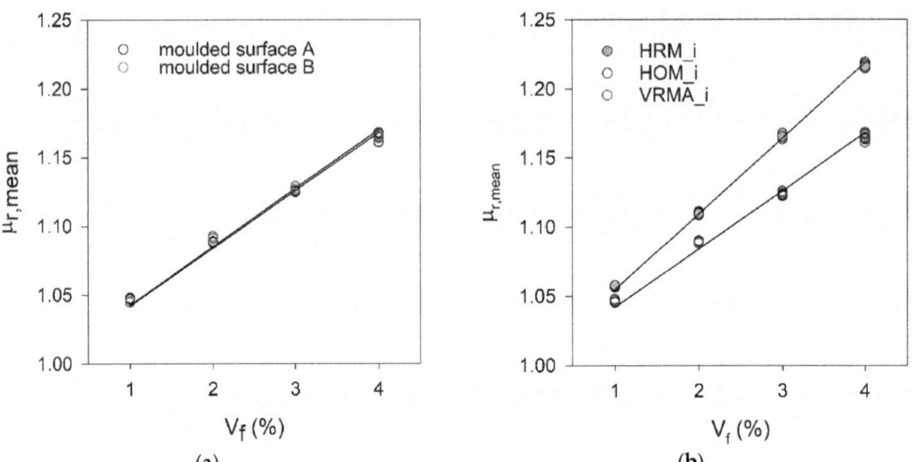


**Figure 16.** Increase of $\mu_{r,mean}$ results as a function of $V_f$ and corresponding fitting lines for (**a**) VR_i specimens (**b**) all moulded surfaces.

Figures 15 and 16 show the positioning of the mould during casting and the choice of the surface to measure inductance can influence the value of the calibrating constant $k_V$. No difference was found in results obtained in both faces of VR_i specimens. However, in both HR_i and HO_i specimens, lower $\mu_{r,mean}$ results were obtained in the casting (upper) surface compared to the moulded surface, attributed to the surface roughness and/or fibres segregation (in-depth).

Table 6. Calibrating constant $k_V$ and $R^2$ of fitting lines.

| Mould Position | Fibres Orientation | Reference | $k_V$ | $R^2$ |
|---|---|---|---|---|
| Horizontal | Random | HRM_i | 5.5 | 0.999 |
| | | HRC_i | 3.1 | 0.965 |
| | | HRS_i | 3.5 | 0.965 |
| | Oriented | HOM_i | 4.2 | 0.992 |
| | | HOC_i | 3.4 | 0.980 |
| | | HOS_i | 3.6 | 0.979 |
| Vertical | Random | VRMA_i | 4.2 | 0.991 |
| | | VRMB_i | 4.3 | 0.990 |

A rough surface does not allow for good contact between the probe and the UHPFRC material, and air gaps are introduced between the probe and the material, significantly reducing the measured inductance. When this occurs, a possible solution is to apply mechanical polishing to the surface to smooth the surface before performing the inductance measurements (see Figure 8). Figure 15 shows $\mu_{r,mean}$ results increased after polishing the casting surfaces, proving that a better contact was achieved between the probe and the material. Similar surface treatment was adopted in the work of Shen and Brühwiler [12].

### 4.5. Effect of Fibres Segregation (In-Depth)

When comparing the results obtained on all moulded surfaces (see Figure 16b), from both vertical and horizontal moulds, it can be observed that the fitting lines from VRMA/B_i and HOM_i specimens are overlapped (similar $k_V$ results), but are below the fitting line corresponding to HRM_i specimens. In order to understand this difference, further studies were carried out to investigate the occurrence of fibres segregation (in-depth) in HRM_i. For this purpose, small cubic specimens of 30 × 30 × 30 $mm^3$ were cut from the centre of plate specimens so that the fibres intersecting a chosen cross-section normal to the X- or Y-axis could be counted ($N_f$). The images of each cross-section were subdivided into 3 mm horizontal layers and the number of fibres per layer was manually counted using ImageJ software [31] (see Figures 17b and 18b).

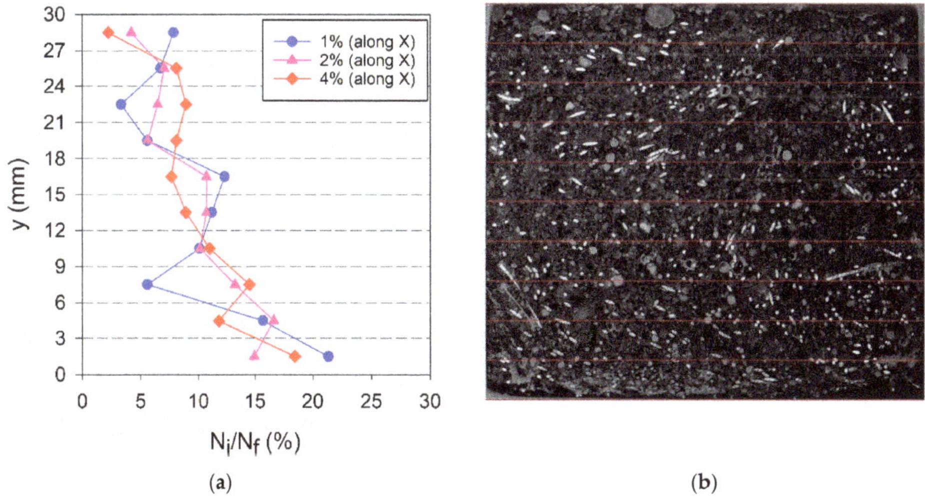


Figure 17. (a) Fibre distribution along the plate thickness for HR_i specimens (b) image of HR_4% along X-direction.

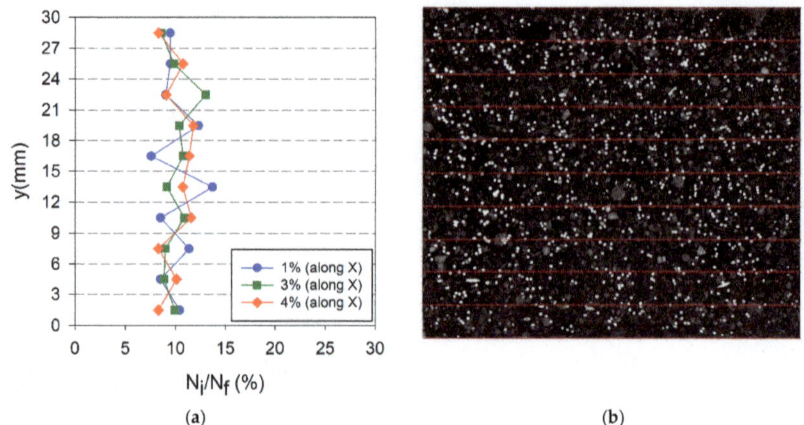


**Figure 18.** (**a**) Fibre distribution along the plate thickness (X-direction) for HO_i specimens; (**b**) image of HO_4% specimen along X-direction.

Table 7 summarises the total number of fibres counted in each cross-section. As expected, for the same orientation, $N_f$ increases with the fibre content. On the other hand keeping the fibre content, $N_f$ increases when fibres are more aligned with the direction normal to the cross-section.

**Table 7.** Total number of fibres ($N_f$) counted in a square area of 30 mm in size.

| Fibre Content | HR_i | | HO_i | | VR_i | |
|---|---|---|---|---|---|---|
| $V_f$ | Along X | Along Y | Along X | Along Y | Along X | Along Y |
| 1% | 89 | – | 212 | – | 217 | – |
| 2% | 354 | | – | 158 | | 152 |
| 3% | – | 639 | 747 | | 713 | – |
| 4% | 481 | – | 953 | – | | 583 |

The percentage of fibres per layer found is represented in Figures 17a, 18a and 19a for each type of specimen. In these graphs, y = 0 corresponds to a moulded surface which is the bottom face for horizontally cast specimens and a lateral face for vertically cast specimens. The fibres distribution in Figure 17 in comparison to Figure 18 confirms the occurrence of fibres segregation in HR_i specimens. More fibres were found in the bottom part (moulded surface) of HR_i specimens compared to the upper part (casting surface), and this explains the higher $\mu_{r,mean}$ results of HRM_i in Figure 16b. In-depth fibres segregation in HR_i specimens was promoted by the higher fluidity of the UHPFRC mixes used in this study (see Section 4.1) and, probably, some agitation of the material during casting, which reduces the matrix yield stress causing fibres settlement.

The wall effect is evidenced in the distribution of fibres of VR_i specimens presented in Figure 19. A more symmetric distribution of the fibres along the thickness is observed and significantly fewer fibres were found in the layers closer to the mould walls compared to the centre of the plate.

It is interesting to notice that in HO_i specimens if we analyse a cross-section parallel to the direction of preferential orientation, a much smaller number of fibres is found, and thus the fibres distribution along the plate thickness has a larger scatter, as shown in Figure 20a This did not occur with other cross-sections normal to the Y-direction, represented in Figures 17a and 18a, because these specimens have a random fibres distribution and thus not differing significantly in X- and Y-directions.

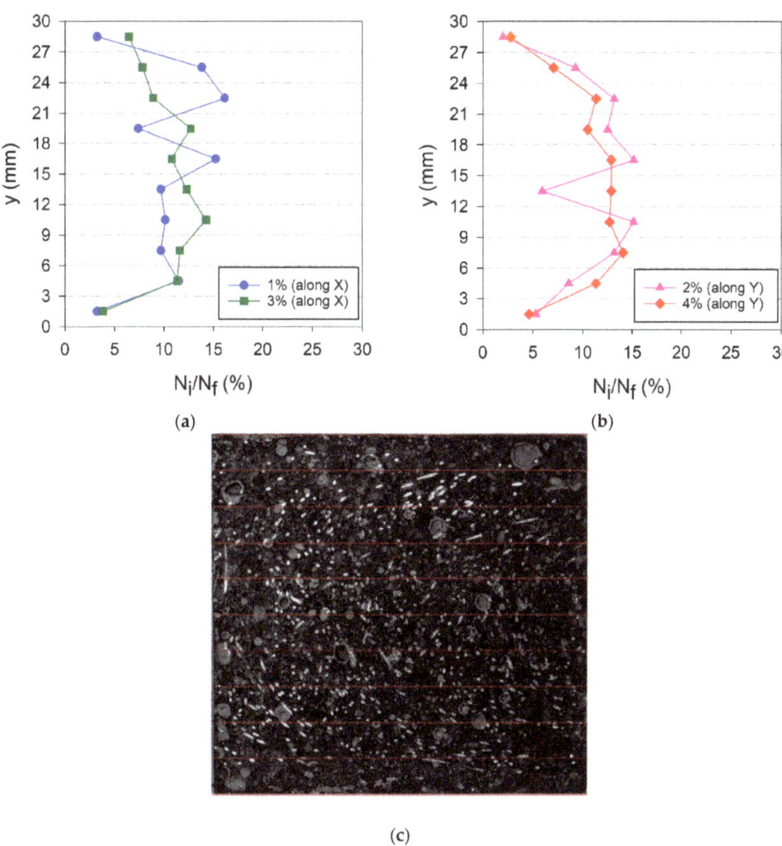


**Figure 19.** Fibre distribution along the plate thickness for VR_i specimens: (**a**) along X-direction, (**b**) along Y-direction; (**c**) image of VR_4% along X-direction.

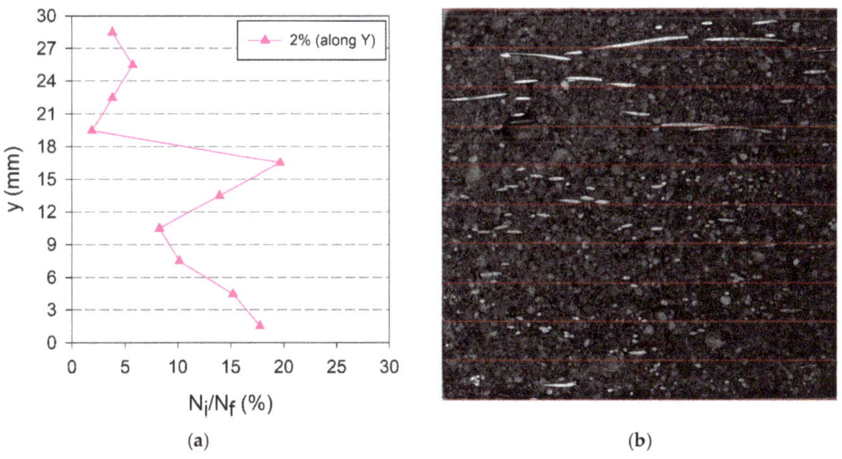


**Figure 20.** (**a**) Fibre distribution along the plate thickness (Y-direction) for HO_i specimens; (**b**) image of HO_2% along Y-direction.

## 5. Conclusions

This paper focused on the effect of key factors for successfully implementing the magnetic NDT method previously developed by the authors [18]. The following conclusions can be drawn from the results presented in this paper:

1. The proportionality constant $k_V$ that allows to predict $V_f$ from the $\mu_{r,mean}$ estimated using the magnetic probe changes with the thickness of the UHPFRC layer ($h_U$). A correction function is proposed for $k_V$ as a function of $h_U$.
2. The magnetic probe used in this study is capable of evaluating the fibre content and fibre orientation in the UHPFRC up to a depth of 70 mm measured from the contact surface. In UHPFRC elements where it is possible to take measurements on two opposite sides the assessed material thickness can be doubled.
3. To perform measurements with a probe similar to the one used in this study, the measuring points (the point at which the probe is centred) are recommended to be at a minimum distance of 100 mm from the sides of the test specimens/elements.
4. The probe measurements should be performed preferably over a moulded and smooth surface since perfect contact between the probe and the UHPFRC material is necessary to obtain representative measurements. Unevenness, for example, in the casting surfaces, should be removed by grinding.
5. The magnetic probe is sensitive to the occurrence of in-depth fibres segregation. Even in thin UHPFRC layers, fibres settlement can occur when casting with more fluid mixtures due to gravity action and/or when the material is agitated during casting.

This study raised the awareness that the thickness, area and roughness of the UHPFRC specimen/element under study, as well as the occurrence of in-depth fibres segregation, can have a significant influence on the inductance results. With the recommendations presented in this paper, one expects to contribute to more efficient use of this NDT method for quality assurance of UHPFRC elements at an industrial scale; being able to provide valuable information on the casting quality (fibres distribution and orientation) and the in-structure tensile strength in any direction.

**Author Contributions:** Conceptualization, S.N. and M.P.; methodology, S.N. and M.P.; software, A.S.; validation, A.S., S.N., and M.P.; formal analysis, A.S.; investigation, P.M. and A.S.; resources, S.N.; data curation, A.S. and P.M.; writing—original draft preparation, S.N.; writing—review and editing, M.P.; visualization, S.N.; supervision, S.N. and M.P.; project administration, M.P.; funding acquisition, M.P. All authors have read and agreed to the published version of the manuscript.

**Funding:** This work was financially supported by: Base Funding—UIDB/04708/2020 and Programmatic Funding—UIDP/04708/2020 of the CONSTRUCT—Instituto de I&D em Estruturas e Construções—funded by national funds through the FCT/MCTES (PIDDAC); by the project POCI-01-0145-FEDER-031777—"UHPGRADE—Next generation of ultra-high performance fibre-reinforced cement-based composites for rehabilitation and strengthening of the existing infrastructure" funded by FEDER funds through COMPETE2020—Programa Operacional Competitividade e Internacionalização (POCI) and by national funds (PIDDAC) through FCT/MCTES; by the Fundação Calouste Gulbenkian through the PhD grant n°144945; and by FCT—Fundação para a Ciência e a Tecnologia through the PhD scholarship UID/04708/2020.

**Institutional Review Board Statement:** Not applicable.

**Informed Consent Statement:** Not applicable.

**Acknowledgments:** Collaboration and materials supply by Concremat, Secil, Omya Comital, Sika, MC-Bauchemie and KrampeHarex is gratefully acknowledged. Lastly, the third author is grateful for the support of the Laboratório de Engenharia de Moçambique (LEM).

**Conflicts of Interest:** The authors declare no conflict of interest.

## References


1. Graybeal, B.; Brühwiler, E.; Kim, B.-S.; Toutlemonde, F.; Voo, Y.L.; Zaghi, A. International Perspective on UHPC in Bridge Engineering. *J. Bridge Eng.* **2020**, *25*, 04020094. [CrossRef]
2. Toutlemonde, F.; Bernadi, S.; Brugeaud, Y.; Simon, A. Twenty Years-Long French Experience in UHPFRC Application and Paths Opened from the Completion of the Standards for UHPFRC. 2018. Available online: https://hal.archives-ouvertes.fr/hal-01955204 (accessed on 26 July 2021).
3. Marek, J.; Kolisko, J.; Tej, P.; Čítek, D.; Komanec, J.; Kalný, M.; Vráblík, L. New UHPFRC bridges in the Czech Republic. *IOP Conf. Ser. Mater. Sci. Eng.* **2019**, *596*. [CrossRef]
4. López, J.Á.; Serna, P.; Navarro-Gregori, J.; Camacho, E. Construction of the U-Shaped Truss Footbridge over the Ovejas Ravine in Alicante. In Proceedings of the 2° International Symposium on UHPFRC. Designing and Building with UHPFRC, Marseille France, 1–3 October 2013.
5. Brühwiler, E.; Denarié, E. Rehabilitation and Strengthening of Concrete Structures Using Ultra-High Performance Fibre Reinforced Concrete. *Struct. Eng. Int.* **2013**, *23*, 450–457. [CrossRef]
6. Alberti, M.G.; Enfedaque, A.; Galvez, J. A review on the assessment and prediction of the orientation and distribution of fibres for concrete. *Compos. Part. B Eng.* **2018**, *151*, 274–290. [CrossRef]
7. Huang, H.; Gao, X.; Teng, L. Fiber alignment and its effect on mechanical properties of UHPC: An overview. *Constr. Build. Mater.* **2021**, *296*, 123741. [CrossRef]
8. Pastor, F.; Hajar, Z.; Palu, P.D. UHPFRC Footbridge in le CANNET des MAURES. In Proceedings of the AFGC-ACI-fib-RILEM Int. Symposium on Ultra-High Performance Fibre-Reinforced Concrete, UHPFRC, Montpellier, France, 2–4 October 2017.
9. Brühwiler, E.; Bastien-Masse, M.; Mühlberg, H.; Houriet, B.; Fleury, B.; Cuennet, S.; Schär, P.; Boudry, F.; Maurer, M. Strengthening the Chillon viaducts deck slabs with reinforced UHPFRC. In Proceedings of the IABSE Conference Geneva 2015 'Structural Engineering: Providing Solutions to Global Challenges', Geneva, Switzerland, 23–25 September 2015. [CrossRef]
10. Bastien-Masse, M.; Denarié, E.; Brühwiler, E. Effect of fiber orientation on the in-plane tensile response of UHPFRC reinforcement layers. *Cem. Concr. Compos.* **2016**, *67*, 111–125. [CrossRef]
11. Abrishambaf, A.; Pimentel, M.; Nunes, S. Influence of fibre orientation on the tensile behaviour of ultra-high performance fibre reinforced cementitious composites. *Cem. Concr. Res.* **2017**, *97*, 28–40. [CrossRef]
12. Shen, X.; Brühwiler, E. Influence of local fiber distribution on tensile behavior of strain hardening UHPFRC using NDT and DIC. *Cem. Concr. Res.* **2020**, *132*, 106042. [CrossRef]
13. Miletić, M.; Kumar, L.M.; Arns, J.-Y.; Agarwal, A.; Foster, S.; Arns, C.; Perić, D. Gradient-based fibre detection method on 3D micro-CT tomographic image for defining fibre orientation bias in ultra-high-performance concrete. *Cem. Concr. Res.* **2020**, *129*, 105962. [CrossRef]
14. Krause, M.S.; Hausherr, J.M.; Burgeth, B.; Herrmann, C.; Krenkel, W. Determination of the fibre orientation in composites using the structure tensor and local X-ray transform. *J. Mater. Sci.* **2010**, *45*, 888–896. [CrossRef]
15. Ferrara, L.; Faifer, M.; Toscani, S. A magnetic method for non destructive monitoring of fiber dispersion and orientation in steel fiber reinforced cementitious composites—Part 1: Method calibration. *Mater. Struct.* **2011**, *45*, 575–589. [CrossRef]
16. Ferrara, L.; Faifer, M.; Muhaxheri, M.; Toscani, S. A magnetic method for non destructive monitoring of fiber dispersion and orientation in steel fiber reinforced cementitious composites. Part 2: Correlation to tensile fracture toughness. *Mater. Struct.* **2011**, *45*, 591–598. [CrossRef]
17. Cavalaro, S.H.P.; López-Carreño, R.; Torrents, J.M.; Aguado, A.; Juan-García, P. Assessment of fibre content and 3D profile in cylindrical SFRC specimens. *Mater. Struct.* **2015**, *49*, 577–595. [CrossRef]
18. Nunes, S.; Pimentel, M.; Carvalho, A. Non-destructive assessment of fibre content and orientation in UHPFRC layers based on a magnetic method. *Cem. Concr. Compos.* **2016**, *72*, 66–79. [CrossRef]
19. Ozyurt, N.; Mason, T.O.; Shah, S.P. Non-destructive monitoring of fiber orientation using AC-IS: An industrial-scale application. *Cem. Concr. Res.* **2006**, *36*, 1653–1660. [CrossRef]
20. Lataste, J.; Behloul, M.; Breysse, D. Characterisation of fibres distribution in a steel fibre reinforced concrete with electrical resistivity measurements. *NDT E Int.* **2008**, *41*, 638–647. [CrossRef]
21. Nunes, S.; Pimentel, M.; Ribeiro, F.; Milheiro-Oliveira, P.; Carvalho, A. Estimation of the tensile strength of UHPFRC layers based on non-destructive assessment of the fibre content and orientation. *Cem. Concr. Compos.* **2017**, *83*, 222–238. [CrossRef]
22. Li, L.; Xia, J.; Chin, C.; Jones, S. Fibre Distribution Characterization of Ultra-High Performance Fibre-Reinforced Concrete (UHPFRC) Plates using Magnetic Probes. *Materials* **2020**, *13*, 5064. [CrossRef]
23. Li, L.; Xia, J.; Galobardes, I. Magnetic probe to test spatial distribution of steel fibres in UHPFRC prisms. In Proceedings of the 5th International fib Congress: Better-Smarter-Stronger, Melbourne, Australia, 7–11 October 2018.
24. Nunes, S.; Ribeiro, F.; Carvalho, A.; Pimentel, M.; Brühwiler, E.; Bastien-Masse, M. Non-destructive measurements to evaluate fiber dispersion and content in UHPFRC reinforcement layers. In Proceedings of the Multi-Span Large Bridges Conference, Porto, Portugal, 1–3 July 2015.
25. Davis, J.; Huang, Y.; Millard, S.G.; Bungey, J. Determination of Dielectric Properties of Insitu Concrete at Radar Frequencies. *Non-Destr. Test. Civ. Eng.* **2003**. Available online: https://www.ndt.net/article/ndtce03/papers/v078/v078.htm (accessed on 23 July 2021).

26. Sine, A.G. *Strengthening of Reinforced Concrete Elements with UHPFRC*; Faculty of Civil Engineering, Porto University: Porto, Portugal, 2021.
27. Krenchel, H. Fibre spacing and specific fibre surface. In *Fibre Reinforced Cement and Concrete*; Construction Press: London, UK, 1975; pp. 69–79.
28. Abrishambaf, A.; Pimentel, M.; Nunes, S. A meso-mechanical model to simulate the tensile behaviour of ultra-high performance fibre-reinforced cementitious composites. *Compos. Struct.* **2019**, *222*, 110911. [CrossRef]
29. BIBM; Cembureau; ERMCO; EFCA; EFNARC. *The European Guidelines for Self-Compacting Concrete Specification, Production and Use.* 2005. Available online: https://www.theconcreteinitiative.eu/images/ECP_Documents/EuropeanGuidelinesSelfCompactingConcrete.pdf (accessed on 2 August 2021).
30. Wang, R.; Gao, X.; Huang, H.; Han, G. Influence of rheological properties of cement mortar on steel fiber distribution in UHPC. *Constr. Build. Mater.* **2017**, *144*, 65–73. [CrossRef]
31. Rasband, W.S. *ImageJ*; U.S. National Institutes of Health: Bethesda, MD, USA. Available online: https://imagej.nih.gov/ij/ (accessed on 26 July 2021).

*Article*

# Numerical Modelling of Concrete-to-UHPC Bond Strength

**Alireza Valikhani [1,*], Azadeh Jaberi Jahromi [2], Islam M. Mantawy [1] and Atorod Azizinamini [1]**

[1] Department of Civil and Environmental Engineering Department, Florida International University, Miami, FL 33172, USA; imantawy@fiu.edu (I.M.M.); aazizina@fiu.edu (A.A.)

[2] Washington State Department of Transportation, Olympia, WA 98505, USA; Jabera@wsdot.wa.gov

* Correspondence: avali023@fiu.edu; Tel.: +1-305-877-0315

Received: 11 February 2020; Accepted: 16 March 2020; Published: 18 March 2020

**Abstract:** Ultra-High Performance Concrete (UHPC) has been a material of interest for retrofitting reinforced concrete elements because of its pioneer mechanical and material properties. Numerous experimental studies for retrofitting concrete structures have shown an improvement in durability performance and structural behaviour. However, conservative and sometimes erroneous estimates for bond strength are used for numerically calculating the strength of the composite members. In addition, different roughening methods have been used to improve the bond mechanism; however, there is a lack of numerical simulation for the force transfer mechanism between the concrete substrate and UHPC as a repair material. This paper presents an experimental and numerical programme designed to characterize the interfacial properties of concrete substrate and its effect on the bond strength between the two materials. The experimental programme evaluates the bond strength between the concrete substrates and UHPC with two different surface preparations while using bi-surface test and additional material tests, including cylinder and cube tests for compression property, direct tension test, and flexural test to complement UHPC tensile properties. Non-linear finite element analysis was conducted, which uses a numerical zero thickness volume model to define the interface bond instead of a traditional fixed contact model. The numerical results from the zero thickness volume model show good agreement with the experimental results with a reduction in error by 181% and 24% for smooth and rough interface surfaces if compared to the results from the model with a fixed contact.

**Keywords:** UHPC; interface; bond strength; numerical analysis

---

## 1. Introduction

Ultra-high performance concrete (UHPC) has been in the interest of research with different range of applications for bridge construction, repair and rehabilitation, overlays, building, petroleum industry, hydraulic structures, and architectural components [1,2]. UHPC is developed by the inspiration of three concrete technologies: self-compacting concrete (SCC), high-performance concrete (HPC), and fibre reinforced concrete (FRC) with a high compressive strength that is higher than 126 MPa (18 ksi) and post-cracking tensile strength of more than 5 MPa (0.7 ksi) [1,3–7]. UHPC with a different mix design model is a combination of portland cement, sand, quartz powder, silica fume, superplasticizer, water, and steel fibres [4,8–11]. In the UHPC mixture, a higher portion of cement is used compared to the HPC and normal strength concrete (NSC) [12,13] and the key factor of UHPC production is the enhancement of material density, mechanical homogeneity, and particle packing by improving macro and micro properties [12–20], which help the durability of UHPC against chloride diffusion. The water-to-cement ratio (W/C) of UHPC is typically less than 0.25 due to the replacement of the portion of un-hydrated cement with blast furnace slag, fly ash, or crushed quartz [21–25]. Superplasticizer is added to the mixture to improve the workability of UHPC, which is low due to the low W/C ratio [15,16,26–29].

Adding silica fume increases the workability of the mixture and fills the voids between aggregate particles, therefore increasing the compressive strength [15,30–32]. Steel fibres are added to change the brittle behaviour of the mixture to a ductile behaviour. The most common size of steel fibres is 13 mm (0.5 in.) in length and 0.20 mm (0.008 in.) in diameter with a recommended ratio of 2% by volume [3,7,18,28,33]. The outstanding characteristics of UHPC enable it as a practical option for repairing and retrofitting damaged structural elements [2,34] connecting the precast elements [35,36] or as an overlay for bridge decks [37,38]. In all of these applications, sufficient interfacial bond strength between normal concrete substrates and UHPC is required to ensure the resulting section is sufficiently composite. Although the bond strength between normal concrete and UHPC has been experimentally investigated by many researchers while using bi-surface, slant shear, push off, and pull-off tests [39–44], there is a gap knowledge that is related to the numerical modelling of the interfacial bond strength for repair design and evaluation. Harris et al., [45] and Azizinamini et al., [46] presented an approach to solve this challenge using simple fix contact surface or tie model between two layers. Such an approach might cause the overestimation of bond strength and global structural performance. This paper presents a realistic approach to evaluate the interfacial bond strength between UHPC and NSC that is numerically based on experimental data using ATENA finite element (FE) software (version 5, Červenka Consulting s.r.o., Czchia, Czech Republic). Mechanical properties of UHPC are investigated to achieve this goal, and the test results are used to simulate the behaviour of UHPC in ATENA software. These three-dimensional FE models can be used to simulate the response of structures that are made of UHPC or repaired and retrofitted while using UHPC and solve the challenge of interfacial bond modelling between UHPC and NSC.

## 2. Materials and Methods

The first phase of this research included the testing of ten bi-surface shear specimens to experimentally quantify the interfacial bond between UHPC as a repair material and substrates that are made of NSC with two surface different preparations for NSC substrates (as cast and sand-blasted) [39,47]. The second phase included the development of numerical models to calibrate the interfacial bond strength between UHPC and NSC. To develop numerical models for the interfacial bond between UHPC and NSC, mechanical properties of UHPC were tested and calibrated numerically, including compressive strength, tensile strength, and flexural strength. For all the experimental tests, a universal testing machine (UTM) with 2224 kN (500 kips) maximum capacity and an Mesure Test Simulte (MTS) machine with 111 kN (25 kips) maximum capacity were used to apply the loads as testing apparatus. For the numerical modelling, a commercial finite element software, ATENA, was used because of its accurate concrete material models for fibre reinforced concrete and interface model.

### 2.1. Material and Mixing

This research utilized Ductal® JS1000 (LafargeHolcim, Clamart, France), which is a proprietary UHPC mix that is manufactured by Lafarge. This premix includes Portland cement, fine sand, ground quartz, silica fume, and accelerator. However, the rest of the UHPC components, such as superplasticizer and steel fibres, were shipped in different packaging by the same manufacturer, as shown in Figure 1a. The superplasticizer was Chryso® Fluid Premia 150 (manufacturer, city, country), and the steel fibres were straight fibres with a radius of 0.1 mm (0.004 in.) and a length of 13 mm (0.5 in.). The concentration of steel fibers in the mixture was chosen to be 2% by volume. UHPC was proportioned by weight while using a 136-kg (300-lbs) scale with an accuracy of ±0.01 kg (0.022 lbs). Water and steel fibres were individually batched using 19-L (5-gallon) buckets, and the superplasticizer was batched in a smaller plastic cup. Water and the superplasticizer were batched 20 min. before the mixing process to reduce the potential of evaporation. Table 1 lists the UHPC weight proportions.

(a)
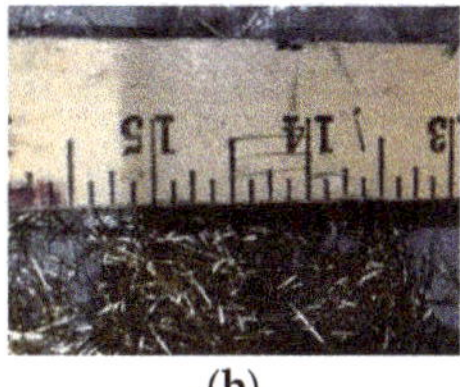
(b)

(c)

**Figure 1.** UHPC mixing process: (**a**) UHPC components; (**b**) steel fibers used in the mixture; The large pan-style mixture used for mix UHPC; and, (**c**) orbital pan-style mixer used.

**Table 1.** UHPC mixture proportions. (1 kg = 2.2 lb) [3,39].

| Constituent | | Portion Based on Each Premix Bag (Kg) | Percentage by Weight (%) |
|---|---|---|---|
| Ductal® JS1000 | Portland cement | 7.43 | 28.5 |
| | Fine sand | 10.64 | 40.8 |
| | Ground quartz | 2.2 | 8.4 |
| | Silica fume | 2.41 | 9.3 |
| | Accelerator | 0.31 | 1.2 |
| | Total weight of premix | 23 | 88.2 |
| Water | | 1.2 | 4.4 |
| Steel fiber 2% | | 1.6 | 6.2 |
| Superplasticizer | | 0.32 | 1.2 |

A large orbital pan-style mixer was used to mix UHPC specimens, as shown in Figure 1c. The mixing procedure of UHPC components started by dispatching UHPC dry premix into the mixture for four minutes mixing time. The required water and half of the superplasticizer were added to the mixer for 15 min. mixing time. Subsequently, the other half of the superplasticizer was added, and, after two minutes of mixing, the dry mix turned to a concrete paste. Afterwards, the steel fibres were added to the mixture. The UHPC mixture was mixed for a further five to six minutes to have a uniform mix. Table 2 shows the mixing procedure.

**Table 2.** UHPC mixing procedure [3,39].

| Procedure | Start Time (min) |
|---|---|
| Mixing UHPC dry premix | 0 |
| Adding water | 4 |
| Adding half superplasticizer | 4 |
| Adding the other half superplasticizer | 19 |
| Adding steel fibers | 21 |
| Mixing until complete uniformity | ≃30 |

### 2.2. *Quality Control and Curing*

ASTM C1437 [48] and Cortes et al., [49] were used to evaluate the rheology of the fresh UHPC. In this test, the fresh UHPC was discharged in a brass cone mould, which is placed over a standard flow table with a diameter of 254 mm (10 in.). The mould was removed straightly upward to allow the fresh UHPC to flow out and settle. Subsequently, the diameter of UHPC was measured along the four perpendicular lines on the flow table, as shown in Figure 2a. The average of these diameter measurements is called static flow. The flow table was manually dripped in height for 13-mm (0.5-in.) interval 20 times, and then the average of the diameter measurement in four perpendicular directions, after 20 drops, was calculated to obtain the dynamic flow. The static flow and dynamic flow of UHPC were measured at 216 mm (8.5 in.) and 228 mm (9 in.), respectively. The rheological property of the mix was categorized as fluid based on Table 3.

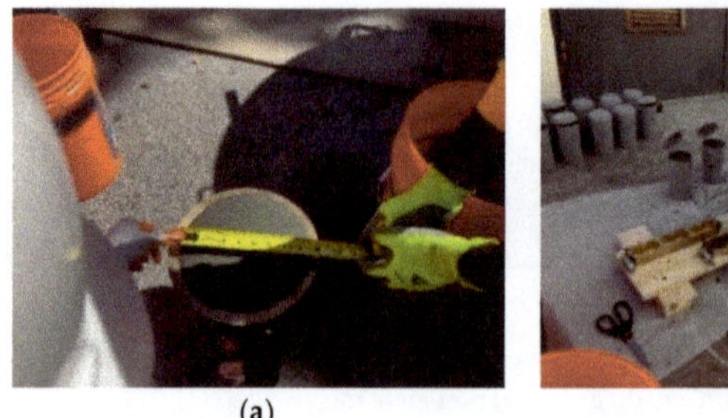
(a)
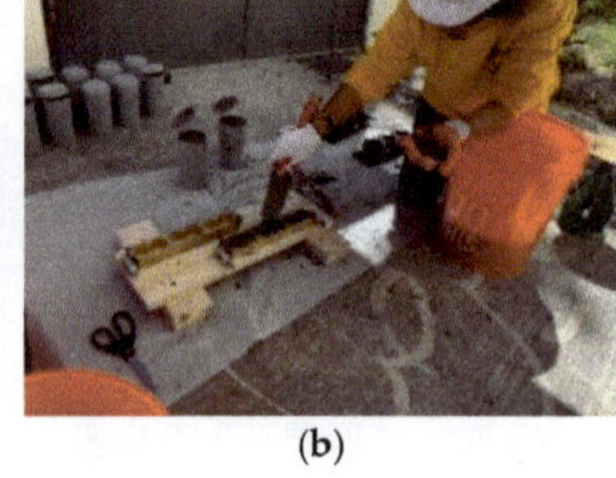
(b)

**Figure 2.** UHPC quality control: (**a**) rheological property measurement, (**b**) sampling process.

**Table 3.** Rheological property measurement based on ASTM C1437 [48]. (1 in. = 25.4 mm.).

| Spread Diameter after 20 Drops (mm) | Mix Rheology |
|---|---|
| <200 | Stiff |
| 200–250 | Fluid |
| >250 | Highly Fluid |

The fresh UHPC was cast in moulds that were based on ASTM C1856 [50] with no need for compaction due to the high flowability and self-consolidating characteristics of UHPC [51]. The sample cylinders and cubes for compressive strength test, dog bone specimens for tensile test, beams for the flexural test, and the portion of large cubes for bond strength test were cast, as shown in Figure 2b, and as described hereafter. The sample moulds were removed after 48 h of casting and were then left in an ambient condition at a temperature of 23 ± 2 °C (74 ± 3 °F) and humidity of 50% ± 5% inside the laboratory. It should be noted that all test specimens were untreated to mimic UHPC conditions in the field [52].

## 3. Bond Strength Test

For different structural applications, such as repairing and strengthening old concrete structures or connecting full-depth deck panels using closure joints [53,54], casting UHPC next to concrete at different ages or even casting UHPC next to steel [55] highlights the challenge of bond strength between these two materials. A bi-surface shear test setup was selected to measure the bond strength for smooth and rough interface surfaces between the two materials to quantify the interfacial bond strength between UHPC and NSC [56,57].

In this paper, ten cubical specimens of 153 mm (6 in.) sides were cast. The concrete substrate portion of the cube occupies two-third of the volume; however, UHPC occupies the other third, as shown in Figure 3. It should be noted that in the first stage of casting, NSC was cast and, in the second stage, UHPC was added like an overlay after 56 days. These specimens were divided into two groups that were based on interface surface preparation. In the first group, the concrete surface was kept as cast, hereafter referred to as "Smooth". In the second group, the concrete surface was roughened while using sandblasting with an average surface roughness of 1.72 mm (0.068 in.), hereafter referred to as "Rough".

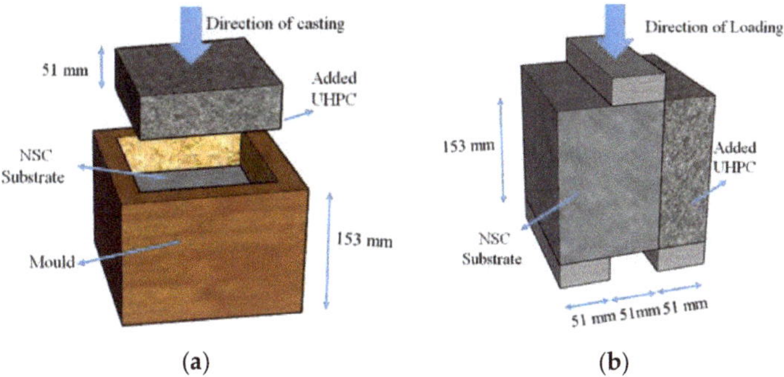


(a) (b)

**Figure 3.** The dimension of the bi-surface shear test specimen: (**a**) casting process, (**b**) test setup (1 in. = 25.4 mm).

Loading plate of 38 mm × 51 mm × 153 mm (1.5 in. × 2 in. × 6 in.) was used in the bi-surface shear test setup, which results in two shear planes, as shown in Figure 3. One shear plane is located at the interface between NSC and UHPC. The other shear plane is located inside the concrete substrate. The universal testing machine (UTM) was used with 935 N/s (210 lb/s) load rate, which is equal to 0.02 MPa/sec (2.92 psi/s) bond strength. The experimental bond strength is calculated using Equation (1).

$$\tau = \frac{P}{2 \times b \times d} \tag{1}$$

where $\tau$: bond strength; $P$: load at failure; $b$: the width of the cube cross-section; and, $d$: the depth of the cube cross-section.

The compressive strength of NSC was measured based on the ASTM C39 [58] for six concrete cylinders of 75 mm (3 in.) in diameter and 150 mm (6 in) in height and it was measured at 43 MPa (6.2 ksi) [39]. The failure modes of each bi-surface shear specimen are divided into three categories: (1) concrete crushing, (2) debonding at the interface, and (3) concurrent failure in bond and concrete, hereafter referred to as cohesive failure, adhesive failure, and mixed failure, respectively, as shown in Figure 4. Figure 5 shows the results of the bi-surface shear test. The average bond strengths are 2.8 MPa (406 psi) for specimens with a smooth interface and 6.3 MPa (914 psi) for the specimens with a rough interface [39].

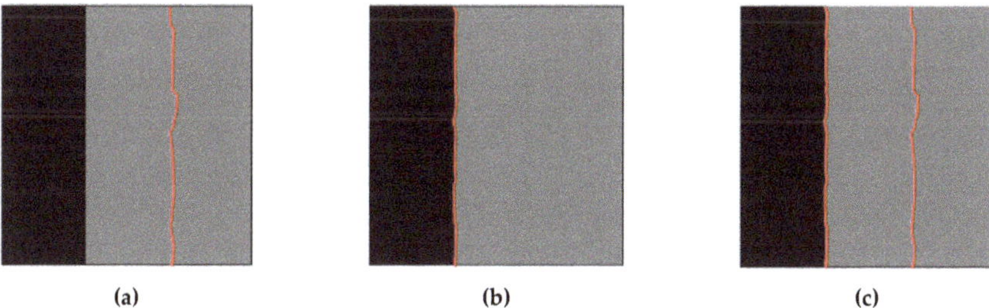

(a) (b) (c)

**Figure 4.** Failure modes for bi-surface shear specimens: (**a**) cohesive failure; (**b**) adhesive failure; and, (**c**) mixed failure.

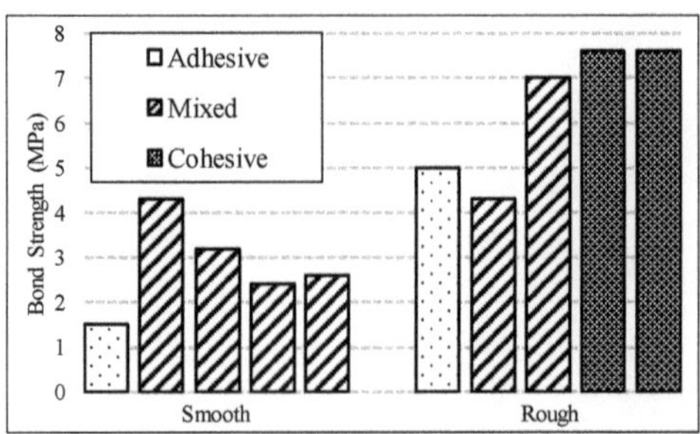


**Figure 5.** Bond strength results (1 MPa = 145 psi).

## 4. Modelling Assumptions

The ATENA software considers three-dimensional constitutive material models for simulating concrete behaviour with a combination of plasticity and fracture models [59]. Rankine tensile criterion is the base of the orthotropic smeared crack model that is used to model fracture. Menétrey and Willam (1995) [60] suggested a hardening/softening plasticity model that is used to simulate concrete crushing with a three-parameter failure surface [61] in ATENA. In this study, NSC compressive strength is used as the concrete class and all of the parameters are calculated by the software based on a fracture-plastic model. For UHPC, compression and tensile behaviours differ from NSC in tensile strength and fracture energy values and in the tensile and compression softening branch behaviour [61]. However, in ATENA software, user-defined material models with constitutive laws can be used, such as "CC3DNonLinCementitious2user". These constitutive laws are tensile and post-cracking softening behaviour, compression behaviour, the effect of lateral compression on tensile strength, the effect of lateral tensile strain on the compression capacity, post-cracking shear strength, and post-cracking shear stiffness [62]. Readers are advised to consult the ATENA manual for more detailed information about material constitutive laws.

Different modelling parameters are defined in this research. These parameters include tensile strength, modulus of elasticity, Poisson ratio, compressive strength, UHPC behaviour after elastic zoom, and UHPC compressive behaviour after elastic zoon. It should be noted that, in ATENA software, two parameters are defined as "characteristic length" and "localization onset", which are defined to reduce the mesh dependency. The characteristic length is the length of strain guage that is used in the experimental test or the element size, which is used to calibrate the material [62], and the localization onset is defined as strain at maximum stress. In this research, the characteristic length is chosen as the dimension of the mesh element.

After defining the nonlinear parameters of both NSC and UHPC in the fracture-plastic model, the modelling of the UHPC-to-concrete interface for the bond test specimens can be conducted. Generally, in numerical simulation, interfaces between two layers of concrete are modelled as a fixed contact for surface or using tie models that cause the overestimation of the interfacial bond strength that might result in eliminating the sliding between the substrate and repair material. Numerically, the interface can be idealized as a zero thickness volume model that can transfer the tangential shear and normal tractions. These transfer tractions are a function of tangential displacement ($\delta_t$) and normal displacement ($\delta_n$) [63].

In this study, the interface constitutive law is formulated based on the Moher-coulomb failure criterion (Figure 6), with a zero thickness volume and post-failure traction-separation laws in shear and tension [62,63].

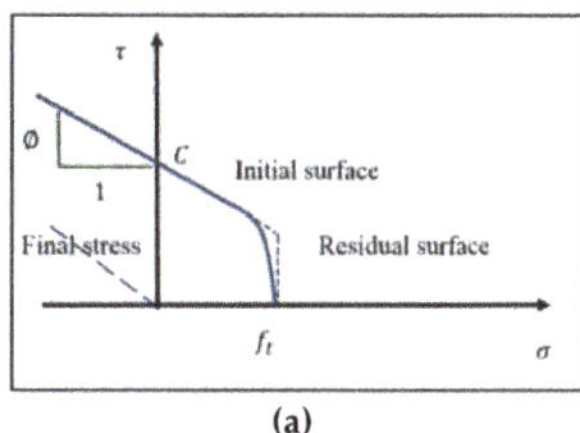


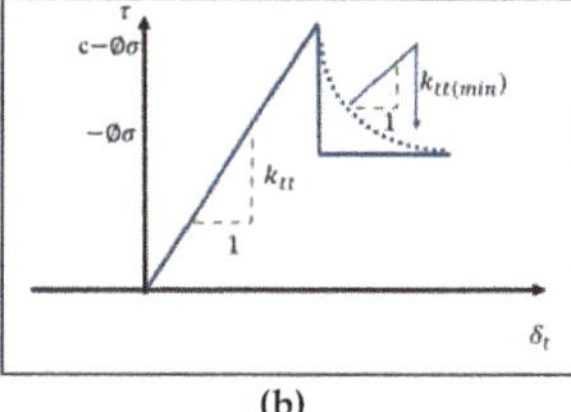


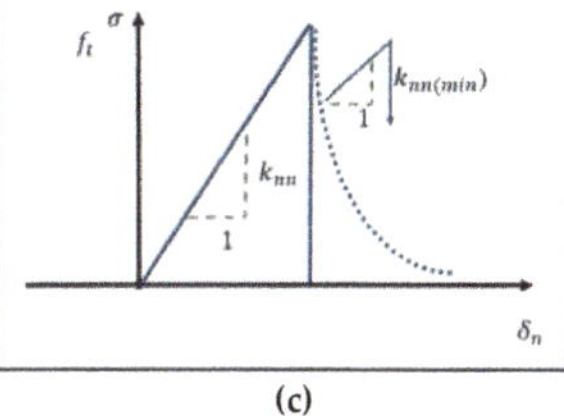


(a) (b) (c)

**Figure 6.** Interface modelling parameters: (**a**) failure surface for interface material based on Moher-column model; (**b**) the interface model behavior in shear; and, (**c**) the interface model behavior in tension [62].

The parameters that are shown in Figure 6 are defined, as follows: $f_t$ is the tensile strength of the bond from direct pull-off test; C is the bond cohesion measured from bond shear test; Ø is the coefficient of friction; σ is normal stress; τ is shear stress; $K_{nn}$ is tangent stiffness that correlates the normal displacement to the normal tractions (calibrated based on experimental results); $K_{tt}$ is tangent stiffness which correlates tangential displacement to the tangent tractions (calibrated based on experimental results); $K_{nn(min)}$ is minimum normal stiffness; and, $K_{tt(min)}$ is the minimum tangential stiffness.

## 5. Material Modelling Calibration

Two initial steps were conducted to simulate the interfacial bond strength between UHPC and NSC. In the first step, the fundamental characteristics of UHPC (compressive, tensile, and flexural behaviours) were experimentally tested. In the second step, the experimental results from the first step were used to calibrate and define the fracture-plastic model parameters for UHPC.

### *5.1. Compression Test*

A total of nine cylindrical specimens and nine cubical specimens were cast to evaluate UHPC compressive strength and modulus of elasticity. All of the cylindrical specimens were cast in moulds of 76-mm (3 in.) in diameter and 150 mm (6 in) in height and then tested based on ASTM C39 [58]. All of the cubical specimens were cast in moulds of 51 mm (2 in.) each side and then tested based on ASTM C109 [64,65]. Both sides of cylinders' surfaces were smooth and out of air bubbles by grinding to create a uniform pressure on the specimen's surface during testing [66]. The length and radius of the cylinders and cube side length were measured to calculate the true stress, true strain, and density. The load rate was chosen to be 1.0 MPa/s (150 psi/s) based on federal highway administration (FHWA) recommendations [3]. Figure 7 shows the test specimens before and after testing.

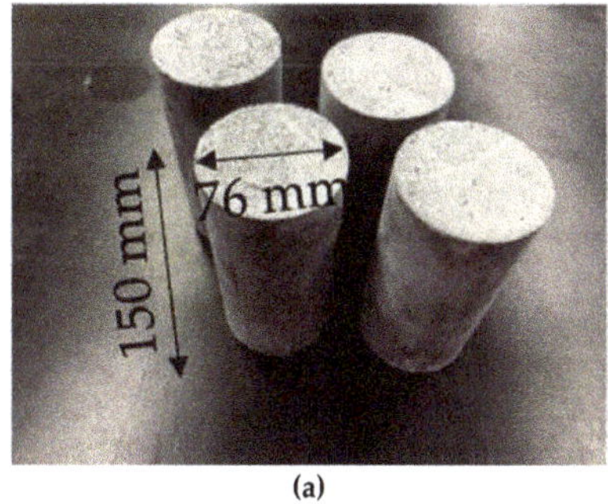


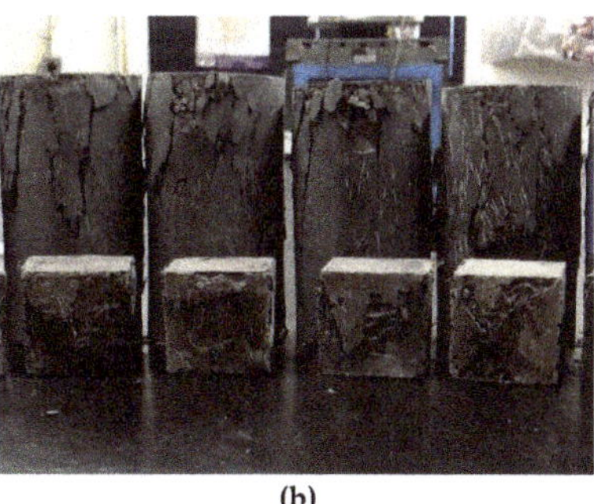

(a) (b)

**Figure 7.** Compressive test specimens: (**a**) cylindrical specimens before the test; (**b**) cylindrical and cubical specimens after the test.

Figure 8a,b, respectively, shows the stress-strain responses from the tested specimens for the untreated cylindrical and cubical specimens at the age of 28-day. UHPC shows a ductile behaviour for both tests due to the interaction between fibres and UHPC mix components. Table 4 shows the compressive strength and strain at peak stress for both cylindrical and cubical specimens.

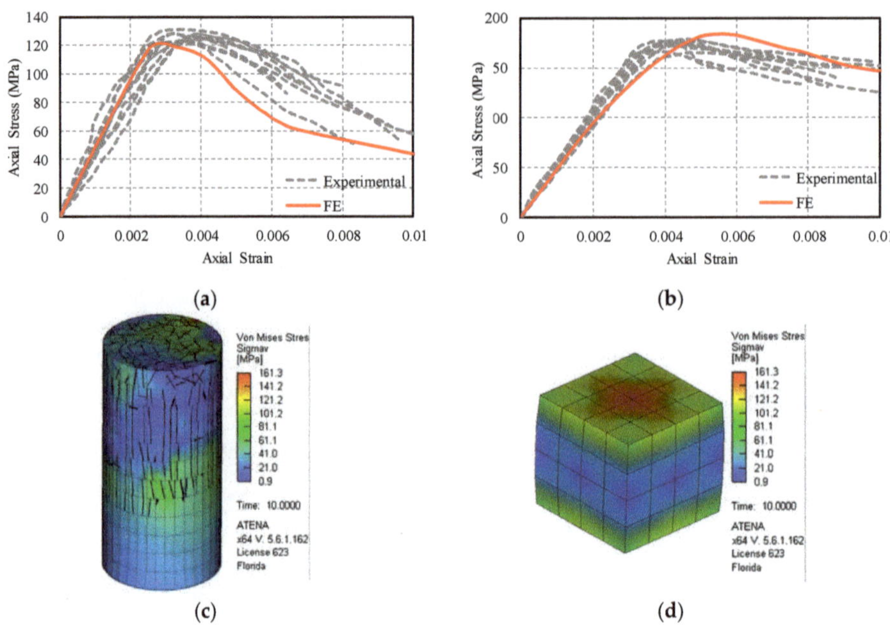


**Figure 8.** The stress-strain comparison between finite element and experimental tests: (**a**) cylindrical specimens; (**b**) cubical specimens and finite element stress distribution; (**c**) cylindrical specimen; and, (**d**) cubical specimen. (1 MPa= 145 psi).

**Table 4.** Experimental stress-strain results for the test specimens under compressive test (1 MPa = 145 psi).

| Specimen Type | Property | Average (MPa) | Standard Deviation (MPa) | Coefficient of Variance % |
|---|---|---|---|---|
| Cylinderical Specimen after 28 days | Compressive strength | 126 | 3 | 7.3 |
| | Strain at peak stress | 0.00353 | 0.000510 | 0.007 |
| | Secant elastic modulus, $E_0$ | 36,016 | 3153.5 | 8.8 |
| | Tangent elastic modulus, $E$ | 52,081 | 4136.9 | 7.9 |
| Cubical Specimen after 28 days | Compressive strength | 173 | 5.0 | 12.45 |
| | Strain at peak stress | 0.00408 | 0.000258 | 0.002 |
| | Secant elastic modulus, $E_0$ | 42,560 | 4095.0 | 9.6 |
| | Tangent elastic modulus, $E$ | 61,191 | 3909.3 | 8.1 |

Furthermore, the modulus of elasticity for UHPC was calculated based on two methods. The first method is based on scant modulus ($E_0$), which is calculated based on the peak strength (maximum) and the corresponding strain. The second is based on the tangent modulus of elasticity ($E_0$), which is calculated based on the stress and corresponding strain between 10% and 30% of the maximum compressive strength [3]. The cubical specimens show a higher modulus of elasticity and compressive strength when compared to the cylindrical specimen due to the shorter aspect ratio and larger lateral confinement provided by the machine plates, the same trend can be noticed in NSC [3]. Table 4 shows the results for both the secant and tangent modulus of elasticity for both cylindrical and cubical specimens.

The results from the experimental tests are used to calibrate the parameters that are needed for modelling UHPC. Table 5 lists the calibrated parameters for the fracture-plastic model. Figure 8 shows comparison between the experimental and numerical results. It should be noted that, for all numerical models, the mesh size of 13 mm (0.5 in.) was used. The mesh elements were hexahedra for cubical specimens, however, for cylindrical specimens, prism mesh element was used, except for the top and bottom portion, where quadrilateral mesh elements were required for geometrical requirements. The numerical models were run in displacement control steps of 0.127 mm (0.005 in.) with a total number of running steps equal to 100 steps. The Newton-Raphson method [62] was used as the solution method by setting the displacement error and residual error equal to 1% and number of iteration limit to 30. No convergence issues were observed for any model. All of the supports and loading plates were attached to the specimens with a fix-contact surface; in addition, the vertical and horizontal displacements of the supports were restricted.

**Table 5.** Calibrated UHPC plastic-fracture model parameters based on experimental test results in ATENA software. (1 MPa = 145 psi, 1 in. = 25.4 mm).

| | **Elastic Zone** | | |
|---|---|---|---|
| Modulus of elasticity | 52,081 MPa | Poisson's ratio | 0.2 |
| Compressive strength | −126 MPa | Tensile strength | 5.8 MPa |
| | **Plastic Zone** | | |
| Compression characteristic size | 1.27 mm | Tension characteristic size | 1.27 mm |
| Compression localization onset | −0.001 | Tension localization onset | 0.002 |
| **Compressive behavior** | | **Tensile behavior** | |
| Yield strain | Compressive stress | Crack strain | Tensile stress |
| 0 | −126 MPa | 0 | 1.1 MPa |
| −0.001 | −126 MPa | 0.002 | 6 MPa |
| −0.01 | −38 MPa | 0.1 | 1 MPa |

## *5.2. Direct Tension and Flexural Tests*

The high compressive strength and high tensile strength [3,67,68], shorter reinforcement development length [69], and shorter lap splice length [70] are the main advantages of UHPC when compared to NSC. The tensile behaviour of UHPC before and after cracking was investigated under direct tension and flexural tests. For the direct tension test, six dog bone shape briquettes were tested according to AASHTO T132 [71]. The dimension of test specimens is 76.2 mm (3 in.) in length, 25.4 mm × 25.4 mm (1 in. × 1 in.) in cross-section at the middle, and 25.4 mm (1 in.) in thickness. The loading rate that was used in this test was 0.0254 mm/sec (0.001 in/sec.) according to [3]. Figure 9a shows one of the test specimens after testing. The testing of briquette specimens shows that UHPC behaved linear elastic before the first crack and then stress hardening occurred because the post-cracking resistance is higher than the resistance of the mixture. In this case, when the initial crack happened, the fibres would carry the tensile stress, which is known as the "bridge effect". The average tensile cracking strength that was measured for UHPC in this series of tests was 3.8 kN (0.85 kips), with standard deviation and coefficient of variance of 0.2 kN (0.045 kips) and 0.06, respectively, with post cracking peak strength of 4.2 kN (0.94 kips) with a standard deviation and coefficient of variance of 0.2 kN (0.045 kips) and 0.05, respectively (Figure 9b). In Figure 9b, the results were only plotted with an offset of 1 mm (0.0394 in.) to better represent the results; however, the results, in reality, are not with an offset.

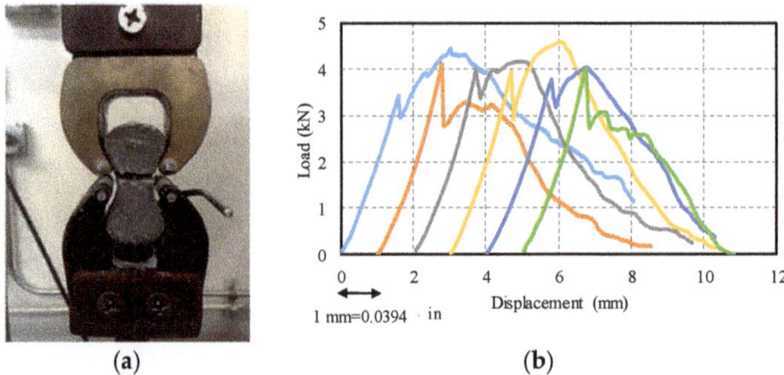


**(a)** **(b)**

**Figure 9.** Direct tension test: (**a**) AASHTO T132 test; (**b**) load-displacement for briquettes. (1 kN = 0.22 kips).

It should be noted that this test is not considered directly in the simulation because of the dimension of the briquette and effect of boundary condition, and it was just used as the preliminary data for simulating the flexural test.

Although the uniaxial test can be considered to be the most realistic method for determining the tensile post-cracking behaviour, it has some difficulties, such as the boundary condition of the testing machine, the complicacy in the test setup and data collection [4], difficulties in obtaining evenly distributed stresses through the section, and controlling the stable load versus displacement/crack opening [5]. In this series of tests, the flexural test was conducted according to ASTM C1018 [72] on three small scale beams with a cross-section area of 153 mm × 153 mm (6 in. × 6 in.), a total length of 612 mm (24 in.), and an effective span length of 459 mm (18 in.). The small scale beams were supported over roller supports and they were tested using a three-point load test setup while using a hydraulic jack with a loading rate of 110 N/s (24.7 (lb/s)), as shown in Figure 10a. From the experimental results, the UHPC beam specimens show linear behaviour before the occurrence of the first crack and then the beam deformation was localized in the first crack with a nonlinear increase in deformation until failure, as shown in Figure 10b. The steel fibres could resist the tensile forces from the external load after the growth of the first crack that kept the beam specimen intact. The results from flexural tests were used to define the fracture-plastic model parameters of UHPC in tension, as shown in Table 5. A comparison between the experimental results and finite element results shows good correlation in force-displacement curves, as shown in Figure 10c. Figure 10d shows the numerical stress distribution showing the first crack propagation, which is comparable to the mode of failure in Figure 10b.

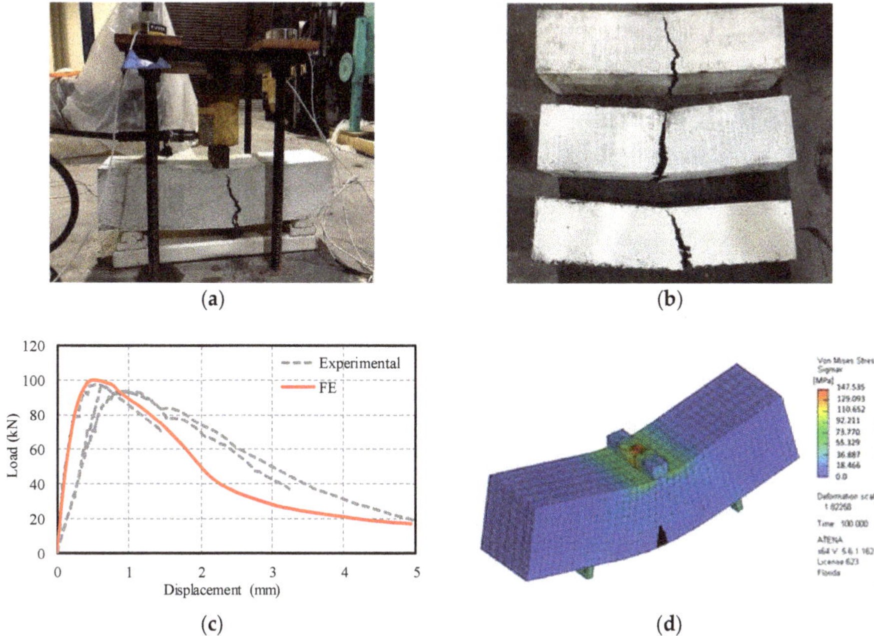


**Figure 10.** Flexural test: (**a**) test setup; (**b**) specimens after testing; (**c**) comparison between experimental and finite element results; and, (**d**) stress distribution in finite element model (1 kN = 0.22 kips, 1 in = 25.4 mm).

## 6. Results of Bond Strength Modelling

In the last step, the interface model was used to simulate the interfacial bond strength between UHPC and NSC. Table 6 shows the parameters that were related to the bond model for smooth and rough surfaces. These parameters are calculated by calibrating the experimental results of bi-surface shear tests and from literature. For this series of tests, the average surface roughness of sandblasted and smooth surfaces was measured as 1.2–2.2 mm (0.05-0.08 in.) and 0.17–0.28 mm (0.0067–0.110), respectively [39].

**Table 6.** The parameters of the interface model for different surface preparation calibrated from bi-surface shear test. (1 MN/m$^3$ = 3.68 lb/in$^3$).

| | $C$ MPa | $f_t$ MPa | Friction Coefficient | $K_{tt}$ (MN/m$^3$) | $K_{nn}$ (MN/m$^3$) | $K_{tt(min)}$ (MN/m$^3$) | $K_{nn(min)}$ (MN/m$^3$) |
|---|---|---|---|---|---|---|---|
| Sand blasted surface | 6.28 | 2 | 1 | $2.2 \times 10^8$ | $2.2 \times 10^8$ | $2.2 \times 10^6$ | $2.2 \times 10^6$ |
| Smooth surface | 2.8 | 0.5 | 0.5 | $10^6$ | $10^6$ | $10^4$ | $10^4$ |

The bond tension strength ($f_t$) was assumed from the direct pull-off test result from literature [37], bond cohesion (C) was calculated from a test with pure shear stresses and no normal stresses condition; however, in this model, the value, as calculated from the bi-surface shear test, was input directly into software as a reasonable approximation. The recommended value from AASHTO-LRFD was used for the coefficient of friction (μ) [73]. Normal stiffness $K_{nn}$ and tangential stiffness $K_{tt}$ are calibrated based on the experimental test results. Additionally, it should be noted that the minimum normal stiffness $K_{nn(min)}$ and minimum tangential stiffness $K_{tt(min)}$ that represent the interface stiffness after failure are chosen as 0.1% of the initial values to eliminate the numerical errors [63].

The results of the model with experimental and the model with a fixed contact for surface between the two layers of UHPC and NSC are compared in Figure 11a to highlight the importance of simulating the interface. Figure 11c shows the stress distribution from the finite element and its corresponding actual mode of failure from experimental testing (Figure 11b).

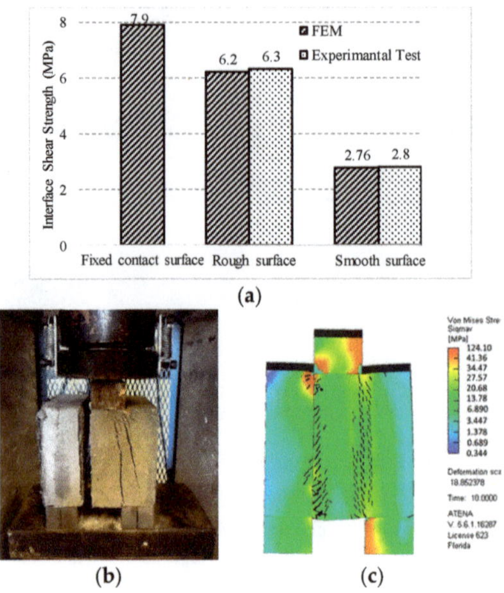


**Figure 11.** Bond strength numerical results: (**a**) Comparison between experimental and numerical model results of bond strength; (**b**) specimen failure; and, (**c**) stress distribution in the finite element model. (1MPa = 145 psi).

## 7. Summary and Conclusions

In this paper, the interfacial bond strength between normal strength concrete and ultra-high performance concrete with smooth and rough interface surfaces was experimentally and numerically investigated. First, the interfacial bond strength was evaluated experimentally by testing 10 cubical specimens using a bi-surface shear test setup. Second, 18 different test specimens, including cylinders, cubes, briquettes, and flexural beams, were cast and tested under compression, direct tension, and flexural tests to calibrate the UHPC material model in ATENA FE software. In the end, the results from both experimental and numerical results were used to calibrate the parameters of a zero thickness volume interface model in FE software. The numerical results were compared with the experimental results and the conventional fixed contact model approach. The following conclusions and observations can be drawn based on the conducted research:

- In ambient conditions and after 28-day, the compressive strength and tensile strength of UHPC reached 126 MPa (18 ksi) and 6.5 Mpa (0.95 kips), respectively, which nominates UHPC as an efficient repair material for damaged structures.
- The bi-surface shear test results showed an average bond strength of 2.9 MPa (420 psi) for specimens with smooth interface surfaces, whereas this value increased by 134% for specimens with rough interface surfaces by sandblasting with an average surface roughness between 1.2–2.2 mm (0.05–0.08 in.).
- The plastic-fracture model could predict the tensile and compressive behaviours of UHPC with acceptable accuracy, which makes it a practical tool for modelling structures, including UHPC.

- For modelling the interface between UHPC and normal strength concrete, the result from the bi-surface test could be directly used as the interface cohesion parameter; however, the only calibrated parameters were the normal stiffness $K_{nn}$ and tangential stiffness $K_{tt}$.
- Modelling of the interface using a fixed contact for the surface model cannot distinguish the effect of surface preparation on bond strength between normal strength concrete and UHPC, which might lead to erroneous numerical results.
- By comparing the fixed contact model and the zero thickness volume model with experimental results, the error of simulation for smooth and rough surface dropped from 182% and 25% to around 1%, respectively.

**Author Contributions:** Conceptualization, A.V.; methodology, A.V.; software, A.V.; validation, A.V., and A.J.J.; investigation, A.V. and A.J.J.; data curation, A.V. and I.M.M..; Visualization, A.V. and I.M.M.; writing—original draft preparation, A.V. and A.J.J.; writing—review and editing, I.M.M.; supervision, A.A.; project administration, A.A.; funding acquisition, A.A. All authors have read and agreed to the published version of the manuscript.

**Funding:** This project was supported by the Accelerated Bridge Construction University Transportation Center (ABC-UTC at www.abc-utc.fiu.edu) at Florida International University (FIU).

**Acknowledgments:** The authors would like to acknowledge the ABC-UTC support.

**Conflicts of Interest:** The authors declare no conflict of interest.

## References

1. Azmee, N.M.; Shafiq, N. Ultra-high performance concrete: From fundamental to applications. *Case Stud. Constr. Mater.* **2018**, *9*, e00109. [CrossRef]
2. Valikhani, A.; Azizinamini, A. Experimental Investigation of High-Performing Protective Shell Used for Retrofitting Bridge Elements. Presented at Transportation research board 97th Annual meeting, Washington, DC, USA, 7–11 January 2018.
3. Graybeal, B.A. *Material Property Characterization of Ultra-High Performance Concrete*; Office of Infrastructure Research and Development: McLean, VA, USA, 2006.
4. Russell, H.G.; Graybeal, B.A.; Russell, H.G. *Ultra-High Performance Concrete: AS-of-the-Art Report for the Bridge Community*; Office of Infrastructure Research and Development: McLean, VA, USA, 2013.
5. De Larrard, F.; Sedran, T. Optimization of ultra-high-performance concrete by the use of a packing model. *Cem. Concr. Res.* **1994**, *24*, 997–1009. [CrossRef]
6. Graybeal, B. *Ultra-High Performance Concrete*; FHWA-HRT-11-038; U.S. Department of Transpotation Federal Highway Administration: McLean, VA, USA, 2011.
7. Richard, P.; Cheyrezy, M. Composition of reactive powder concretes. *Cem. Concr. Res.* **1995**, *25*, 1501–1511. [CrossRef]
8. Yu, R.; Przemek, S.; Brouwers, H.J.H. Mix design and properties assessment of ultra-high performance fibre reinforced concrete (UHPFRC). *Cem. Concr. Res.* **2014**, *56*, 29–39. [CrossRef]
9. Lohaus, L.; Peter, R. Robustness of UHPC-A new approach for mixture proportioning. In Proceedings of the 2nd International Symposium on Ultra High Performance Concrete, Kassel, Germany, 5–7 March 2008; pp. 113–120.
10. Fennis, S.A.; Walraven, J.C.; Den Uijl, J.A. The use of particle packing models to design ecological concrete. *Heron* **2009**, *54*, 185–204.
11. Geisenhansluke, C. Methods for modeling and calculation of high density packing for cement and filler in UHPC. In Proceedings of the International Symposium on Ultra-High Performance Concrete, Kassel, Germany, 13–15 September 2004; pp. 303–312.
12. Schmidt, M.; Ekkehard, F. Ultra-high-performance concrete: Research, development and application in Europe. *ACI Spec. Publ.* **2005**, *228*, 51–78.
13. Ghafari, E.; Hugo, C.; Eduardo, J. Critical review on eco-efficient ultra-high performance concrete enhanced with nano-materials. *Constr. Build. Mater.* **2015**, *101*, 201–208. [CrossRef]
14. Vernet, C.P. Ultra-durable concretes: Structure at the micro-and nanoscale. *MRS Bull.* **2004**, *29*, 324–327. [CrossRef]

15. Wille, K.; Antoine, N.E.; Parra-Montesinos, G.J. Ultra-High Performance Concrete with Compressive Strength Exceeding 150 MPa (22 ksi): A Simpler Way. *ACI Mater. J.* **2011**, *108*, 1.
16. Shi, C.; Wu, Z.; Xiao, J.; Wang, D.; Huang, Z.; Fang, Z. A review on ultra-high performance concrete: Part I. Raw materials and mixture design. *Constr. Build. Mater.* **2015**, *101*, 741–751. [CrossRef]
17. Abbas, S.; Soliman, A.M.; Nehdi, M.L. Exploring mechanical and durability properties of ultra-high performance concrete incorporating various steel fiber lengths and dosages. *Constr. Build. Mater.* **2015**, *75*, 429–441. [CrossRef]
18. Schmidt, M.; Ekkehard, F.; Geisenhanslüke, C. Ultra high performance concrete (UHPC). In Proceedings of the International Symposium on Ultra High Performance Concrete, Kassel, Germany, 13–15 September 2004.
19. Talebinejad, I.; Seyed, A.B.; Amirhossein, I.; Mohammad, S. Optimizing mix proportions of normal weight reactive powder concrete with strengths of 200–350 MPa. In Proceedings of the International Symposium on UHPC, Kassel, Germany, 13–15 September 2004; pp. 133–141.
20. Ghoddousi, P.; Armin, M.B.; Esmail, S.; Mohammad, A. Prediction of Plastic Shrinkage Cracking of Self-Compacting Concrete. *Adv. Civ. Eng.* **2019**, *2019*, 1–7. [CrossRef]
21. Yazıcı, H. The effect of curing conditions on compressive strength of ultra-high strength concrete with high volume mineral admixtures. *Build. Environ.* **2007**, *42*, 2083–2089. [CrossRef]
22. Abbas, S.M.L.N.; Saleem, M.A. Ultra-high performance concrete: Mechanical performance, durability, sustainability and implementation challenges. *Int. J. Concr. Struct. Mater.* **2016**, *10*, 271. [CrossRef]
23. Nicolaides, D.; Antonios, K.; Michael, F.P.; Marios, S. Mix design, mechanical properties and impact resistance of UHPFRCCs. In Proceedings of the 3rd International Conference on Concrete Repair, Rehabilitation and Retrofitting, ICCRRR-3, Alexander, MG, USA, 3–5 September 2012; pp. 181–186.
24. Gao, R.; Liu, Z.M.; Zhang, L.Q.; Stroeven, P. Static properties of plain reactive powder concrete beams. *Key Eng. Mater.* **2006**, *302*, 521–527. [CrossRef]
25. Wen-yu, J.; Ming-zhe, A.; Gui-ping, Y.; Jun-min, W. Study on reactive powder concrete used in the sidewalk system of the Qinghai-Tibet railway bridge. In Proceedings of the International Workshop on Sustainable Development and Concrete Technology, Beijing, China, 20–21 May 2004; pp. 333–338.
26. Rougeau, P.; Béatrice, B. Ultra high performance concrete with ultrafine particles other than silica fume. In Proceedings of the International Symposium on Ultra High Performance Concrete, Kassel, Germany, 13–15 September 2004; Volume 932, pp. 213–225.
27. Tue, N.V.; Ma, J.; Marko, O. Influence of addition method of superplasticizer on the properties of fresh UHPC. In Proceedings of the 2nd International Symposium on Ultra-High Performance Concrete, Kassel, Germany, 5–7 March 2008; pp. 93–100.
28. Schmidt, M.; Ekkehard, F.; Christoph, G.; Susanne, F.; Siemon, P. Ultra-High Performance Concrete and Nanotechnology in Construction. In Proceedings of the Hipermat 2012 3rd International Symposium on UHPC and Nanotechnology for High Performance Construction Materials, Kassel, Germany, 7–9 March 2012.
29. Droll, K. Influence of additions on ultra high performance concretes–grain size optimisation. In Proceedings of the International Symposium on Ultra-High Performance Concrete, Kassel, Germany, 13–15 September 2004; Volume 915, pp. 285–301.
30. Xing, F.; Li, D.H.; Zheng, L.C.; Liang, P.D. Study on preparation technique for low-cost green reactive powder concrete. *Key Eng. Mater.* **2006**, *302*, 405–410. [CrossRef]
31. Chan, Y.-W.; Chu, S.-H. Effect of silica fume on steel fiber bond characteristics in reactive powder concrete. *Cem. Concr. Res.* **2004**, *34*, 1167–1172. [CrossRef]
32. Matte, V.; Moranville, M. Durability of reactive powder composites influence of silica fume on the leaching properties of very low water/binder pastes. *Cem. Concr. Compos.* **1999**, *21*, 1–9. [CrossRef]
33. Bayard, O.; Plé, O. Fracture mechanics of reactive powder concrete: Material modelling and experimental investigations. *Eng. Fract. Mech.* **2003**, *70*, 839–851. [CrossRef]
34. Momayez, A.; Ehsani, M.R.; Ramezanianpour, A.A.; Rajaie, H. Comparison of methods for evaluating bond strength between concrete substrate and repair materials. *Cem. Concr. Res.* **2005**, *35*, 748–757. [CrossRef]
35. Jaberi, J.; Alireza, V.; Islam, M.; Atorod, A. Service Life Design of Deck Closure Joints in ABC Bridges: Guidelines and Practical Implementation. *Front. Built Environ.* **2019**, *5*, 152. [CrossRef]
36. Graybeal, B. *Design and Construction of Field-Cast UHPC Connections*; Federal Highway Administration: Washington, DC, USA, 2014.

37. Haber, Z.B.; Jose, F.M.; Igor, D.L.V.; Benjamin, A.G. Bond characterization of UHPC overlays for concrete bridge decks: Laboratory and field testing. *Constr. Build. Mater.* **2018**, *190*, 1056–1068. [CrossRef]
38. Khayat, K.H.; Mahdi, V. *Design of Ultra High Performance Concrete as an Overlay in Pavements and Bridge Decks*; Missouri University of Science and Technology, Center for Transportation Infrastructure and Safety: Rolla, MO, USA, 2014.
39. Valikhani, A.; Azadeh, J.J.; Islam, M.M.; Atorod, A. Experimental evaluation of concrete-to-UHPC bond strength with correlation to surface roughness for repair application. *Constr. Build. Mater.* **2020**, *238*, 117753. [CrossRef]
40. Tayeh, B.A.; Abu Bakar, B.H.; Megat Johari, M.A.; Yen, L.O. Mechanical and permeability properties of the interface between normal concrete substrate and ultra high performance fiber concrete overlay. *Constr. Build. Mater.* **2012**, *36*, 538–548. [CrossRef]
41. Harris, D.K.; Jayeeta, S.; Theresa, M.A. Characterization of interface bond of ultra-high-performance concrete bridge deck overlays. *Transp. Res. Rec.* **2011**, *2240*, 40–49. [CrossRef]
42. Banta, T.E. Horizontal Shear Transfer between Ultra High Performance Concrete and Lightweight Concrete. Ph.D. Thesis, Virginia Tech, Blacksburg, VA, USA, 2005.
43. Crane, C.K. Shear and Shear Friction of Ultra-High Performance Concrete Bridge Girders. Ph.D. Thesis, Georgia Institute of Technology, Atlanta, GA, USA, 2010.
44. Aaleti, S.; Sri, S. Quantifying bonding characteristics between UHPC and normal-strength concrete for bridge deck application. *J. Bridge Eng.* **2019**, *24*, 04019041. [CrossRef]
45. Harris, D.K.; Miguel, A.C.M.; Amir, G.; Theresa, M.A.; Sarah, V.R. The challenges related to interface bond characterization of ultra-high-performance concrete with implications for bridge rehabilitation practices. *Adv. Civ. Eng. Mater.* **2015**, *4*, 75–101. [CrossRef]
46. Azizinamini, A.; Sheharyar, R.; Sadeghnejad, A. Enhancing resiliency and delivery of bridge elements using ultra-high performance concrete as formwork. *Transp. Res. Rec.* **2019**, *2673*, 443–453. [CrossRef]
47. Valikhani, A.; Azadeh, J.J.; Azizinamini, A. Retrofitting Damaged Bridge Elements Using Thin Ultra High Performance Shell Elements. Presented at Transportation Research Board 96th Annual Meeting, Washington, DC, USA, 8–12 January 2017.
48. ASTM International. *C1437-15 Standard Test Method for Flow of Hydraulic Cement Mortar*; ASTM International: West Conshohocken, PA, USA, 2015.
49. Cortes, D.D.; Kim, H.-K.; Palomino, A.M.; Santamarina, J.C. Rheological and mechanical properties of mortars prepared with natural and manufactured sands. *Cem. Concr. Res.* **2008**, *38*, 1142–1147. [CrossRef]
50. ASTM International. *C1856/C1856M-17 Standard Practice for Fabricating and Testing Specimens of Ultra-High Performance Concrete*; ASTM International: West Conshohocken, PA, USA, 2017.
51. Habel, K.; Charron, J.-P.; Shadi, B.; Douglas, R.H.; Paul, G.; Bruno, M. Ultra-high performance fibre reinforced concrete mix design in central Canada. *Can. J. Civ. Eng.* **2008**, *35*, 217–224. [CrossRef]
52. Park, J.-S.; Young, J.K.; Jeong, R.C.; Jeon, S.-J. Early-age strength of ultra-high performance concrete in various curing conditions. *Materials* **2015**, *8*, 5537–5553. [CrossRef] [PubMed]
53. Jahromi, A.J.; Alireza, V.; Azizinamini, A. *Toward Development of Best Practices for Closure Joints in ABC Projects*; paper numbers 18-05330, Association Number: 01657436; Transportation Research Board 97th Annual Meeting: Washington, DC, USA, January 2018.
54. Jahromi, A.J.; Morgan, D.; Alireza, V.; Azizinamini, A. *Assessing Structural Integrity of Closure Pours in ABC Projects*; paper numbers 18-05307, Association Number: 01657435; Transportation Research Board 97th Annual Meeting: Washington, DC, USA, January 2018.
55. Zhang, Y.; Shukun, C.; Zhu, Y.; Liang, F.; Shao, X. Flexural responses of steel-UHPC composite beams under hogging moment. *Eng. Struct.* **2020**, *206*, 110134. [CrossRef]
56. Momayez, A.; Ramezanianpour, A.A.; Rajaie, H.; Ehsani, M.R. Bi-surface shear test for evaluating bond between existing and new concrete. *Mater. J.* **2004**, *101*, 99–106.
57. Lee, H.S.; Jang, H.-O.; Cho, K.-H. Evaluation of bonding shear performance of ultra-high-performance concrete with increase in delay in formation of cold joints. *Materials* **2016**, *9*, 362. [CrossRef] [PubMed]
58. ASTM International. *C39/C39M-18 Standard Test Method for Compressive Strength of Cylindrical Concrete Specimens*; ASTM International: West Conshohocken, PA, USA, 2018.
59. Verre, S.; Luciano, O. Numerical modelling approaches of FRCMs/SRG confined masonry columns. *Front. Built Environ.* **2019**, *5*, 143.

60. Menetrey, P.; Willam, K.J. Triaxial failure criterion for concrete and its generalization. *Struct. J.* **1995**, *92*, 311–318.
61. Cervenka, V.; Jan, C.; Radomir, P. ATENA—A tool for engineering analysis of fracture in concrete. *Sadhana* **2002**, *27*, 485–492. [CrossRef]
62. Cervenka, V.; Libor, J.; Jan, C. ATENA program documentation part 1 theory. *Cervenka Consult. Prague* **2007**, *231*, 43–65.
63. Sajdlová, T.; Kabele, P. *Finite Element Analyasis of Test Configuration Identification of Interface Parameters in Layered FRCC Systems*; High performance Fiber Reinforced Cement Composits (HPFRCC7): Stuttgart, Germany, 2015.
64. ASTM International. *C109/C109M-13 Standard Test Method for Compressive Strength of Hydraulic Cement Mortars (Using 2-in. or [50-mm] Cube Specimens)*; ASTM International: West Conshohocken, PA, USA, 2013.
65. Li, C.; Zheng, F.; Ke, L.; Pan, R.; Nie, J. Experimental Study on Shear Performance of Cast-In-Place Ultra-High Performance Concrete Structures. *Materials* **2019**, *12*, 3254. [CrossRef]
66. Aboukifa, M.; Mohamed, A.M.; Ahmad, M.I.; Negar, N. *Durable UHPC Columns with High-Strength Steel*; Accelerated Bridge Construction University Transportation Center (ABC-UTC), University of Nevada Reno: Reno, NV, USA, 2019.
67. Khosravani, M.R.; Peter, W.; Dirk, F.; Kerstin, W. Dynamic fracture investigations of ultra-high performance concrete by spalling tests. *Eng. Struct.* **2019**, *201*, 109844. [CrossRef]
68. Fan, L.; Weina, M.; Le, T.; Kamal, H.K. Effects of lightweight sand and steel fiber contents on the corrosion performance of steel rebar embedded in UHPC. *Constr. Build. Mater.* **2020**, *238*, 117709. [CrossRef]
69. Tazarv, M.; Saiid, M. Design and construction of UHPC-filled duct connections for precast bridge columns in high seismic zones. *Struct. Infrastruct. Eng.* **2017**, *13*, 743–753. [CrossRef]
70. Haber, Z.B.; Benjamin, A.G. Lap-spliced rebar connections with UHPC closures. *J. Bridge Eng.* **2018**, *23*, 04018028. [CrossRef]
71. AASHTO T132. *Standard Method of Test for Tensile Strength of Hydraulic Cement Mortars*; America Association of State Highway and Transportation Officials (AASHTO): Washington, DC, USA, 2013.
72. ASTM International. *Standard Test Method for Flexural Toughness and First-Crack Strength of Fiber-Reinforced Concrete (Using Beam with Third-Point Loading), ASTM C1018*; ASTM International: West Conshohocken, PA, USA, 1997.
73. AASHTO. *LRFD Bridge Design Specifications, Part I: Sections 1–6*, 8th ed.; American Association of State Highway and Transportation Officials: Washington, DC, USA, 2017.

*Article*

# Effects of MgO Expansive Agent and Steel Fiber on Crack Resistance of a Bridge Deck

**Feifei Jiang [1,2,*], Min Deng [1], Liwu Mo [1] and Wenqing Wu [3]**

[1] College of Materials Science and Engineering, Nanjing Tech University, Nanjing 211800, China; dengmin@njtech.edu.cn (M.D.); andymoliwu@njtech.edu.cn (L.M.)
[2] College of Naval Architecture Civil Engineering, Jiangsu University of Science and Technology, Zhangjiagang Campus, Suzhou 215600, China
[3] School of Transportation, Southeast University, Nanjing 210089, China; wuwenqing@seu.edu.cn
* Correspondence: 999620140019@just.edu.cn

Received: 31 May 2020; Accepted: 8 July 2020; Published: 9 July 2020

**Abstract:** To prevent cracks caused by shrinkage of the deck of the Xiaoqing River Bridge, MgO concrete (MC) and steel fiber reinforced MgO concrete (SMC) were used. The deformation and strength of the deck were measured in the field, the resistance to chloride penetration of the concrete was measured in the laboratory, and the pore structure of the concrete was analyzed by a mercury intrusion porosimeter (MIP). The results showed that the expansion caused by the hydration of MgO could suppress the shrinkage of the bridge deck, and the deformation of the deck changed from $-88.3 \times 10^{-6}$ to $24.9 \times 10^{-6}$, effectively preventing shrinkage cracks. At the same time, due to the restriction of the expansion of MgO by the steel bars, the expansion of the bridge deck in the later stage gradually stabilized, and no harmful expansion was produced. When steel fiber and MgO were used at the same time, the three-dimensional distribution of steel fiber further limited the expansion of MgO. The hydration expansion of MgO in confined space reduced the porosity of concrete, optimized the pore structure, and improved the strength and durability of concrete. The research on the performance of concrete in the in-situ test section showed that MgO and steel fiber were safe for the bridge deck, which not only solved the problem of shrinkage cracking of the bridge deck but also further improved the mechanical properties of the bridge deck.

**Keywords:** MgO expansive agent; steel fiber reinforced concrete; bridge deck; laminated slab; shrinkage crack

## 1. Introduction

Cracks are the direct cause of concrete penetration, which seriously affect the durability of concrete. Due to shrinkage and temperature changes, concrete volume will be deformed, which is the main reason for early cracks. From the perspective of structural calculation, deformation under free conditions will not produce tensile stress, and the concrete will not crack. However, once the deformation is restrained, tensile stress will be generated, and since the tensile strength of concrete is very low, it is easy to crack. In thin-walled structures such as bridge decks, due to the large surface area, the volume deformation caused by shrinkage and temperature is greater, and the cracking caused by the constraints of steel bars and bearings is also more serious. [1,2]. Early deformation cracks do not necessarily affect the safety of the structure immediately, but will seriously affect the durability of the bridge, so it is necessary to pay attention to the damage caused by cracks [3–5].

The cracks of cement-based materials are closely related to the non-load deformation of the materials. American academic Bazant pointed out that the early deformation of concrete can be divided into drying shrinkage, chemical shrinkage, and thermal shrinkage [6]. Early non-load cracks easily make concrete more vulnerable to the invasion of potentially harmful substances, and this kind of crack is more serious in bridge engineering for three main reasons. The first reason is that the concrete for bridges has high strength and low water-to-binder ratio (w/b), so it shrinks greatly [7–9]. It is reported that chemical shrinkage occurs in concrete when the w/b ratio is less than about 0.42, and increases with the decrease of w/b [10]. In addition, the bridge deck with a thin-walled structure has a large contact area with air, resulting in large drying shrinkage. Furthermore, due to direct sunlight, the moisture loss of the concrete of the bridge deck is quicker, which also aggravates the drying shrinkage. The second reason is that, compared to the concrete of houses, there are more steel bars in bridges. These steel bars severely restrict shrinkage and produce greater tensile stress. The third reason is large temperature cracks. Bridges are in field environments without any other shelter from buildings. The temperature affected by solar radiation is high in the daytime, reaching 40–50 °C, but at night, due to the small thickness of the bridge deck, the heat will be quickly distributed to the air, and the temperature will be rapidly reduced to 20 °C. The strength of fresh concrete is low, and temperature cracks easily occur with such large temperature differences.

The bridge deck is an important part of bridge. It not only needs to have high strength to bear automobile loads but also needs to have good crack resistance and waterproof performance to ensure that the rainwater will not penetrate into concrete and lead to corrosion of steel bars [11–13]. For example, only two months after the Shantou Bay Bridge was put into use in China, a large number of cracks appeared on the bridge deck. What is more, two years later, the cracking area reached 75% of the total area of the deck, causing great economic losses. This kind of accident reminds researchers of the need to develop new high-performance concrete to reduce shrinkage and improve the crack resistance of bridge decks [14–17].

The Changshen Expressway is one of the "three vertical" lines of China's highway network, connecting the Northeast, the Yellow River Delta, the Yangtze River Delta, and the Pearl River Delta and is a major north–south corridor through eastern China. As shown in Figure 1, the Xiaoqing River Bridge is a steel–concrete composite beam bridge on the Changshen Expressway, with a total length of 260 m (70 m + 120 m + 70 m). As shown in Figure 2, the bridge deck is a laminated slab structure. The lower part is a precast concrete slab with a thickness of 80 mm, and the upper part is cast-in-place concrete with a thickness of 240 mm.

**Figure 1.** Photo of Xiaoqing River Bridge.

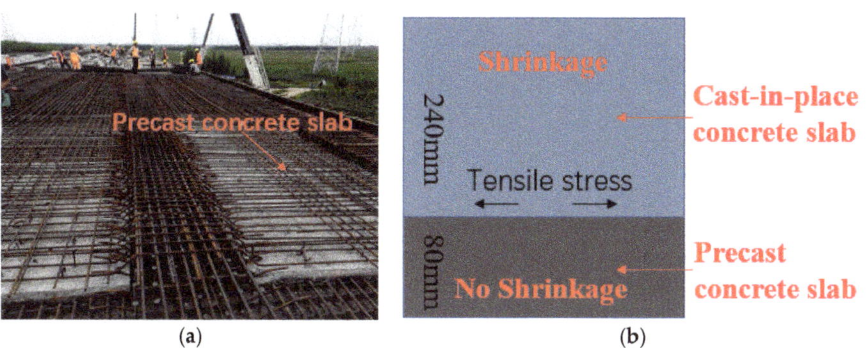


**Figure 2.** Structure of bridge deck: (**a**) construction site; (**b**) sketch map.

In general, the precast concrete slab has been stacked in the construction site for about one month, the shrinkage has been almost completed, and there will be no obvious shrinkage in the later stage. However, the cast-in-place concrete in the upper part will shrink, and the shrinkage is constrained by the precast slabs below, resulting in tensile stress [18–21]. When the tensile stress exceeds the tensile strength, cracks will be produced in the deck [22].

Shrinkage of concrete, as a material property, has been extensively studied in the past decade, and researchers have proposed many measures to reduce shrinkage cracks. Huang added 8% MgO to the concrete, which caused the concrete to expand by $60 \times 10^{-6}$, preventing shrinkage cracks in airport pavement. However, at the same time, MgO reduced the compressive strength of concrete by 27% [23]. Yousefieh studied the effect of fibers on drying shrinkage and crack resistance of concrete, and he found that steel fibers could reduce the width and length of cracks [24]. From the above research, we can find that when MgO is used alone, the expansion of MgO can compensate the shrinkage of the concrete and prevent cracking, but it reduces the strength. When steel fiber is used alone, the steel fiber can only prevent the further expansion of cracks after cracks occur, but cannot prevent the generation of cracks. In order to prevent shrinkage cracks without reducing the strength of concrete, MgO and steel fibers are used simultaneously in this research.

MgO has primarily been used in dam and airport pavement to reduce shrinkage cracks. Many researchers have proved that MgO concrete is superior to ordinary concrete in terms of carbonization resistance, chloride ion penetration resistance, sulfate corrosion resistance, and freeze-thaw cycle, which means that MgO can also be used in bridges to achieve the purpose of inhibiting shrinkage cracks [25–31]. In the previous study, our research team invented steel fiber reinforced MgO concrete (SMC) by using both MgO and steel fiber [32]. In addition, the deformation of SMC in the laboratory environment was studied, and it was found that SMC can produce micro expansion and prevent early shrinkage cracks. However, this kind of new material has remained in laboratory research and has not been applied in practical engineering. In the laboratory environment, the temperature and humidity have remained stable, but in the field environment, the temperature and humidity are constantly changing, and even the temperature change within a day reaches 20 °C. In the field environment, the deformation and strength of SMC have not been tested, which hinders the application and promotion of SMC. In this research, SMC was used in the cast-in-place deck of the Xiaoqing River Bridge, and the effect of SMC was evaluated by analyzing temperature, deformation, and strength. The study proved that this specially designed concrete of the bridge deck was able to maintain good performance under the harsh field environment. In addition, the deck is required to have good durability, because it not only bears vehicle loads but also is eroded by rain water. In this study, the durability of the bridge deck was measured by chloride permeability. Thanks to the positive results of this study, the combined application of MgO and steel fiber in concrete should be highly promoted for building bridge decks of long span bridges.

## 2. Materials and Methods

### 2.1. Materials

In this research, the cement was Class 52.5 ordinary Portland cement (Shan Aluminum Cement Co., Ltd., Shandong, China). Fly ash was Class I fly ash (Shenhua Guohua Shoudian Electric Co., Ltd., Beijing, China). The chemical composition of cement and fly ash was tested according to GBT 176, 2017 "Chemical Analysis Method of Cement" [33], and the test results are shown in Table 1. The coarse aggregate was limestone macadam with 5–20 mm continuous grading. Its apparent density was 2.73 g/cm$^3$, its bulk density was 1.48 g/cm$^3$, its porosity was 46.7%, and its water absorption was 0.45%. The fine aggregate was river sand, with a fineness modulus of 2.94, apparent density of 2.63 g/cm$^3$, bulk density of 1.54 g/cm$^3$, porosity of 41.7%, and water absorption of 0.40%. The MgO expansion agent was produced by Wuhan Sanyuan Special Building Materials Co. LTD in China, and its activity was 115 s. The steel fiber was wave shaped steel fiber (SF, Shuanglian Building Materials Co., Ltd., Shandong, China), and its detailed parameters are shown in Table 2. The mix proportions are listed in Table 3.

**Table 1.** Chemical composition of cement and fly ash.

| No. | Chemical Composition/% | | | | | | | | | |
|---|---|---|---|---|---|---|---|---|---|---|
| | CaO | MgO | $Al_2O_3$ | $SiO_2$ | $Fe_2O_3$ | $SO_3$ | $K_2O$ | $Na_2O$ | Loss | Total |
| Cement | 60.51 | 2.18 | 6.34 | 22.02 | 3.05 | 1.86 | 0.47 | 0.23 | 1.96 | 98.62 |
| Fly ash | 5.01 | 1.03 | 34.18 | 48.91 | 5.22 | 1.20 | 0.89 | 0.62 | 1.50 | 98.56 |

**Table 2.** Basic properties of steel fiber.

| Length/mm | Diameter/mm | Tensile Strength/MPa |
|---|---|---|
| 38 (±5%) | 0.58 (±5%) | 520 |

**Table 3.** Mix proportion of concrete.

| No. | Composition/kg·m$^{-3}$ | | | | | | | |
|---|---|---|---|---|---|---|---|---|
| | Cement | Fly Ash | Fine Aggregate | Coarse Aggregate | Water | Water Reducer | Steel Fiber | MgO |
| Ref | 450 | 50 | 713 | 1025 | 160 | 6 | 0 | 0 |
| 8%MgO(MC) | 450 | 50 | 713 | 1025 | 160 | 7.5 | 0 | 40 |
| 1%SF + 8%MgO(SMC) | 450 | 50 | 713 | 1025 | 160 | 7.6 | 78 | 40 |

### 2.2. Experimental

#### 2.2.1. Temperature and Deformation Test

As shown in Figure 3, a VWS-10 vibrating wire strain gauge was used to measure the temperature and deformation of concrete. The temperature measurement range of the strain gauge was from −40 °C to 80 °C with an accuracy of 0.5 °C. *F* was changed with the change of concrete length, and the deformation of concrete could be calculated by calculating the change value of *F* [34]. The strain was calculated by Formula (1) as follows:

$$\varepsilon = k\Delta F + (b - a)\Delta T \tag{1}$$

where $\varepsilon$ is expansion or contraction of concrete, in $10^{-6}$; $k$ is the constant coefficient of the strain gauge, in $10^{-6}$/F; $\Delta F$ is the change of F; $b$ is the temperature correction factor, in $10^{-6}$/°C; $a$ is the thermal expansion coefficient of concrete, in $10^{-6}$/°C; and $\Delta T$ is the change of the temperature of concrete, in °C.

(a) (b)

**Figure 3.** Field arrangement of strain gauge: (**a**) placement of strain gauge; (**b**) data acquisition. Where strain gauge 1 (S1) was used to measure the temperature and deformation of Ref, and S2 was used to measure the surface temperature of the deck.

As shown in Figure 4, in order to compare the effects of different MgO and steel fiber on the performance of the bridge deck, the deck was divided into two parts. The area of the first part was 9 $m^2$. The material used was SMC, and S3 was used to measure its temperature and deformation. The area of the second part was 519 $m^2$. The material used was MC, and S4 was used to measure the temperature and deformation of MC.

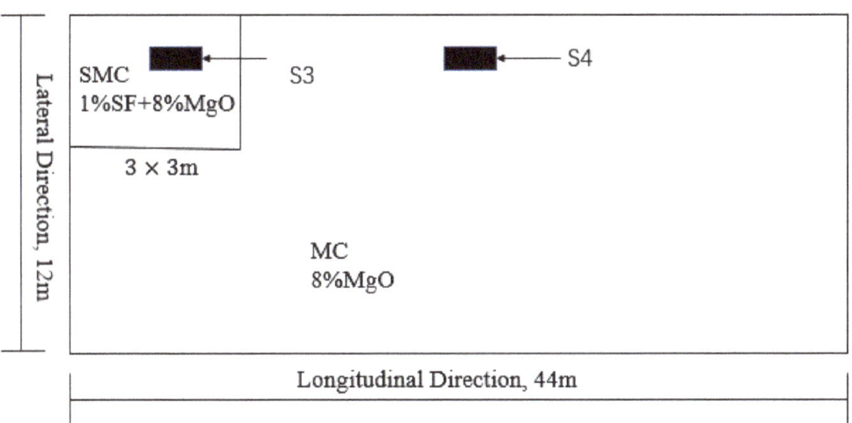


**Figure 4.** Location of testing sections. SMC: steel fiber reinforced MgO concrete; SF: steel fiber; MC: MgO concrete.

2.2.2. Strength Test

The strength of concrete was tested by a concrete rebound instrument (Gantan Instrument Co., Ltd., Shanghai, China). As shown in Figure 5, in order to evaluate the influence of MgO and steel fiber on the strength, the strength of the bridge deck was tested at 3 days, 7 days, and 21 days. To ensure the accuracy of the experiment, the partial deck was polished before the test. In order to ensure the effectiveness of the rebound strength, before the test the rebound meter was calibrated in accordance with the "Technical Specification for Testing the Compressive Strength of Concrete by the Rebound Method" JTJ/T 23-2011 [35]. The bouncing rod rotated in four directions, each rotation being 90°, and the rebound values were 80.6, 80.5, 80.4, and 80.5, respectively, which met the requirements of the range of 80 ± 2 in the specifications.

**Figure 5.** Test of the concrete strength by rebound instrument.

2.2.3. Permeability Test

Considering that in winter, in order to prevent the deck from freezing, deicing salt is often sprinkled on the deck. When chloride ions diffuse into the bridge deck, they cause corrosion and expansion of steel bars, which lead to cracking and affect the safety of the structure. In order to analyze the chloride permeability of SMC, a rapid chloride permeability test was carried out according to the procedure described in ASTM C1202 [36–39]. The test device is shown in Figure 6.

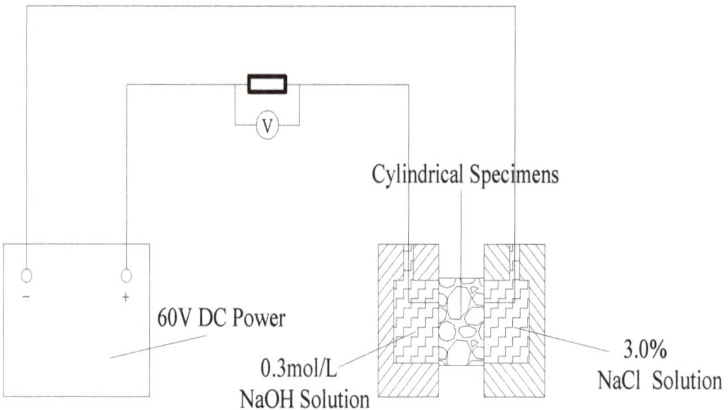


**Figure 6.** Setup for the chloride diffusion test.

A cylindrical concrete specimen with a height of 50 mm and a length of 100 mm was used to test the chloride diffusion coefficient. Two different solutions, namely 3% NaCl solution and 0.3 mol/L NaOH solution, were selected in the experiment. A concrete chloride flow meter was used to measure the flux through the sample within the specified time. The total electric flux of the concrete test block for 6 h was calculated as follows according to Formula (2):

$$Q = 900 \times (I_0 + 2I_{30} + 2I_{60} + \cdots + 2I_t + \cdots + 2I_{300} + 2I_{330} + I_{360}) \quad (2)$$

where $Q$ is total flux through the specimen within 6 h (C); $I_0$ is initial flux (A), to 0.001A; $I_t$ is flux (A) at specified time t (min), to 0.001A.

According to the Nernst–Planck equation, the chloride diffusion coefficient (CDC) can be calculated by Q according to Formula (3) as follows:

$$CDC = 2.57765 + 0.00492 \times Q \quad (3)$$

2.2.4. Pore Structure Test

The influence of MgO and steel fiber on the pore structure of concrete was analyzed by mercury intrusion porosimeter (MIP). First, the concrete was broken into small pieces with a length of 2 mm, and the coarse aggregate was removed. Then in order to stop the hydration of cement and MgO, the samples were immersed in anhydrous ethanol for 24 h. After that, the samples were vacuum dried for another 12 h. By comparing the changes of the pore structure before and after adding MgO and steel fiber into the concrete, the reasons why the deformation, strength, and durability of bridge deck changed can be explained from the microscopic point of view.

## 3. Results

### *3.1. Temperature Variation*

Figure 7 shows the temperature variation on the surface and interior of deck. At the beginning of pouring, the external environment temperature was 25.2 °C, and the data collection was started four hours after pouring. It can be seen from Figure 7 that due to the solar radiation on the surface of the deck, the temperature changed greatly in the daytime and evening, and the biggest range of temperature in a day increased from 20.8 °C to 46.9 °C. On the other hand, due to the poor thermal conductivity of concrete, there was little temperature change inside deck, controlled within 6 °C [40–42].

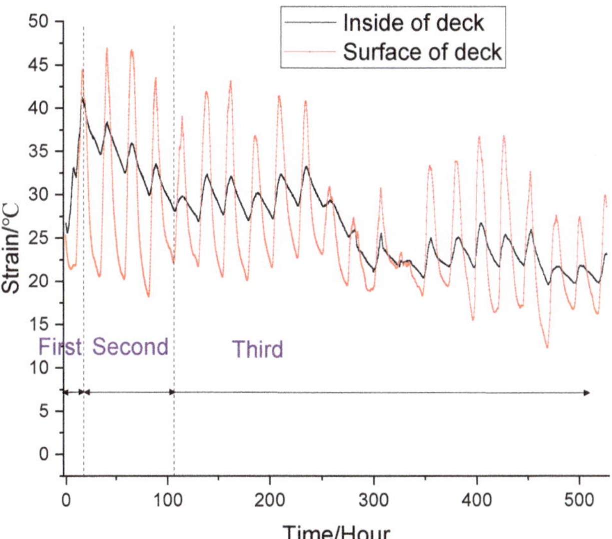


**Figure 7.** Curve of temperature variation.

The internal temperature variation of deck could be divided into three stages. The first stage was the rapid temperature rise stage. Due to the exothermic hydration of cement and MgO, the temperature increased rapidly to 40.9 °C at 17 h and reached 46.9 °C at 41 h. On the other hand, the heat on the concrete surface dissipated rapidly, and the temperature dropped rapidly to 37.9 °C, resulting in a temperature difference of 13 °C between the inside and outside of the bridge deck. According to the stress calculation formula (Formula (4)) proposed by Bradbury, the tensile stress increases with the increase of temperature difference [43]. Therefore, the bridge deck easily produced temperature cracks in the early stage when the concrete strength was low. However, the addition of steel fibers significantly reduced cracks due to two main factors. On the one hand, steel fibers increased the thermal conductivity of concrete and reduced the temperature difference; on the other hand, the combined effects of steel fibers and MgO improved the strength of concrete and limited the development of cracks.

$$\sigma_t = \frac{1}{2}\Delta T \times E\alpha \times \left(\frac{C_x + \mu C_y}{1-\mu^2}\right) \tag{4}$$

where $\Delta T$ is the temperature difference between the inside and the surface of the bridge deck (°C); $E$ is the elastic modulus of concrete (MPa); $\alpha$ is the thermal expansion coefficient of concrete (°C$^{-1}$); $\mu$ is Poisson's ratio of concrete; and $C_x$ and $C_y$ are the stress coefficients.

The second stage was the cooling stage. In this stage, the hydration rate slowed down noticeably. As the heat diffused into the air, the internal temperature decreased gradually, and it decreased to 28.1 °C at 106 h. The third stage was the natural temperature stage. After 106 h, most of the hydration of cement and MgO was completed, and the temperature changed with the external environment temperature. Compared with the three different stages, the following conclusions can be reached: (1) In the temperature rise stage, the temperature difference between the inside and outside of the deck was very small, and the concrete was not fully hardened during this time, so it was not easy to produce cracks. (2) In the temperature drop stage, the surface temperature was higher than the internal temperature in the daytime. However at night, the surface temperature was less than the internal temperature. The alternating action of plus–minus temperature stress made it easy to produce temperature cracks in the deck. (3) What is more, in the early stage, the strength of the concrete was low, and the large temperature stress made it easy to produce through-cracks in the bridge deck. Therefore, the maintenance of concrete needed to be strengthened in the temperature drop stage. As shown in Figure 8a, when the membrane curing method was used, it could not only ensure the early hydration of cement, but it also ensured the water demand during the hydration of MgO, which made the early strength of MgO concrete develop stably and improve the early performance of MgO concrete. As shown in Figure 8b, during the construction of the Xiaoqing River Bridge, after 17 h of pouring, the surface of the bridge deck was covered with geotextile, and water was sprayed every 6 h. After 21 days, the bridge deck surface was intact and there were no temperature cracks.

(a)

(b)

**Figure 8.** Photos of bridge deck: (**a**) curing; (**b**) 21 days after pouring.

### *3.2. Volume Deformation*

Figure 9 shows the change of deformation of concrete with different proportions over time. The deformation of concrete changed periodically with the temperature change of the external environment. Similar to the law of temperature change, deformation can be divided into three stages.

The first stage was rapid expansion stage, lasting for 21 h. The REF also expanded at this stage, but its expansion value was smaller than that of the other two kinds of concrete, which indicated that MgO had begun to hydrate and had produced effective expansion at the early stage of temperature rise. For example, at 21 h, the expansion of REF was $48.4 \times 10^{-6}$, which was only 34.9% and 45.2% of MC and SMC, respectively. The second stage was the shrinkage stage, lasting for 100 h. The third stage was the stable stage, at which point the shrinkage or expansion no longer changed dramatically.

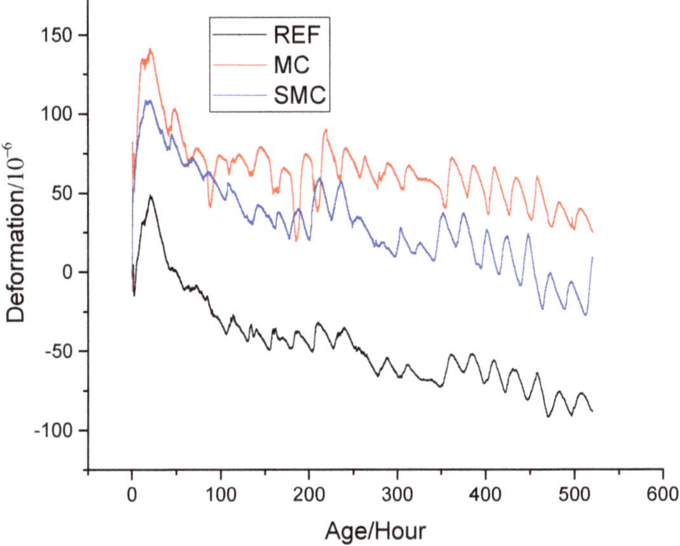


**Figure 9.** Deformation of concrete with different mix proportions.

It can be seen from Figure 9 that the deformation of REF decreased with time due to drying shrinkage and chemical shrinkage, and the shrinkage reached $-88.3 \times 10^{-6}$ at 520 h. In this stage, if no measures are taken to reduce shrinkage, the shrinkage of the upper cast-in-place slab will be restrained by the lower precast slab, resulting in tensile stress. When the tensile stress exceeds the tensile strength of the concrete, it will lead to cracking, which was also one of the main reasons for the cracking of many bridge decks in the past. According to Formula (5), the width of the crack increases with the length of concrete slab and the shrinkage deformation [44]. Moreover, in this project, there was no expansion joint in the 44 m long bridge deck. Therefore, if the shrinkage of the bridge deck is not limited, there will be wide cracks in such a long concrete slab.

$$d = C \times L \times (\Delta T \alpha + \varepsilon) \quad (5)$$

where $d$ is the width of the crack; $C$ is the constant coefficient; $L$ is the length of the concrete slab; $\Delta T$ is the change in temperature; $\alpha$ is the thermal expansion coefficient of concrete; and $\varepsilon$ is the shrinkage deformation of concrete.

For MC, the shrinkage of concrete was compensated by the expansion of MgO hydration, and the deformation was $24.9 \times 10^{-6}$ at 520 h, attributed to the generation of $Mg(OH)_2$. MgO made the concrete from original shrinkage to expansion and fundamentally prevented the shrinkage cracking of the bridge deck. For SMC, by contrast, due to the steel fiber restricted the expansion of MgO, part of the energy generated by MgO was used to tension the steel fiber and generated self-stress, which reduced the porosity of concrete and improve the strength of concrete [45–48]. Furthermore, the expansion of SMC was smaller than that of MC, which was $9.1 \times 10^{-6}$ at 520 h. During the whole time, both MC and SMC were always in the expansion state, so there was no shrinkage cracking, which proved that MgO and steel fiber could be used to solve the early cracks of decks.

### *3.3. Expansion Caused by MgO Hydration*

The deformation measured by strain gauge was the comprehensive result of concrete shrinkage and MgO expansion, which was the sum of shrinkage ($\varepsilon_c$) and expansion ($\varepsilon_M$). As shown in Figure 10, in order to analyze the influence of steel fiber constraint on MgO expansion, the deformation of MC and SMC simultaneously subtracted the deformation of REF so as to obtain the deformation caused by MgO hydration. The expansion process of MgO could be divided into two stages. The first stage was the rapid expansion stage. In this stage, the expansion caused by rapid hydration of MgO was obviously greater than the shrinkage of concrete. The second stage was the stable stage. In this stage, the hydration rate of MgO slowed down significantly, the expansion and shrinkage were approximately the same, and the total deformation remained stable. By comparing the different expansions of MC and SMC, the following conclusions can be found: (1) In the early days, for MC, the expansion of MgO was able to produce an expansion of $60.7 \times 10^{-6}$, which completely compensated the shrinkage of the concrete. In the later period, the expansion did not increase without limit, but gradually tended to a stable value. Therefore, the use of MgO in the bridge deck is safe and will not cause stability problems due to harmful expansion. (2) Due to the restraint of steel fiber, the expansion of SMC was smaller than that of MC. In the early stage (21 h), the expansion of SMC was $60.7 \times 10^{-6}$, only 65.8% of MC. In the later stage (520 h), the expansion of SMC was 55.8 μ ε, only 48.9% of MC. In contrast, for SMC, the energy generated by the hydration of MgO was not completely used to expand the concrete, but was partially used to tension the steel fibers, which was helpful to improve the density of the concrete.

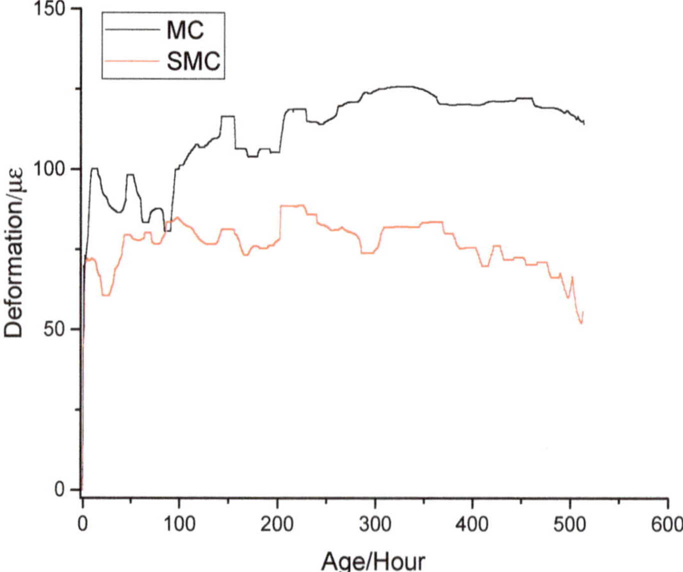


**Figure 10.** MgO expansion test.

### *3.4. Strength Test*

Figure 11 shows the strength variation of concrete with different proportions. With the increase of curing age, the strength of all concrete gradually increased. Compared with REF, the strength of MC increased by −0.4%, 1.2%, and 1.7% at 3 d, 7 d, and 21 d, respectively, which means that MgO had little effect on the strength of deck. On the other hand, the strength of SMC increased significantly. Compared with REF, the strength of SMC increased by 2.1%, 10.3%, and 17.5% at 3 d, 7 d, and 21 d, respectively. The test results showed that the combined action of MgO and steel fiber reduced the internal defects of concrete and improved the density of concrete, which improved the strength of the deck. The results of the two tests, strength and deformation, showed that the combination of MgO and steel fiber not only prevented the shrinkage crack of the deck, but also improved the strength of the deck, which provided a new method for preventing the early cracks in the deck.

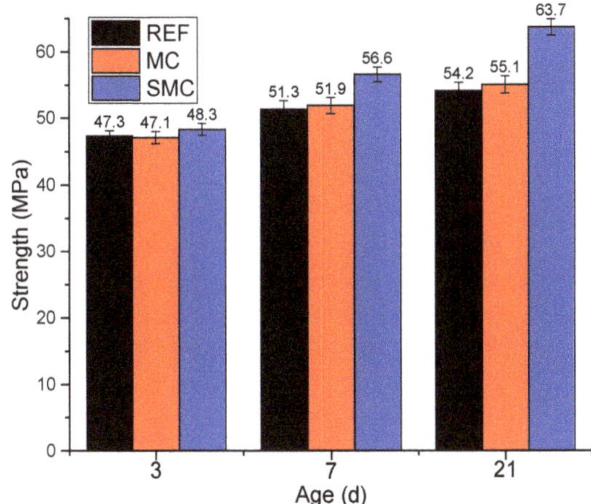


**Figure 11.** Strength variation of concrete with different proportions.

The strength test results showed that after 8% MgO was added to the bridge deck, the strength of the deck at 21 d was improved due to the increase of the total amount of cementing material. However, in the early days, the strength of the matrix was small, and the constraint on expansion of MgO was small. The hydration of MgO produced $Mg(OH)_2$, which caused the concrete to expand outward, which increased the porosity and reduced the early strength. Therefore, at 3 d, the strength of MC was 0.4% lower than that of REF. On the other hand, when MgO and steel fiber were used at the same time, the steel fiber restricted the expansion of MgO, and $Mg(OH)_2$ grew in the restricted space, which reduced the porosity of the concrete, improved the density, and significantly increased the strength.

### *3.5. Chloride Permeability of Concrete*

The results of the rapid chloride permeability test of concrete at different ages are shown in Table 4. It can be seen that MgO and steel fibers enhanced the resistance to chloride penetration of concrete, and the concrete with both MgO and steel fibers had a higher resistance than that of MgO alone.

**Table 4.** Chloride permeability of concrete at different ages.

| Type | Chloride Diffusion Coefficient ($10^{-9}$ $cm^2/s$) | | | | Reduced Extent (%) | | | |
|---|---|---|---|---|---|---|---|---|
| | 3 d | 7 d | 28 d | 60 d | 3 d | 7 d | 28 d | 60 d |
| REF | 15.6 | 15.4 | 10.8 | 8.8 | 0 | 0 | 0 | 0 |
| MC | 15.4 | 15.2 | 10.7 | 8.4 | 1.3 | 1.3 | 0.9 | 4.6 |
| SMC | 13.7 | 11.2 | 6.4 | 6.2 | 12.2 | 27.3 | 40.7 | 29.6 |

It can be seen from Table 4 that MgO not only did not reduce the chloride penetration resistance of concrete, but it also improved the chloride penetration resistance of concrete. The increase within 60 days was 4.6%–0.9%, which showed that it was safe to use MgO in the bridge deck. When MgO and steel fiber were used at the same time, the resistance to chloride penetration was further improved, the increase within 60 days was from 12.2% to 40.7%, indicating that the combined effect of MgO and steel fiber made the concrete more dense, and the same conclusion can be obtained in the pore structure test. Therefore, in summary, it was found that MgO could be used alone to suppress the shrinkage cracks of bridge decks in the southern regions with high winter temperatures, and MgO and steel fibers could be used to suppress the shrinkage cracks of bridge decks in the northern regions with low winter temperatures.

### 3.6. Performance Enhancement Mechanism of Concrete

#### 3.6.1. The Relationship between Deformation and Stress

Figure 12 shows the stresses due to the deformation of three types of concrete under the constraint conditions. Under the restraint of reinforcement and precast slab, tensile stress is generated due to shrinkage in REF, and the bridge deck will crack when the tensile stress exceeds the tensile strength of concrete (Figure 12a). In contrast, for MC and SMC, the expansion of MgO compensates for the contracting of the concrete, causing a slight expansion of the bridge deck. Under restraint conditions, compressive stress is generated due to expansion inside the concrete, which not only prevents cracking, but also makes the concrete denser and improves the strength of the concrete (Figure 12b). Furthermore, the greater the constraint, the greater the compressive stress and the denser the concrete. For SMC, under the combined constraint of steel fiber and reinforcement, the concrete produced greater compressive stress, which also explains why the strength of SMC was greater than that of REF and MC.

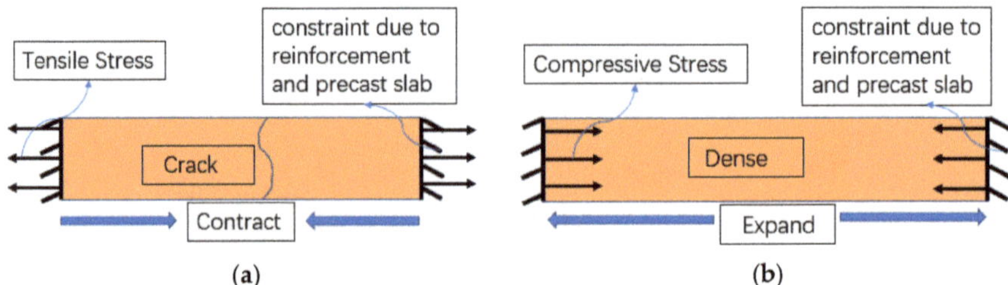


**Figure 12.** Stress produced by deformation under constraint: (**a**) REF; (**b**) MC and SMC.

#### 3.6.2. Pore Structure of Concrete

The porosity of mortar with two different mix ratios is given in Table 5. According to the literature [49–53], pores less than 50 nm, 50–500 nm, and greater than 500 nm are defined as harmless pores, less harmful pores, and harmful pores, respectively.

**Table 5.** Pore structure of mortar.

| No. | Total Porosity/% | Pore Distribution/% | | |
|---|---|---|---|---|
| | | 0–50 nm | 50–500 nm | >500 nm |
| REF | 15.76 | 8.51 | 2.42 | 4.83 |
| SMC | 10.69 | 4.38 | 3.98 | 2.38 |

Table 5 shows the pore structure of REF and SMC. Due to the existence of compressive stress, the total porosity of SMC was 10.69%, which was 32.2% less than that of REF. Moreover, the harmful pores of SMC totaled 2.38%, which was 50.7% less than that of REF. The above data showed that the combined action of MgO and steel fiber optimized the pore structure of concrete.

When steel fiber and MgO were added to concrete at the same time, MgO was used as an expansion element, and steel fiber simultaneously played a role in limiting expansion and preventing cracks. Both of them had a compound effect and produced chemical compressive stress in concrete, reduced the porosity of concrete, and optimized the pore structure, which was the key to improving the performance of SMC.

## 4. Conclusions

In this research, the effect of MC and SMC on the temperature, deformation, and strength of a bridge deck is studied. Through detailed experimental study, the following conclusions can be drawn:

(1) With the increase of bridge spans, the cement content of modern concrete is increasing, which significantly increases the early shrinkage. The shrinkage of REF measured at the Xiaoqing River Bridge in 520 h reached −88.3 μ ε. If no measures are taken to reduce shrinkage, the shrinkage under the constraint of the precast slab below will produce tensile stress, which will lead to shrinkage cracks.

(2) Due to the radiation of the sun, the temperature of the bridge deck is very high during the daytime, but at night the temperature will quickly drop. In this process, the bridge deck is subjected to repeated tension and compression. Because the concrete is not completely hardened in the early stage, the strength is low, and it is easy to produce temperature cracks. Therefore, in the early stage of construction, it is necessary to strengthen the maintenance of the bridge deck.

(3) After adding MgO to concrete, the expansion of MgO hydration compensated the shrinkage, and the expansion of MC in 520 h was $24.9 \times 10^{-6}$, preventing shrinkage cracks of the deck. The later expansion of MgO tended to a stable value, and there was no backward shrinkage. In addition, the strength test showed that MgO did not reduce the strength of the deck.

(4) Due to the restraint of steel fibers, $Ca(OH)_2$ was generated in the restricted space, resulting in pre-pressure, which improved the performance of concrete. When MgO and steel fiber were used at the same time, the expansion of SMC decreased to $9.1 \times 10^{-6}$ in 520 h, and the strength increased by 17.5% in 21 days. The test results showed that the combined action not only prevented shrinkage cracking of the slab, but also improved the strength of the concrete. In this way, temperature cracks and shrinkage cracks could be effectively prevented.

(5) The use of MgO alone had little effect on the resistance to chloride penetration of concrete, with a maximum increase of only 4.6%. On the other hand, the combined effect of MgO and steel fibers at the same time significantly improved the resistance to chloride penetration of concrete, with a maximum increase of 40.7%.

(6) The combined effect of MgO and steel fiber significantly improved the pore structure of concrete, which reduced the total porosity of SMC by 32.2% and the harmful pore by 50.7%, which also explained the reason for the increase of rebound strength of SMC from the micro level.

(7) Through in-situ tests at the construction site, this study clearly shows that the combined application of MgO and steel fiber can still maintain good performance in outdoor environments. Therefore, SMC, a new type of cement-based material, should be widely promoted and applied to bridge decks of long-span bridges, which has great significance in improving the durability of bridges.

**Author Contributions:** F.J. designed and conducted the experimental program. M.D. gave many suggestions about the experiment and helped solve the testing problems. W.W. and L.M. assisted in pouring the concrete. All authors assisted in the analysis. All authors have read and agreed to the published version of the manuscript.

**Funding:** This research was funded by Science and Technology Development Plan of Suzhou (SNG201904), National Key Research and Development Plan of China (2017YFB0309903-01) and Transportation Science and Technology Planning Project of Shandong Province (2018B37-02).

**Acknowledgments:** The authors would like to thank Zhongyang Mao from Nanjing Tech University for his precious contribution in the experiments.

**Conflicts of Interest:** The authors declare no conflict of interest.

## References


1. Shan, J.H.; Zhou, M.; Li, B.X.; Cai, J.W.; Xu, F. Application of polypropylene fiber concrete in bridge deck pavement. *Key Eng. Mater.* **2006**, 418–423. [CrossRef]
2. Su, L.; Wang, S.; Gao, Y.; Liu, J.; Shao, X. In Situ experimental study on the behavior of UHPC composite orthotropic steel bridge deck. *Materials* **2020**, *13*, 253. [CrossRef] [PubMed]
3. Khotbehsara, M.M.; Manalo, A.; Aravinthan, T.; Ferdous, W.; Nguyen, K.T.Q.; Hota, G. Ageing of particulate-filled epoxy resin under hygrothermal conditions. *Constr. Build. Mater.* **2020**, *249*, 118846. [CrossRef]
4. Manalo, A.; Maranan, G.; Benmokrane, B.; Cousin, P.; Alajarmeh, O.; Ferdous, W.; Liang, R.; Hota, G. Comparative durability of GFRP composite reinforcing bars in concrete and in simulated concrete environments. *Cem. Concr. Compos.* **2020**, *109*, 103564. [CrossRef]
5. Abousnina, R.; Manalo, A.; Ferdous, W.; Lokuge, W.; Al-Jabri, K.S. Characteristics, strength development and microstructure of cement mortar containing oil-contaminated sand. *Constr. Build. Mater.* **2020**, *252*, 13. [CrossRef]
6. Bazant, Z. Advances in material modelling of concrete. In Proceedings of the Tenth International Conference on Structural Mechanics in Reactor Technology (SMiRT 10), Anaheim, CA, USA, 14–18 August 1989; pp. 301–330.
7. Craeye, B.; Geirnaert, M.; Schutter, G.D. Super absorbing polymers as an internal curing agent for mitigation of early-age cracking of high-performance concrete bridge decks. *Constr. Build. Mater.* **2011**, *25*, 1–13. [CrossRef]
8. Shen, D.; Liu, X.; Zeng, X.; Zhao, X.; Jiang, G. Effect of polypropylene plastic fibers length on cracking resistance of high-performance concrete at early age. *Constr. Build. Mater.* **2020**, *244*, 117874. [CrossRef]
9. Yang, L.; Shi, C.; Wu, Z. Mitigation techniques for autogenous shrinkage of ultra-high-performance concrete—A review. *Compos. Part B Eng.* **2019**, *178*, 107456. [CrossRef]
10. Shen, D.; Jiang, J.; Wang, W.; Shen, J.; Jiang, G. Tensile creep and cracking resistance of concrete with different water-to-cement ratios at early age. *Constr. Build. Mater.* **2017**, *146*, 410–418. [CrossRef]
11. Zhu, J.; Wang, Y.; Yan, J.; Guo, X. Shear behaviour of steel-UHPC composite beams in waffle bridge deck. *Compos. Struct.* **2020**, *234*, 111678. [CrossRef]
12. Hajibabaee, A.; Moradllo, M.K.; Behravan, A.; Ley, M.T. Quantitative measurements of curing methods for concrete bridge decks. *Constr. Build. Mater.* **2018**, *162*, 306–313. [CrossRef]
13. You, J.; Park, J.; Lee, C.; Park, S.; Hong, S. Analysis of expressway bridge deck slab deterioration. In Proceedings of the AIP Conference Proceedings, Okinawa, Japan, 20 April 2018.
14. Su, Q.; Dai, C.; Jiang, X. Bending performance of composite bridge deck with T-shaped ribs. *Front. Struct. Civil. Eng.* **2019**, *13*, 990–997. [CrossRef]
15. Alizadeh, E. Efficient composite bridge deck consisting of GFRP, steel, and concrete. *J. Sandw. Struct. Mater.* **2019**, *21*, 154–174. [CrossRef]
16. Pei, B.; Li, L.; Shao, X.; Wang, L.; Zeng, Y. Field Measurement and practical design of a lightweight composite bridge deck. *J. Constr. Steel Res.* **2018**, *147*, 564–574. [CrossRef]
17. Bajzecerováa, V.; Kanóczb, J. The Effect of environment on timber-concrete composite bridge deck. In Proceedings of the Bridges in Danube Basin 2016—New Trends in Bridge Engineering and Efficient Solution for Large and Medium Span Bridges, Zilina, Slovakia, 30 September–1 October 2016; pp. 32–39.
18. Elsanadedy, H.M.; Abbas, H.; Al-Salloum, Y.A.; Almusallam, T.H. Prediction of intermediate crack debonding strain of externally bonded FRP laminates in RC beams and one-way slabs. *J. Compos. Constr.* **2014**, *18*, 04014008. [CrossRef]
19. Mayencourt, P.; Mueller, C. Structural optimization of cross-laminated timber panels in one-way bending. *Structures* **2019**, *18*, 48–59. [CrossRef]
20. Rabinovitch, O.; Frostig, Y. High-order analysis of refine concrete slabs strengthened with circular composite laminated patches of general layup. *J. Eng. Mech. ASCE* **2004**, *127*, 1334–1345. [CrossRef]
21. Teguedy, M.C.; Joly-Lapalice, C.; Sorelli, L.; Conciatori, D. Optical fiber sensors implementation for monitoring the early-age behavior of full-scale timber-concrete composite slabs. *Constr. Build. Mater.* **2019**, *226*, 564–578. [CrossRef]

22. Zhang, J.Z.J.; Hou, D.H.D.; Gao, Y.G.Y. Calculation of shrinkage stress in early-age concrete pavements. II: Calculation of shrinkage stress. *J. Transp. Eng.* **2013**, *139*, 971–980. [CrossRef]
23. Huang, K.; Shi, X.; Zollinger, D.; Mirsayar, M.; Wang, A.; Mo, L. Use of MgO expansion agent to compensate concrete shrinkage in jointed reinforced concrete pavement under high-altitude environmental conditions. *Constr. Build. Mater.* **2019**, *202*, 528–536. [CrossRef]
24. Yousefieh, N.Y.N.; Joshaghani, A.J.A.; Hajibandeh, E.H.E.; Shekarchi, M.S.M. Influence of fibers on drying shrinkage in restrained concrete. *Constr. Build. Mater.* **2017**, *148*, 833–845. [CrossRef]
25. Hu, Y.; Li, H.; Wang, Q.; Zhang, J.; Song, Q. Characterization of LDHs prepared with different activity MgO and resisting $Cl^-$ attack of concrete in salt lake brine. *Constr. Build. Mater.* **2019**, *229*, 116921. [CrossRef]
26. Heidari, A.; Hashempour, M.; Chermahini, M.D. Influence of reactive MgO hydration and cement content on C&DW aggregate concrete characteristics. *Int. J. Civ. Eng.* **2019**, *17*, 1095–1106.
27. Dung, N.T.; Unluer, C. Performance of reactive MgO concrete under increased $CO_2$ dissolution. *Cem. Concr. Res.* **2019**, *118*, 92–101. [CrossRef]
28. Pu, L.; Unluer, C. Durability of carbonated MgO concrete containing fly ash and ground granulated blast-furnace slag. *Constr. Build. Mater.* **2018**, *192*, 403–415. [CrossRef]
29. Zhang, R.; Panesar, D.K. Mechanical properties and rapid chloride permeability of carbonated concrete containing reactive MgO. *Constr. Build. Mater.* **2018**, *172*, 77–85. [CrossRef]
30. Ruan, S.; Unluer, C. Influence of mix design on the carbonation, mechanical properties and microstructure of reactive MgO cement-based concrete. *Cem. Concr. Compos.* **2017**, *80*, 104–114. [CrossRef]
31. Dung, N.T.; Unluer, C. Development of MgO concrete with enhanced hydration and carbonation mechanisms. *Cem. Concr. Res.* **2018**, *103*, 160–169. [CrossRef]
32. Jiang, F.; Mao, Z.; Deng, M.; Li, D. Deformation and compressive strength of steel fiber reinforced MgO concrete. *Materials* **2019**, *12*, 3617. [CrossRef]
33. *Chemical Analysis Method of Cement*; GB/T 176-2017; Standards Press: Beijing, China, 29 December 2017.
34. Huang, K.; Deng, M.; Mo, L.; Wang, Y. Early Age stability of concrete pavement by using hybrid fiber together with MgO expansion agent in high altitude locality. *Constr. Build. Mater.* **2013**, *48*, 685–690. [CrossRef]
35. *Technical Specification for Testing the Compressive Strength of Concrete by the Rebound Method*; JTJ/T 23-2011; Architecture & Building Press: Beijing, China, 1 December 2011.
36. *Standard Test Method for Electrical Indication of Concrete's Ability to Resist Chloride Ion Penetration, ASTM C1202*; ASTM International: West Conshohocken, PA, USA, 2012.
37. Salmasi, F.; Mostofinejad, D. Investigating the effects of bacterial activity on compressive strength and durability of natural lightweight aggregate concrete reinforced with steel fibers. *Constr. Build. Mater.* **2020**, *251*, 119032. [CrossRef]
38. Provete Vincler, J.; Sanchez, T.; Turgeon, V.; Conciatori, D.; Sorelli, L. A modified accelerated chloride migration tests for UHPC and UHPFRC with PVA and steel fibers. *Cem. Concr. Res.* **2019**, *117*, 38–44. [CrossRef]
39. Lehner, P.; Konečný, P.; Ponikiewski, T. Comparison of material properties of SCC concrete with steel fibres related to ingress of chlorides. *Crystals* **2020**, *10*, 220. [CrossRef]
40. Lee, C.; Kim, S.; Kim, Y.; Kim, S.; Hwang, J.; Park, J. Experimental study on thermal conductivity of concrete using ferronickel slag powder. *KSCE J. Civ. Eng.* **2020**, *24*, 219–227. [CrossRef]
41. Wang, W.; Wang, H.; Chang, K.; Wang, S. Effect of high temperature on the strength and thermal conductivity of glass fiber concrete. *Constr. Build. Mater.* **2020**, *245*, 118387. [CrossRef]
42. Marie, I.M.I. Thermal conductivity of hybrid recycled aggregate—Rubberized concrete. *Constr. Build. Mater.* **2017**, *133*, 516–524. [CrossRef]
43. Bradbury, R.D. *Reinforced Concrete Pavement*; Wire Reinforcement Institute: Washington, DC, USA, 1938.
44. Darter, M.I. *Design of Zero-Maintenance Plain Jointed Concrete Pavement, Volume 1: Development of Design Procedures*; Report No. FHWA-RD-77-III; Federal Highway Administration: Washington, DC, USA, 1977.
45. Chidighikaobi, P.C. Thermal effect on the flexural strength of expanded clay lightweight basalt fiber reinforced concrete. *Mater. Today Proc.* **2019**, *19*, 2467–2470. [CrossRef]
46. Xiaoke, L.; Changyong, L.; Minglei, Z.; Hui, Y.; Siyi, Z. Testing and prediction of shear performance for steel fiber reinforced expanded-shale lightweight concrete beams without web reinforcements. *Materials* **2019**, *12*, 1594.

47. Zhao, M.; Zhang, B.; Shang, P.; Fu, Y.; Zhang, X.; Zhao, S. Complete stress-strain curves of self-compacting steel fiber reinforced expanded-shale lightweight concrete under uniaxial compression. *Materials* **2019**, *12*, 2979. [CrossRef]
48. Zhao, M.; Zhang, X.; Song, W.; Li, C.; Zhao, S. Development of steel fiber-reinforced expanded-shale lightweight concrete with high freeze-thaw resistance. *Adv. Mater. Sci. Eng.* **2018**, *2018*, 1–8. [CrossRef]
49. Bharadwaj, K.; Glosser, D.; Moradllo, M.K.; Isgor, O.B.; Weiss, W.J. Toward the prediction of pore volumes and freeze-thaw performance of concrete using thermodynamic modelling. *Cem. Concr. Res.* **2019**, *124*, 105820. [CrossRef]
50. Pacheco, F.; de Souza, R.P.; Christ, R.; Rocha, C.A.; Silva, L.; Tutikian, B. Determination of volume and distribution of pores of concretes according to different exposure classes through 3D microtomography and mercury intrusion porosimetry. *Struct. Concr.* **2018**, *19*, 1419–1427. [CrossRef]
51. Mehta, P.K. Studies on blended Portland cements containing Santorin earth. *Cem. Concr. Res.* **1981**, *11*, 507–518. [CrossRef]
52. Chen, L.; Wang, Y.; Yin, X. Influence of pore size of concrete on its impermeability. *J. Silicate* **2005**, *33*, 97–102. (In Chinese)
53. Chen, S.; Hu, C. Research trends of cement concrete durability of highway structures. *Highway* **2003**, *5*, 124–129. (In Chinese)

*Article*

# Effects of Traffic Vibrations on the Flexural Properties of Newly Placed PVA-ECC Bridge Repairs

**Xiaodong Zhang [1], Shuguang Liu [1,2,*], Changwang Yan [2], Xiaoxiao Wang [1,*] and Huiwen Wang [3]**

[1] School of Materials Science and Engineering, Inner Mongolia University of Technology, Hohhot 010051, China; zhangxd8808@126.com
[2] School of Mining and Technology, Inner Mongolia University of Technology, Hohhot 010051, China; yancw20031013@126.com
[3] Beijing Tuowei Times Architectural Design Co., Ltd., Beijing 100020, China; wanghw0717@126.com
* Correspondence: liusg6011@126.com (S.L.); wxiaoxiao.good@163.com (X.W.)

Received: 21 September 2019; Accepted: 11 October 2019; Published: 13 October 2019

**Abstract:** Polyvinyl alcohol fiber reinforced engineering cementitious composites (PVA-ECCs) exhibit excellent tight-cracking and super-high toughness behaviors and have been widely used in bridge repair projects. In reality, the conventional method in bridge repair is that a portion of the bridge is closed and repaired while the other portion is left open to traffic. Consequently, newly placed PVA-ECC bridge repairs (NP-ECC-BRs) are exposed to continuous traffic vibrations (TRVs), even during the setting periods. However, whether or not TRVs affect the expected flexural properties of NP-ECC-BRs remains unknown. The purpose of this investigation was to determine the effects of TRVs on the attainable flexural properties of NP-ECC-BRs. For this purpose, a total of 324 newly fabricated thin-plate specimens were exposed to different vibration variables using self-designed vibration equipment. After vibration, a four-point flexural test was conducted to determine the flexural properties of the specimens. The results indicate that the effects of TRVs on the strengths of NP-ECC-BRs was significantly negative, but insignificantly positive for flexural deformation. We concluded that in the design of PVA-ECC bridge repairs, effects of TRVs on the flexural deformation capacity of NP-ECC-BRs are not a cause for concern, but serious consideration should be given to the associated reduction of flexural load-bearing capacity.

**Keywords:** PVA-ECCs; bridge repairs; setting periods; traffic vibrations; strain-hardening characteristics; flexural properties

## 1. Introduction

Concrete was the most-used solid material over the past century. To date, a huge number of concrete structures have needed repair, for which enormous sums are spent annually [1–3]. Polyvinyl alcohol fiber reinforced engineering cementitious composites (PVA-ECCs) possess unique tight-cracking, strain-hardening, and super-high toughness behaviors under flexural loading [4], hence are becoming an attractive choice for the repair of existing concrete structures. Recently, PVA-ECCs have been widely used to repair concrete bridge deck slabs [5], pavements [6], overlays [7], and expansion or non-expansion joints [8].

In reality, completely closing traffic during bridge repair is unrealistic, so the conventional method is that a portion of the bridge is closed and repaired while the other portion is left open to traffic. As a result, newly placed PVA-ECC bridge repairs (NP-ECC-BRs) are exposed to continuous traffic vibrations (TRVs), even during the curing times. However, whether or not TRVs affect the expected flexural properties of NP-ECC-BRs remains unknown.

Similar to ordinary concrete and other cement-based composites bridge repairs, NP-ECC-BRs feature different rheological characteristics [9] during different setting periods and thus present variable physicochemical states. Therefore, it could be speculated that NP-ECC-BRs are vulnerable and sensitive to TRVs, and a series of their physicochemical processes might be affected during different setting periods.

Before the initial set, continuous TRVs will cause some extent of bleeding to the cement motor or concrete mixture [10], and this might be the same for PVA-ECCs under TRVs. During the setting period (between the initial and final set), low-density calcium silicate hydrate (C-S-H) gels in the cement matrix are gradually transferred to high-density C-S-H gels and agglomerate into larger C-S-H particles [11,12]. If TRVs occur during the setting period, this will probably damage or obstruct the connection of the solid skeleton [13] in the PVA-ECCs, as well as the bond of C-S-H particles or even that of C-S-H gels. After the final set, the hydration of cement is mainly restricted by the diffusion of capillary water in the matrix [14,15], any TRVs that occurred during this period would probably result in the accelerated transportation of free water from the capillary pores toward the interface of the anhydrous cement grains, which would be convenient for the consumption of anhydrous cement grains [16] and further increase the degree of hydration of the matrix, thereby affecting the flexural properties of the PVA-ECCs as a consequence.

As there is the potential for TRVs to greatly affect the properties of newly placed bridge repairs, a number of related investigations have been carried out over recent decades. Previous studies mostly focused on newly placed concrete bridge repairs (NP-C-BRs) as the research object, taking the compressive and bond strength, or splitting tensile strength of NP-C-BRs as vibration responses. The majority of these studies showed that the effects of TRVs on the strengths of NP-C-BRs should not be a cause for concern or need to be considered as a serious risk during bridge repair or widening [17–21]. Through a large number of experimental and field investigations, Manning (1981) pointed out in a research report for American association of state highway and transportation officials (AASHTO) that TRVs seem have no substantial effects on the compressive and bond strengths for NP-C-BRs. He also emphasized that both bond and compressive strengths even appeared to increase slightly for NP-C-BRs of high quality and low slump and, in that case, traffic could be maintained on bridge decks undergoing repair [17]. Subsequently, similar conclusions were drawn by Harsh (1986) [18].

Many recent studies [19–21] have also shown that the effects of TRVs on both the compressive and bond strengths of NP-C-BRs are not a cause for concern. Weathere [19] performed an experimental test to simulate the effects of staged bridge deck construction on the bond strength of concrete/reinforcing steel; the results indicate that after being imposed to different displacements, the bond strength of concrete/reinforcing steel of NP-C-BRs was still capable of developing the actual yield strength of the reinforcing bars. Wang [20] performed an experimental test to study the effects of TRVs on the compressive strength of newly placed high performance concrete (HPC) repairs; the results show that after being vibrated under the imposed vibration model of 2 Hz–3 mm and 4 Hz–3 mm, the 28 days' compressive strength of HPC bridge repairs decreased slightly by 3% (which could be ignored). Hong [21] conducted laboratory studies and a field test to investigate the effect of TRVs on the compressive and bond strengths of fresh concrete during bridge widening; the results show that the effect of TRVs on the strengths of NP-C-BRs should not require serious consideration if the durations of vibration are within 6 h and the corresponding peak particle velocities (PPVs) are within 0.3 cm/s.

Some current studies have shown that TRVs not only significantly reduce the compressive and bond strengths, but also result in a considerable reduction in the tensile strength, flexural strength, and elastic coefficient [22–24] of NP-C-BRs. Zhang [22] observed apparent macrocracks on the surface of NP-C-BRs and, moreover, detected serious internal damage after the specimens were vibrated during the setting period (the period between the initial and final set, where the penetration resistance was within the scope of 3.5 to 28 MPa). As this special period was the most vulnerable to TRVs, Zhang [22] described it as a vibration-sensitive stage. Kwan [23] and Ng [24] investigated the effects of TRVs on curing concrete stitch and found that if vibrations began right after pouring, relatively small cracks or slackness caused by threshold curvature would result in a significant reduction (above 20% reduction) in bond and contraflexural strength.

Additionally, over the years, a number of studies have been performed experimentally, methodologically, and numerically to reveal the performance of newly designed concrete structure or newly placed concrete materials subjected to various sources of dynamic loads, including seismic [25,26], pile driving [27], and blasting [28], and constructive research progress has been achieved.

So far, PVA-ECCs appear to have a bright future in the repair and retrofitting of existing constructed facilities, having been widely used especially in maintenance of the upper structures of bridges. Nonetheless, despite the great potential for TRVs to affect the properties of NP-ECC-BRs, to the best of our knowledge, we found that few of the previous studies have revealed the effects of TRVs on the properties of NP-ECC-BRs, except for the authors' previous work that revealed the effects of TRVs on the tensile behaviors of NP-ECC-BRs [29].

The purpose of this study was to study the effects of TRVs on the flexural properties of NP-ECC-BRs. For this purpose, self-designed vibration equipment was adopted to simulate TRVs, and a total of 333 (37 groups of nine) thin-plate PVA-ECC specimens with size of 400 × 100 × 15 mm were fabricated. Each group of thin-plate PVA-ECC specimens was vibrated under three designed variables, including the age at which the specimens were vibrated (AWV), duration of vibration (DV), and vibration frequency. Finally, after being cured for 28 days, all 37 groups of specimens were tested with a four-point flexural test method to determine their flexural properties.

## 2. Materials and Methods

### 2.1. Materials

Materials consisting of ordinary Portland cement, silica sand, fly ash, water, and polyvinyl alcohol (PVA) fiber were used. Additives consisting of viscosity modifying agent (VMA), high-efficiency defoamer (HED), and high-efficiency water reducing agent (HEWRA) were used to modify the properties of the cement matrix. Detailed source information of the materials is listed in Table 1. The basic physical properties of ordinary Portland cement are listed in Table 2, and the chemical compositions of Portland cement are listed in Table 3. The chemical compositions of fly ash with particle sizes of 0.5–2.0 μm are listed in Table 4. The physical properties of the PVA fiber are listed in Table 5. The mixture proportions of the materials are listed in Table 6.

**Table 1.** Source information of materials (Product Model, Manufacturer, City, Country).

| Materials | Product Model | Manufacturer | City | Country |
|---|---|---|---|---|
| Portland cement | P·O 42.5 R [a] | Ji Dong Cement | Hohhot | China |
| Fly ash | Class– [b] | Ordos Thermal Power Plant | Ordos | China |
| Silicon sand | High-quality | Togtoh Silicon Sand | Hohhot | China |
| Polyvinyl alcohol (PVA) fiber | K–II | Kuraray | – | Japan |
| High-efficiency water reducing Agent (HEWRA) | 3301E | Sika Construction and Building Materials | Dalian | China |
| High-efficiency defoamer (HED) | JXPT–1206 | Jinliangbo Technology | Beijing | China |
| Viscosity modifying agent (VMA) | MK–100000S | Chuangyao Biotechnology | Jinan | China |

[a] The "P·O 42.5 R" Portland cement represented the compressive and flexural strengths of tested samples that were higher than 22.0 and 4.0 MPa after 3 days' curing, and were higher than 42.5 and 4.0 MPa after 28 days' curing according to Chinese National Standards GB175–2007 [30]. [b] The main physical property indexes of "Class-I" fly ash must meet the requirements of fineness ≤12.0% (residue after being screened with a 45 μm square mesh sieve), water demand ratio ≤95%, and loss on ignition ≤5% according to Chinese National Standards GB/T1596–2017 [31].

The particle sizes of high-quality silicon sand ranged from 75 to 135 μm. The water-reducing efficiency of PCSP was about 33.0% (provided by the manufacturer). The effects of the VMA were to increase the cohesiveness and water retention of the matrix, as well as to prevent fiber from aggregation. The effects of the HED were to inhibit or eliminate the production and propagation of air in the mixture during mixing.

In this investigation, a constant PVA fiber volume fraction of 2.0% and a constant water-to-binder ratio of 0.24 were used.

**Table 2.** Basic physical indexes of Portland cement.

| Setting Time (h) | | Water Requirement of Normal Consistency (%) | Flexural Strength (MPa) | | Compressive Strength (MPa) | |
|---|---|---|---|---|---|---|
| Initial | Final | | 3 days | 28 days | 3 days | 28 days |
| 1.95 | 2.98 | 26.93 | 5.82 | 8.14 | 28.92 | 47.64 |

**Table 3.** Chemical compositions of Portland cement (%).

| $Al_2O_3$ | $SiO_2$ | CaO | $Fe_2O_3$ | MgO | $SO_3$ | Loss on Ignition |
|---|---|---|---|---|---|---|
| 7.2 | 23.4 | 55.0 | 3.0 | 2.2 | 2.9 | 2.9 |

**Table 4.** Chemical compositions of fly ash (%).

| $SiO_2$ | $Al_2O_3$ | CaO | $Fe_2O_3$ | $CO_2$ | MgO | $SO_3$ | $K_2O$ | $Na_2O$ | $TiO_2$ | SrO | Others |
|---|---|---|---|---|---|---|---|---|---|---|---|
| 40.3 | 18.2 | 18.1 | 8.5 | 5.2 | 2.3 | 2.1 | 1.8 | 1.3 | 1.0 | 0.7 | 0.6 |

**Table 5.** Physical properties of polyvinyl alcohol (PVA) fiber.

| Fineness (dtex) | Density (g/cm$^3$) | Diameter (μm) | Elongation (%) | Tensile Strength (MPa) | Length (mm) | Young's Modulus (GPa) |
|---|---|---|---|---|---|---|
| 15 | 1.3 | 40 | 6 | 1600 | 12 | 40 |

**Table 6.** Mixture proportions of the material (kg/m$^3$).

| Portland Cement | Silicon Sand | Fly Ash | Water | HEWRA | HED | VMA | PVA Fiber |
|---|---|---|---|---|---|---|---|
| 254 | 457 | 1016 | 304 | 15.2 | 2.6 | 0.6 | 26 |

## 2.2. Methods

### 2.2.1. Specimen Preparation and Testing Procedure

First, a penetration test was conducted, and the setting times of PVA-ECC mixtures were obtained for the predetermination of vibration variables. Second, a total of 333 (37 groups of nine) thin-plate PVA-ECC specimens were fabricated to meet the vibration variables. The 37 groups of specimens included one group of control specimens that would not be vibrated, and 36 groups of specimens that would be vibrated under different vibration variables. Third, the 36 groups of specimens were vibrated using self-designed vibration equipment under different vibration variables so that the simulation of TRVs was achieved. Fourth, after vibration, all 37 groups of specimens were cured for 28 days in a few sealable boxes in which a layer of water was reserved, and a stainless-steel frame was placed; the specimens were placed on the stainless-steel and not in contact with water. This curing method ensured that the relative humidity in the curing boxes was higher than 90% and the temperature in the curing boxes was approximately 20 ± 5 °C. Finally, after being cured for 28 days, all 37 groups of PVA-ECC specimens were tested using a four-point flexural test method.

### 2.2.2. Vibration Equipment and Variables

TRVs were simulated using self-designed vibration equipment, as shown in Figure 1. The vibration table of the equipment where specimens were placed was driven by a motor underneath. The rotational frequencies of the motor were controlled by a frequency converter. The scope of vibration frequency of the working table could be adjusted from 1 to 10 Hz through the frequency converter. The vibration model of this equipment was designed to produce simple harmonic vibration.

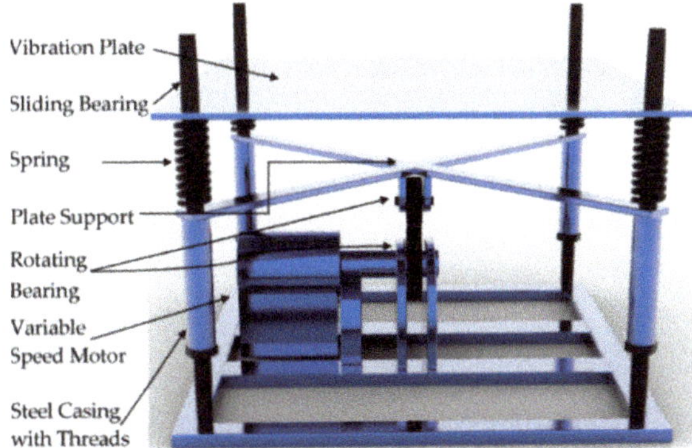


**Figure 1.** Self-designed vibration equipment [29].

In this study, three variables, namely, age when the specimens were vibrated (AWV), magnitude of vibration frequency (VF), and duration of vibration (DV), were considered to be of greatest interest regarding their potential to affect the flexural properties of NP-ECC-BRs.

The first variable, AWV, the age or time when the specimens would be vibrated, was predetermined by conducting a penetration test according to the Chinese National Standard JTG E30–2005 [32]. According to this standard, the hours at which the cement or concrete mixtures reach the stages of the initial and final set are defined as when the penetration resistance reaches 3.5 and 28.0 MPa, respectively. The hours of the initial and final set of the PVA-ECCs in this study were obtained as 7.6 and 23.8 h, respectively, as shown in Figure 2. To investigate the effects of TRVs on the flexural properties of NP-ECC-BRs, the specimens were designed to be vibrated during three setting periods. The selected ages at which the specimens were vibrated, and their corresponding setting periods, are listed in Table 7.

**Table 7.** The ages at which the specimens were vibrated and their corresponding setting periods

| Setting Periods | Before the Initial Set | During the Setting Period | After the Final Set |
|---|---|---|---|
| Penetration Resistance (MPa) | ≤3.5 | 3.5–28.0 | ≥28.0 |
| Setting Time (h) | 0–7.6 | 7.6–23.8 | ≥23.8 |
| Ages When Vibrated (h) | 1.5 | 8, 15, 23 | 36, 48 |

The second variable, DV, was defined as the duration of continuous and uninterrupted vibration disturbance in this study. DVs of 2, 5, 8, and 11 h were chosen to explore whether and to what extent the length of vibration affects the flexural properties of NP-ECC-BRs.

The third variable, vibration frequency, was determined according to previous experimental and field studies on concrete bridges. The vibration frequency was selected to investigate the effects that different magnitudes of TRVs have on the flexural properties of NP-ECC-BRs. Jiang [33] conducted a large number of experimental investigations and field tests and obtained the vibration frequencies of different types of concrete bridges induced by moving vehicles, which ranged from 1.74 to 5.0 Hz, and the corresponding vibration amplitudes, which were in the range of 3.3–9.3 mm The Swiss Federal Laboratories for Materials Science and Technology (EMPA) [34] conducted a series of dynamic tests on 226 concrete girder bridges and obtained a regression empirical equation of the first fundamental frequency and bridge span, as shown in Equation (1).

$$f = 90.4 l^{-0.933} \quad (1)$$

where l is the span (m) of the concrete girder bridge and f is the first fundamental frequency (Hz). According to Equation (1), TRV frequencies of concrete bridges with the commonest spans of 60, 40, 30, and 20 m were calculated as 1.97, 2.88, 3.77, and 5.50 Hz, respectively. Based on these results, vibration frequencies of 2, 3, 4, and 5 Hz were selected in this study, and the corresponding amplitude was maintained as 5 mm according to Jiang [33]. These selected frequencies could cover most of the medium-span and mini-type concrete bridges which are the most common bridge types found.

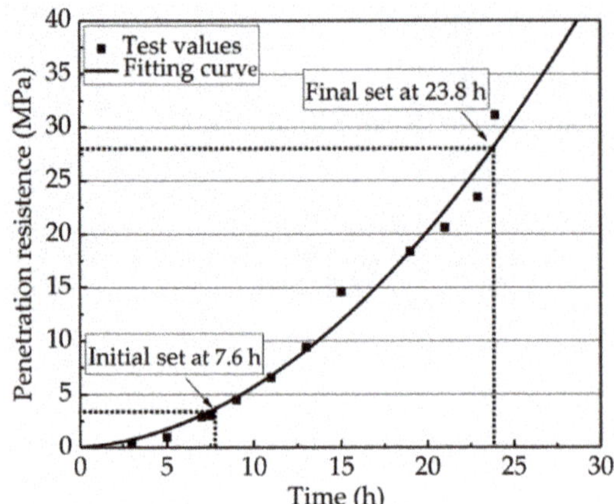


**Figure 2.** Results of the penetration test [29].

2.2.3. Flexural Test Program

A four-point flexural test method was adopted through an MTS exceed universal test system (MTS E43.104, Eden Prairie, MN, USA) with a range of 10 kN. The flexural test was performed on a model with an imposed displacement of 0.06 mm/s during loading. The mid-span deflection was measured by a linear variable differential transformer (LVDT). The loads were collected by a tension-compression sensor (BLR-1/1T, Donghua Electronics, Shanghai, China). The tested data were collected through a Static Strain Collection and Analysis System (DH3820, Huadong Test, Jinjiang, China).

The tested span and force arm of the specimens were designed as 300 and 100 mm, respectively. A sketch of the four-point flexural test is shown in Figure 3.

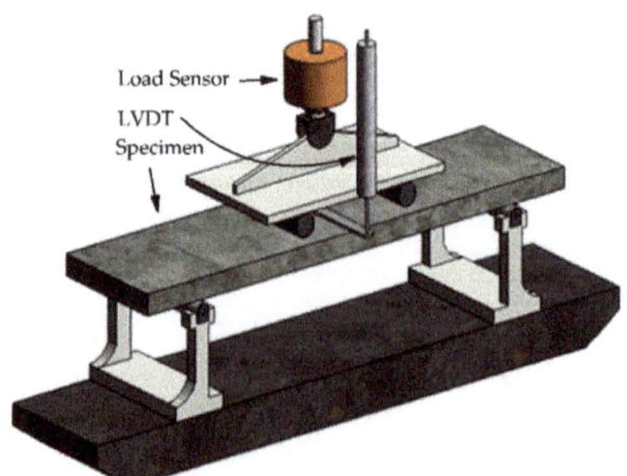


**Figure 3.** Sketch of the four-point flexural test.

### 2.2.4. Specimen Design

The size of the specimens was 400 mm × 100 mm × 15 mm. To meet the designed variables, 36 groups of newly poured PVA-ECC specimens were vibrated under operating conditions for (1) ages of 1.5, 8, 15, 23, 36, and 48 h under the combination of a duration of 5 h and imposed vibration frequencies of 2, 3, 4, and 5 Hz, referred to as Var. 1; and (2) durations of 2, 5, 8, and 11 h at ages of 8 h under imposed vibration frequencies of 2, 3, 4, and 5 Hz, referred to as Var. 2. The detailed numbers, variables, and operating conditions of all the 324 PVA-ECC specimens are shown in Table 8.

**Table 8.** Specimen design.

| Specimen Group | Operating Conditions | Vibration Frequency (Hz) | Age When Vibrated (h) | Duration of Vibration (h) | Vibration Amplitude (mm) | Number of Specimens | Age When Tested (Days) |
|---|---|---|---|---|---|---|---|
| Control | | – | – | – | – | 9 | 28 |
| F2–1.5–5 | Var. 1 | 2 | 1.5 | 5 | 5 | 9 | 28 |
| F2–8–5 | | 2 | 8 | 5 | 5 | 9 | 28 |
| F2–15–5 | | 2 | 15 | 5 | 5 | 9 | 28 |
| F2–23–5 | | 2 | 23 | 5 | 5 | 9 | 28 |
| F2–36–5 | | 2 | 36 | 5 | 5 | 9 | 28 |
| F2–48–5 | | 2 | 48 | 5 | 5 | 9 | 28 |
| F3-1.5-5 | | 3 | 1.5 | 5 | 5 | 9 | 28 |
| F3-8-5 | | 3 | 8 | 5 | 5 | 9 | 28 |
| F3-15-5 | | 3 | 15 | 5 | 5 | 9 | 28 |
| F3-23-5 | | 3 | 23 | 5 | 5 | 9 | 28 |
| F3-36-5 | | 3 | 36 | 5 | 5 | 9 | 28 |
| F3-48-5 | | 3 | 48 | 5 | 5 | 9 | 28 |
| F4–1.5–5 | | 4 | 1.5 | 5 | 5 | 9 | 28 |
| F4–8–5 | | 4 | 8 | 5 | 5 | 9 | 28 |
| F4–15–5 | | 4 | 15 | 5 | 5 | 9 | 28 |
| F4–23–5 | | 4 | 23 | 5 | 5 | 9 | 28 |
| F4–36–5 | | 4 | 36 | 5 | 5 | 9 | 28 |
| F4–48–5 | | 4 | 48 | 5 | 5 | 9 | 28 |
| F5–1.5–5 | | 5 | 1.5 | 5 | 5 | 9 | 28 |
| F5–8–5 | | 5 | 8 | 5 | 5 | 9 | 28 |
| F5-15-5 | | 5 | 15 | 5 | 5 | 9 | 28 |
| F5–23–5 | | 5 | 23 | 5 | 5 | 9 | 28 |
| F5–36–5 | | 5 | 36 | 5 | 5 | 9 | 28 |
| F5–48–5 | | 5 | 48 | 5 | 5 | 9 | 28 |
| F2–8–2 | Var. 2 | 2 | 8 | 2 | 5 | 9 | 28 |
| F2–8–5 | | 2 | 8 | 5 | 5 | 9 | 28 |
| F2–8–8 | | 2 | 8 | 8 | 5 | 9 | 28 |
| F2–8–11 | | 2 | 8 | 11 | 5 | 9 | 28 |
| F3–8–2 | | 3 | 8 | 2 | 5 | 9 | 28 |
| F3–8–5 | | 3 | 8 | 5 | 5 | 9 | 28 |
| F3–8–8 | | 3 | 8 | 8 | 5 | 9 | 28 |
| F3–8–11 | | 3 | 8 | 11 | 5 | 9 | 28 |
| F4–8–2 | | 4 | 8 | 2 | 5 | 9 | 28 |
| F4–8–5 | | 4 | 8 | 5 | 5 | 9 | 28 |
| F4–8–8 | | 4 | 8 | 8 | 5 | 9 | 28 |
| F4–8–11 | | 4 | 8 | 11 | 5 | 9 | 28 |
| F5–8–2 | | 5 | 8 | 2 | 5 | 9 | 28 |
| F5–8–5 | | 5 | 8 | 5 | 5 | 9 | 28 |
| F5–8–8 | | 5 | 8 | 8 | 5 | 9 | 28 |
| F5–8–11 | | 5 | 8 | 11 | 5 | 9 | 28 |

Note: "F2–1.5–5" means that the corresponding group of specimens was vibrated at the age of 1.5 h under the combination of a duration of 5.0 h and a vibration frequency of 2 Hz. A similar naming format is used to indicate the test conditions for the other groups.

### 2.2.5. Data Analysis

To minimize the variance of the data measured in this study, the winsorized mean of the measured data for each group of specimens was taken as the representative value of the corresponding group. Additionally, for the one specimen in each group whose maximum mid-span deflection was the closest to the winsorized mean of the group, its load versus mid-span deflection (P–δ) curve would be taken as the representative P–δ curve of its group.

## 3. Results and Discussion

The results of calculated winsorized means of the cracking load ($P_c$) and the corresponding mid-span deflection ($\delta_c$), extreme flexural load ($P_u$) and the corresponding mid-span deflection ($\delta_u$) of the specimens are shown in Table 9.

**Table 9.** Winsorized means of the cracking load ($P_c$), mid-span deflection ($\delta_c$), extreme flexural load ($P_u$), and mid-span deflection ($\delta_u$) of all 37 groups of polyvinyl alcohol fiber reinforced engineering cementitious composite (PVA-ECC) specimens.

| Specimen Group | Operating Conditions | $P_c$ (kN) | $\Delta P_c$ (%) | $\delta_c$ (mm) | $\Delta\delta_c$ (%) | $P_u$ (MPa) | $\Delta P_u$ (%) | $\delta_u$ (mm) | $\Delta\delta_u$ (%) |
|---|---|---|---|---|---|---|---|---|---|
| Control | | 0.42 | | 0.25 | | 0.98 | | 28.8 | |
| F2–1.5–5 | Var. 1 | 0.41 | −2 | 0.17 | −32 | 0.95 | −3 | 30.8 | 7 |
| F2–8.0–5 | | 0.18 | −57 | 0.15 | −40 | 0.67 | −32 | 30.4 | 6 |
| F2–15.0–5 | | 0.28 | −33 | 0.32 | 28 | 0.89 | −9 | 36.7 | 27 |
| F2–23.0–5 | | 0.41 | −2 | 0.31 | 24 | 0.96 | −2 | 34.6 | 20 |
| F2–36.0–5 | | 0.36 | −14 | 0.31 | 24 | 0.93 | −5 | 31.7 | 10 |
| F2–48.0–5 | | 0.37 | −12 | 0.27 | 8 | 0.95 | −3 | 32.4 | 13 |
| F3–1.5–5 | | 0.32 | −24 | 0.24 | −4 | 0.88 | −10 | 34.7 | 20 |
| F3–8.0–5 | | 0.25 | −40 | 0.22 | −12 | 0.83 | −15 | 35 | 22 |
| F3–15.0–5 | | 0.41 | −2 | 0.28 | 12 | 0.89 | −9 | 39 | 35 |
| F3–23.0–5 | | 0.24 | −43 | 0.44 | 76 | 0.66 | −33 | 26 | −10 |
| F3–36.0–5 | | 0.16 | −62 | 0.32 | 28 | 0.6 | −39 | 29.8 | 3 |
| F3–48.0–5 | | 0.27 | −36 | 0.48 | 92 | 0.7 | −29 | 34 | 18 |
| F4–1.5–5 | | 0.21 | −50 | 0.52 | 108 | 1.01 | 3 | 32.25 | 12 |
| F4–8.0–5 | | 0.27 | −36 | 0.21 | −16 | 0.71 | −28 | 35.5 | 23 |
| F4–15.0–5 | | 0.35 | −17 | 0.28 | 12 | 0.91 | −7 | 33.4 | 16 |
| F4–23.0–5 | | 0.19 | −19 | 0.2 | −20 | 0.83 | −15 | 34.3 | 19 |
| F4–36.0–5 | | 0.32 | −24 | 0.3 | 20 | 0.78 | −20 | 36.6 | 27 |
| F4–48.0–5 | | 0.2 | −55 | 0.25 | 0 | 0.68 | −31 | 38.5 | 34 |
| F5–1.5–5 | | 0.41 | −2 | 0.64 | 156 | 0.97 | −1 | 32.8 | 14 |
| F5–8.0–5 | | 0.28 | −33 | 0.21 | −16 | 0.81 | −17 | 34.2 | 19 |
| F5–15.0–5 | | 0.12 | −71 | 0.35 | 40 | 0.51 | −48 | 27.6 | −4 |
| F5–23.0–5 | | 0.38 | −10 | 0.32 | 28 | 0.96 | −2 | 28.2 | −2 |
| F5–36.0–5 | | 0.37 | −12 | 0.36 | 44 | 0.97 | −1 | 34.2 | 19 |
| F5–48.0–5 | | 0.28 | −33 | 0.3 | 20 | 0.81 | −17 | 33.3 | 16 |
| F2–8.0–2 | Var. 2 | 0.34 | −19 | 0.27 | 8 | 1 | 2 | 33.7 | 17 |
| F2–8.0–5 | | 0.18 | −57 | 0.15 | −40 | 0.67 | −32 | 30.4 | 6 |
| F2–8.0–8 | | 0.32 | −24 | 0.35 | 40 | 0.91 | −7 | 34.3 | 19 |
| F2–8.0–11 | | 0.24 | −43 | 0.19 | −24 | 0.68 | −31 | 44.4 | 54 |
| F3–8.0–2 | | 0.22 | −48 | 0.28 | 12 | 0.63 | −36 | 34.4 | 19 |
| F3–8.0–5 | | 0.25 | −40 | 0.22 | −12 | 0.83 | −15 | 35 | 22 |
| F3–8.0–8 | | 0.34 | −19 | 0.28 | 12 | 0.85 | −13 | 37.5 | 30 |
| F3–8.0–11 | | 0.21 | −50 | 0.18 | −28 | 0.63 | −36 | 38.9 | 35 |
| F4–8.0–2 | | 0.33 | −21 | 0.28 | 12 | 0.84 | −14 | 40.2 | 40 |
| F4–8.0–5 | | 0.27 | −36 | 0.21 | −16 | 0.71 | −28 | 35.5 | 23 |
| F4–8.0–8 | | 0.35 | −15 | 0.24 | −4 | 0.89 | −9 | 29 | 1 |
| F4–8.0–11 | | 0.34 | −19 | 0.36 | 44 | 0.86 | −12 | 33.5 | 16 |

**Table 9.** *Cont.*

| Specimen Group | Operating Conditions | $P_c$ (kN) | $\Delta P_c$ (%) | $\delta_c$ (mm) | $\Delta\delta_c$ (%) | $P_u$ (MPa) | $\Delta P_u$ (%) | $\delta_u$ (mm) | $\Delta\delta_u$ (%) |
|---|---|---|---|---|---|---|---|---|---|
| F5–8.0–2 | | 0.41 | −2 | 0.38 | 52 | 0.97 | −1 | 31.6 | 10 |
| F5–8.0–5 | | 0.28 | −33 | 0.21 | −16 | 0.81 | −17 | 34.2 | 19 |
| F5–8.0–8 | | 0.41 | −2 | 0.15 | −40 | 0.83 | −15 | 32.4 | 13 |
| F5–8.0–11 | | 0.21 | −50 | 0.19 | −24 | 0.64 | −35 | 36.5 | 27 |

Strictly speaking, it should be pointed out that $P_c$ and $\delta_c$ correspond to the load and deflection when the first crack appeared on the specimen. In this study, the end point of the elasticity region in the P–δ curve was defined as the cracking point, as shown in Figure 4, point A. For convenience, the corresponding load and deflection at the cracking point were taken as the $P_c$ and $\delta_c$ of the specimen. Similarly, the corresponding load and deflection at the peak point in the P–δ curve were taken as the $P_u$ and $\delta_u$ of specimens, as shown in Figure 4, point B.

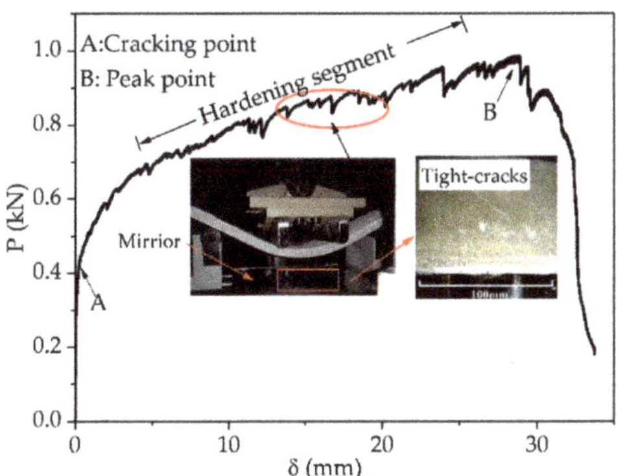


**Figure 4.** Representative load versus mid-span deflection (P–δ) curve of the control specimen, showing the classical strain-hardening and super-toughness characteristics of polyvinyl alcohol fiber reinforced engineering cementitious composites (PVA-ECCs).

### 3.1. *Effects of AWV on the Flexural Properties of NP-ECC-BRs*

In this section, to determine the effects of AWV on the flexural properties of NP-ECC-BRs, a total of 216 (24 groups, and nine specimens in each group) thin-plate PVA-ECC specimens were tested by a four-point flexural test after the specimens were vibrated under the operating conditions for Var. 1.

#### 3.1.1. Effects of AWV on the P–δ curves of NP-ECC-BRs

In a typical P–δ curve of a thin-plate PVA-ECC s specimen, the flexural loads will fluctuate and increase slowly with the increase in deflection after the first crack appears during loading, as shown in Figure 4. Through extensive experimental observation, it was found that the volatility (frequency and value of fluctuation) characteristics of the P–δ curve generally reflected the formation and propagation processes of cracking. That is to say that the number of downward fluctuations of the load at the P–δ curve approximately reflects the number of cracks forming in the specimens, and the levels of load reduction approximately reflect the instantaneous and unstable propagation degree of cracks during the process of cracks from their appearance to maintaining stability.

According to the collected data, the P–δ curves of all the 24 groups of PVA-ECC specimens were obtained, as shown in Figure 5.

It can be seen in Figure 5 that the volatility of the P–δ curves of most of the vibrated groups was not as obvious as that of the control group, indicating that the TRVs that occurred during setting periods affected the multi-cracking characteristics of NP-ECC-BRs to some extent. It can also be seen in Figure 5 that significant hardening segments appeared in the P–δ curves of all the specimens under the operating conditions for Var. 1. In the hardening segments, the flexural loads of the specimens increased slowly with the increase in the corresponding deflection, and the segments presented significant deformation-hardening or strain-hardening characteristics, indicating that vibrations that occurred during different setting periods have no significantly negative impact on the inherent strain-hardening behavior of PVA-ECCs under the operating conditions for Var. 1.The above results are consistent with the tensile stress–strain curves of NP-ECC-BRs that were obtained under the same experimental and parametric conditions in [29].

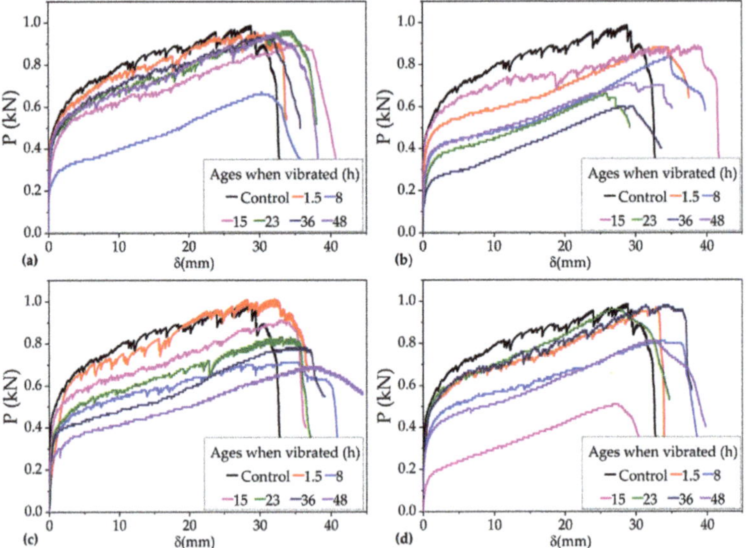


**Figure 5.** P–δ curves of all the 24 groups of specimens under the operating conditions for Var. 1 with frequencies of (**a**) 2 Hz, (**b**) 3 Hz, (**c**) 4 Hz, and (**d**) 5 Hz and that of the control group.

### 3.1.2. Effects of AWV on the Flexural Properties of NP-ECC-BRs

In this subsection, the question of to what extent the TRVs that occurred during different setting periods affected the $P_c$, $\delta_c$, $P_u$, and $\delta_u$ of NP-ECC-BRs was investigated and determined quantitively. The rates of the $P_c$, $\delta_c$, $P_u$, and $\delta_u$ of the groups that were vibrated under the operating conditions for Var. 1 over the corresponding control winsorized means are shown in Figure 6.

Effects of AWV on the cracking load ($P_c$) of NP-ECC-BRs. It can be seen in Figure 6a–d (black lines) that the $P_c$ of each group of specimens decreased by 2%–71% over the control average under the operating conditions for Var. 1, indicating that the effects of AWV on the $P_c$ of NP-ECC-BRs are significantly negative throughout the three setting periods. Furthermore, when the specimens were subjected to frequencies of 2 and 5 Hz, the AWV that resulted in the most negative effects for $P_c$ occurred at ages of 8, 15, or 23 h which were within the setting period (the period between the initial and final set), as shown in Figure 6a,d. However, when the specimens were subjected to the frequencies of 3 and 4 Hz, the AWV that resulted in the most negative effects for $P_c$ occurred at the ages of 36 or 48 h which were after the final set, as shown in Figure 6b,c.

Effects of AWV on the extreme flexural load ($P_u$) of NP-ECC-BRs. It can be seen in Figure 6 that the $P_u$ (blue lines) of NP-ECC-BRs specimens decreased by 1%–39% over the control average under the operating conditions for Var. 1, which shows a similar impact trend of to that of $P_c$. Moreover, it also can be seen in Figure 6 that the rates of $P_u$ and $P_c$ increase or decrease in harmony with the increase in the AWV, and the most negative effect for $P_u$ also occurred at the same AWV as those of $P_c$ under the same vibration frequency (2–5 Hz).

Effects of AWV on the extreme flexural deformation ($\delta_u$) of NP-ECC-BRs. The effects of AWV on the extreme deformation of NP-ECC-BRs specimens were very different to that of the load-bearing properties ($P_c$, $P_u$). The effects of the AWV on the $\delta_u$ of almost all the NP-ECC-BR groups (except for the group F3–23–5, which decreased by 2% and can be ignored) tended to be insignificantly positive, and for most of them, the positive impact degrees were within 20%, as shown in Figure 6a–d (green lines).

Effects of AWV on the cracking deformation ($\delta_c$) of NP-ECC-BRs. Before the initial set and under relatively lower magnitudes of vibration frequency (2 and 3 Hz), it can be seen in Figure 6a,b (red lines) that the effects of AWV on the $\delta_c$ of all the NP-ECC-BR groups tended to be insignificantly negative (within 20% reduction). On the contrary, the effects were significantly positive (more than 20% growth) on the $\delta_c$ of most of the vibrated groups under a relatively higher vibration frequency (4 and 5 Hz) during the same setting period, as shown in Figure 6c,d.

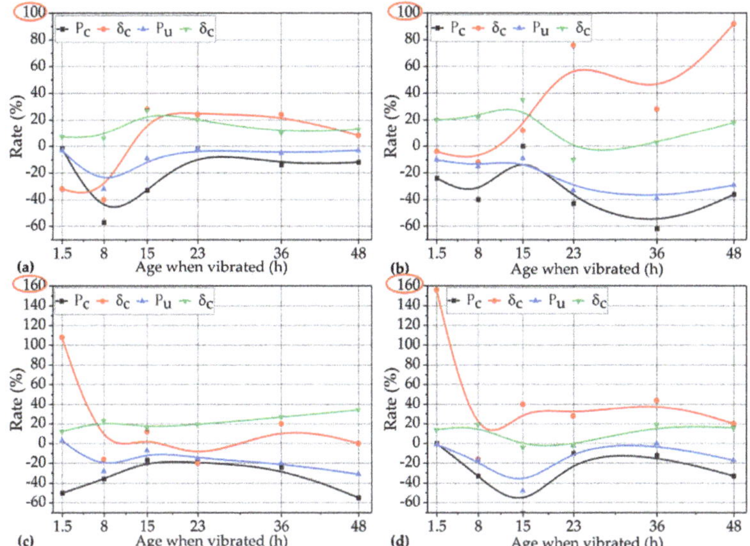


**Figure 6.** Rates of $P_c$, $\delta_c$, $P_u$, and $\delta_u$ for the 24 groups of specimens over the corresponding control averages under the operating conditions for Var. 1 with frequencies of (**a**) 2 Hz, (**b**) 3 Hz, (**c**) 4 Hz, and (**d**) 5 Hz.

In summary, these above results indicate that the effects of TRVs that occurred during different setting periods on the cracking and extreme load-bearing capacity of NP-ECC-BRs tend to be significantly negative under the combination of different frequencies ranging from 2 to 5 Hz and a constant duration of 5 h. On the contrary, the effects of AWV on the extreme flexural deformation of NP-ECC-BRs generally tended to be insignificantly positive. In addition, the effects of AWV on the cracking deformation of newly placed PVA-ECCs varied according to the corresponding vibration frequency that the specimens suffered when TRVs occurred before the initial set.

### 3.2. *Effects of Duration of Vibration (DV) on the Flexural Properties of NP-ECC-BRs*

In this section, to determine the effects of DV on the flexural properties of NP-ECC-BRs when vibrations only occurred during the setting period (the period between the initial and final set, where the penetration resistance was within the scope of 3.5 to 28 MPa), a total of 144 (16 groups) thin-plate

PVA-ECC specimens were tested by a four-point flexural test after the specimens were vibrated under the operating conditions for Var. 2.

### 3.2.1. Effects of DV on the P–δ Curves of NP-ECC-BRs

According to the collected data, the P–δ curves of all the 16 groups of PVA-ECC specimens were obtained, as shown in Figure 7.

It can be seen in Figure 7 that the volatility of the P–δ curves of most of the vibrated groups was not so obvious compared to that of the control group, indicating that there is some extent of negative effect on the multi-cracking characteristics of NP-ECC-BR specimens when the specimens were vibrated during the setting with different DVs. Even so, it also can be seen in Figure 7 that significant hardening segments appeared in the P–δ curves of all the 16 groups of specimens when they were subjected to TRVs with a duration completely covering the setting period. In this section, the longest DV was 11 h, the TRVs began at the age of 8 h, and the time of the final set was 23.8 h. Therefore, for the series of F2,3,4,5–8–11, the duration of 11 h almost covered the whole setting period.

Combined with the results in Subsection 3.1.1 and the results of our previous work in [29], it can be further concluded that the effects of TRVs are not significant, but to some extent there are still negative effects on the strain-hardening behaviors of NP-ECC-BRs within the limits in these studies. On this point, this result was also consistent with the previously found effects of TRVs on the compressive behaviors of NP-C-BRs [17–21].

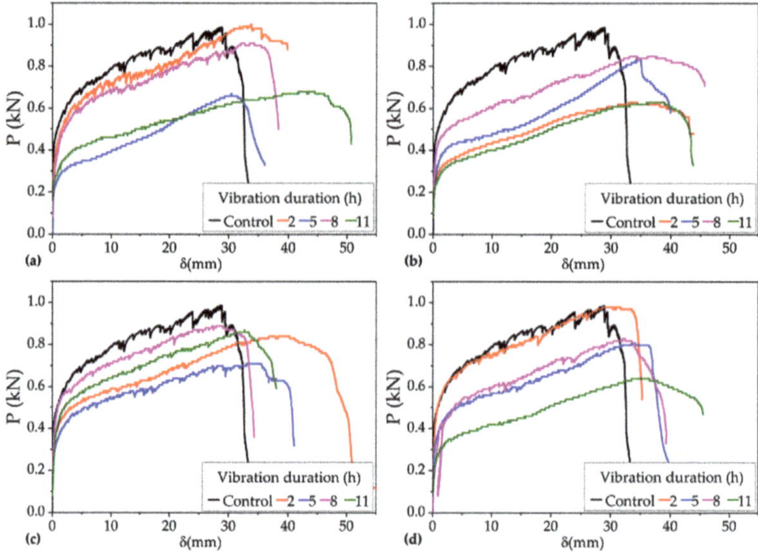


**Figure 7.** P–δ curves of all the 16 groups of specimens under the operating conditions for Var. 2 with frequencies of (**a**) 2 Hz, (**b**) 3 Hz, (**c**) 4 Hz, and (**d**) 5 Hz and that of the control group.

### 3.2.2. Effects of DV on the Flexural Properties of NP-ECC-BRs

In this subsection, we investigate and quantitively characterize the extent to which the DV during the setting period affected the $P_c$, $\delta_c$, $P_u$, and $\delta_u$ of NP-ECC-BRs. The rates of the $P_c$, $\delta_c$, $P_u$, and $\delta_u$ of the groups that were vibrated under the operating conditions for Var. 2 over the corresponding control winsorized means are shown in Figure 8.

The effects of DV on the $P_c$ of NP-ECC-BRs under the operating conditions for Var. 2 are examined. Similar to the $P_c$ obtained under the operating conditions for Var. 1, the effects of DV on the $P_c$ of all the 16 groups of NP-ECC-BRs tended to be significantly negative under the operating conditions for Var. 2, decreasing by 19%–57% over the control average, as shown in Figure 8a–d (black lines). Furthermore, $P_c$ did not change significantly with the increase in DV under the imposed vibration frequencies of 2, 3, and 4 Hz, as shown in Figure 8a–c, while it presented an obvious reduction trend under the imposed vibration frequency of 5 Hz, as shown in Figure 8d. Based on these results and the results obtained in Section 3.1.2, it could be further concluded that the effects of TRVs on the cracking load-bearing capacity of NP-ECC-BRs tended to be significantly negative, while it was not sensitive to increases in the DV when vibrations occurred during the setting period and when the vibration frequencies were lower than 5.0 Hz.

The effects of DV on the $P_u$ of NP-ECC-BRs under the operating conditions for Var. 2 were examined. Likewise, a similar impact trend was presented for $P_u$ compared to $P_c$ under the operating conditions for Var. 2. The difference was that the impact degree of DV on $P_u$ was less obvious than that of $P_c$ in general, as shown in Figure 8a–d (blue lines).

The effects of DV on the $\delta_u$ of NP-ECC-BRs under the operating conditions for Var. 2 were examined. However, the effects of DV on the $\delta_u$ of all the 16 groups of specimens was positive, and for most of them, the positive impact degrees were above 20% under the operating conditions for Var. 2, as shown in Figure 8a–d (green lines). Moreover, $\delta_u$ increased with the increasing of the DV under the imposed vibration frequencies of 2, 3, and 5 Hz, as shown in Figure 8a–c, but it decreased under the imposed vibration frequency of 4 Hz, as shown in Figure 8d.

The effects of DV on the $\delta_c$ of NP-ECC-BRs under the operating conditions for Var. 2 were examined. It can be seen in Figure 8a–d (red lines) that the effects, regardless of the impact trend, polarity, or degree of DV on the $\delta_c$ of the specimens, varied according to the corresponding frequencies that the specimens were subjected to.

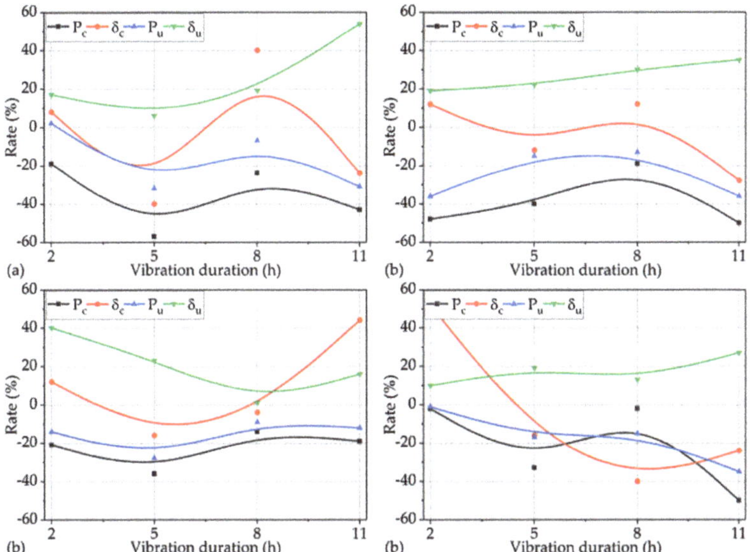


**Figure 8.** Rates of the $P_c$, $\delta_c$, $P_u$, and $\delta_u$ of the 16 groups of specimens over the corresponding control averages under the operating conditions for Var. 2 with frequencies of (**a**) 2 Hz, (**b**) 3 Hz, (**c**) 4 Hz, and (**d**) 5 Hz.

These above results indicate that the effects of DV on the cracking and extreme load-bearing capacity of NP-ECC-BRs were significantly negative (19%–57% reduction for most of them) under the operating conditions for Var. 2. By contrast, the effects of DV on the extreme flexural deformation were significantly positive, increasing by over 20% under the operating conditions for Var. 2. Moreover, the results in Figure 8 also indicate that the longer the DVs, the higher the extreme flexural deformation capacity of NP-ECC-BRs when TRVs occurred only during the setting period. Additionally, the effects of DV on the cracking deformation of NP-ECC-BRs seem to vary according to the corresponding vibration frequencies that the specimens experienced, regardless of the vibration variables of AWV and DV.

These above results regarding the flexural load-bearing capacity obtained in this section was similar to that of the ultimate tensile strength of NP-ECC-BRs in [29], as well as the compressive strength, bond strength, splitting tensile strength, and flexural strength of NP-C-BRs when vibrations occurred only during the setting period [22–24]. The results in [22–24,29] showed that the TRVs that occurred only during the setting periods affected the performance of NP-C-BRs or NP-ECC-BRs to some extent which should not be ignored.

However, these above results concerning the extreme flexural deformation were just opposite to the tensile deformation properties of NP-ECC-BRs when TRVs occurred only during the setting period [29]. The result in [29] showed that the effects of DVs ranging from 2 to 11 h on the extreme tensile deformation capacity of NP-ECC-BRs tended to be negative overall.

### 3.3. *Effects of Vibration Frequency on the Flexural Properties of NP-ECC-BRs*

The results in Sections 3.1 and 3.2 show that effects of TRVs on the flexural properties of NP-ECC-BRs are related to the corresponding vibration frequency. Accordingly, to further quantitatively investigate the flexural properties of NP-ECC-BRs when imposing different vibration frequencies, the effects of vibration frequency on the flexural properties of NP-ECC-BRs are analyzed in this section. The rates of flexural properties of the specimens over the control averages under the operating conditions for Var. 1 and Var. 2 are shown in Figures 9 and 10, respectively.

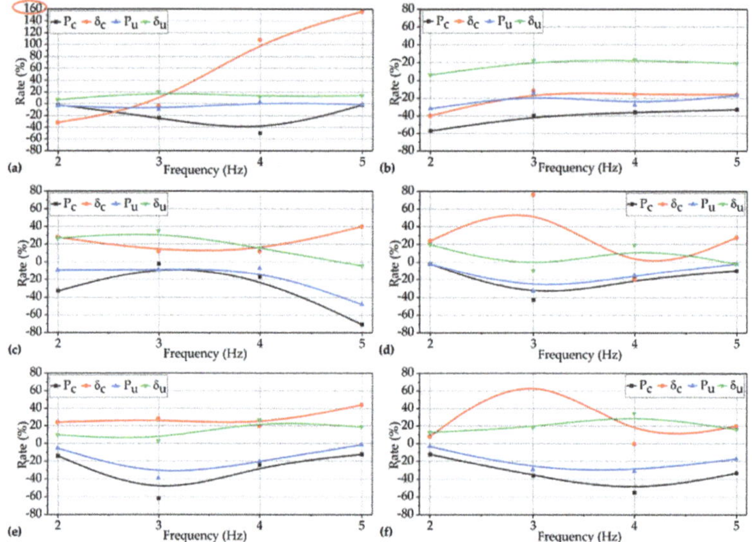


**Figure 9.** The rates of flexural properties with different vibration frequencies over the control averages under the operating conditions for Var. 1 at ages of (**a**) 1.5 h, (**b**) 8 h, (**c**) 15 h, (**d**) 23 h, (**e**) 36 h, and (**f**) 48 h.

### 3.3.1. Effects of Vibration Frequency on the Flexural Properties of NP-ECC-BRs under the Operating Conditions for Var. 1

Effects of Vibration frequency on the flexural properties ($P_c$, $\delta_c$, $P_u$, and $\delta_u$) of NP-ECC-BRs when vibrations occurred before the initial set were examined. It can be seen in Figure 9a that the $P_c$ of the vibrated specimens changed in an approximately linearly fashion from −2% to −24%, and −50% over the control average with an increase in vibration frequency ranging from 2 to 4 Hz, and this changed up to −2% under the vibration frequency of 5 Hz. It also can be seen in Figure 9a that $\delta_c$ presented a roughly opposite trend compared with $P_c$, which changed in an approximately linearly fashion from −32% to 156% over the control average with an increase in vibration frequency ranging from 2 to 5 Hz. Moreover, the effects of vibration frequency on the extreme flexural properties ($P_u$ and $\delta_u$) were not so obvious compared to those of the cracking properties ($P_c$ and $\delta_c$) when vibrations occurred before the initial set.

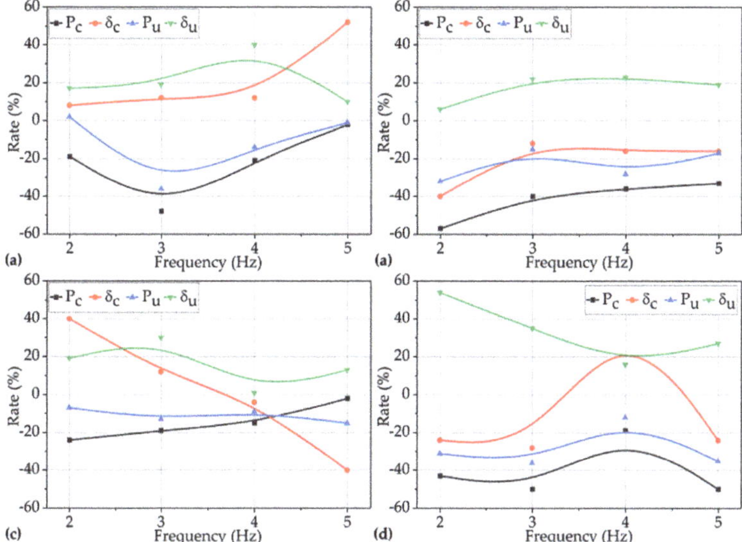


**Figure 10.** The effects of vibration frequency on the flexural properties of newly placed PVA-ECC bridge repairs (NP-ECC-BRs) under the operating conditions for Var. 2 with durations of (**a**) 2 h, (**b**) 5 h, (**c**) 8 h, and (**d**) 11 h.

Effects of Vibration frequency on the flexural properties ($P_c$, $\delta_c$, $P_u$ and $\delta_u$) of NP-ECC-BRs when vibrations began just after the initial set were examined. When the specimens were vibrated at the age of 8 h, it can be seen in Figure 9b that all the flexural properties increase or decrease in harmony and approximately linearly with the increase in vibration frequency. The differences are that the impact polarity was negative on $P_u$, $P_c$, and $\delta_c$, and it was positive on $\delta_u$. Additionally, the impact degree of TRVs on $P_c$ was the largest, indicating that $P_c$ was the most sensitive in its response to TRVs during these periods, followed by $\delta_c$ or $P_u$, and then $\delta_u$.

Effects of Vibration frequency on the flexural properties ($P_c$, $\delta_c$, $P_u$ and $\delta_u$) of NP-ECC-BRs when vibrations occurred after the middle stage of the setting period were examined. When the specimens were vibrated after the age of 15 h, it can be seen in Figure 9c–f that the effects of TRVs on the load-bearing properties ($P_c$ and $P_u$) of the specimens was negative throughout the scope of imposed vibration frequencies in this study. However, it was positive on the deformation properties ($\delta_c$ and $\delta_u$). Figure 9c–f also show that the flexural properties of most of the vibrated groups declined, in general, with the increase in imposed vibration frequency. Moreover, regarding the impact degree (regardless of polarity), the cracking properties (the $P_c$ and $\delta_c$ of most of the specimens presented a reduction or growth above 20%) were more sensitive to the imposed vibration frequency than those of the extreme flexural properties (the $P_u$ and $\delta_u$ of all the specimens presented a reduction or growth within 20%).

#### 3.3.2. Effects of Vibration Frequency on the Flexural Properties of NP-ECC-BRs under the Operating Conditions for Var. 2

As described in Figure 9b and shown in Figure 10b, the effects of imposed vibration frequency on the flexural properties ($P_u$, $\delta_u$, $P_c$, and $\delta_c$) of NP-ECC-BRs specimens is that they increase or decrease in an approximately linearly fashion with the increase in vibration frequency with the constant duration of 5 h during the setting period. However, the results in Figure 10a,c,d show that when subjected to durations of 2, 8, and 11 h during the setting period, the effects of imposed vibration frequency on the deformation properties ($\delta_c$ and $\delta_u$) of some of the vibrated groups were obviously different to those with the constant duration of 5 h. The obvious observed differences can be described as follows: (1) with a duration of 2 h, the effects of TRVs on the deformation properties ($\delta_c$ and $\delta_u$) of the specimens were positive, and they generally increased with an increase in vibration frequency, as shown in Figure 10a; (2) with a duration of 8 h, the rate of the $\delta_c$ of the vibrated PVA-ECC specimens decreased approximately linearly from 40% to − 40% over the control average, as shown in Figure 10c; (3) when subjected to a duration of 11 h, the $\delta_c$ of the vibrated specimens increased by 44% over the control group under the vibration frequency of 4 Hz.

### 3.4. *Explanations and Recommendations*

Regarding the reduction of load-bearing capacity of NP-ECC-BRs under operating conditions for Var. 1 and 2, the following explanations can be made: before the initial set, continuous TRVs will lead to some extent of bleeding to the cement mixture [10], and the stronger the magnitudes of vibration frequency, the greater the amount of bleeding this would lead to; after the initial set, low-density C-S-H gels in the cement matrix gradually are transferred to high-density C-S-H gels and agglomerate into larger C-S-H particles [11,12], which gradually form the solid skeleton of the cement matrix, and TRVs during this period are likely to disturb the connection of the solid skeleton in the cement matrix or even damage the bond of C-S-H particles and C-S-H gels [13]; these above effects demonstrate a great potential for decreasing the cracking and extreme flexural load-bearing capacity of the cement matrix. According to the fiber-bridge constitutive law of engineered cementitious composites (ECCs) and its design principle [35,36], relatively lower cracking strength is beneficial for ECCs to express their bridging effect between PVA fibers and cement matrix and to exhibit larger deformation [37,38]. Therefore, TRVs could result in a negative effect on the load-bearing capacity of NP-ECC-BRs while resulting in a positive effect on the flexural deformation capacity of NP-ECC-BRs.

With regard of the reduction of load-bearing capacity of NP-ECC-BRs under operating conditions for Var. 1 and 2, the authors recommend that (a) in the design of PVA-ECC bridge repairs, engineers should try to improve the design strength of PVA-ECCs, especially its early-age strengths, such as adding an appropriate amount of silica fume and early-strength agents, adopting well-graded silicon sand and high-strength cement; and (b) during the repairing of concrete bridge using PVA-ECCs, constructors should restrict the traffic of heavy-duty trucks and medium-size vehicles in large- and medium-spans of concrete bridges, especially during the period between the initial and final set of PVA-ECC repairs.

## 4. Conclusions

(1) The effects of traffic vibrations are not determinantal, but to some extent, there are negative effects on the strain-hardening behavior of newly placed PVA-ECC bridge repairs within the limits in this investigation.

(2) The effects of traffic vibrations on the cracking and extreme load-bearing capacity of newly placed PVA-ECC bridge repairs were significantly negative (above 20% reduction) for most of the

vibrated groups in this investigation. By contrast, the effects were significantly positive (above 20% growth) on the extreme flexural deformation of newly placed PVA-ECC bridge repairs. Moreover, the longer the durations of vibration, the higher the extreme flexural deformation capacity of newly placed PVA-ECC bridge repairs, generally.

(3) The effects of traffic vibrations on the cracking deformation of newly placed PVA-ECCs varied according to the corresponding vibration frequency that the specimens suffered.

(4) Based on the results obtained, we concluded that the effects of traffic vibrations on the flexural deformation of PVA-ECC bridge repairs should not be a cause for concern, however, serious consideration should be given to the associated reduction of load-bearing capacity.

(5) The authors recommend that (a) in the design of PVA-ECC bridge repairs, engineers should try to improve the design strength of PVA-ECCs, especially its early-age strengths, such as adding an appropriate amount of silica fume and early-strength agents, adopting well-graded silicon sand and high-strength cement; and (b) during the repairing of concrete bridges using PVA-ECCs, constructors should restrict the traffic of heavy-duty trucks and medium-size vehicles in large- and medium-spans of concrete bridge, especially during the period between the initial and final set of PVA-ECC repairs.

**Author Contributions:** Conceptualization, S.L. and C.Y.; Methodology, X.Z., X.W., and H.W.; Project administration, X.W. and C.Y.; Writing—original draft preparation, X.Z.; Writing—review and editing, S.L. and X.W.

**Funding:** This research was funded by the National Natural Science Foundation of China, grant numbers 51768051 and 51968056, Natural Science Foundation of Inner Mongolia, grant number 2017MS0505, and the Science and Technology Innovation Guidance Project of Inner Mongolia, grant number KCBJ2018016.

**Acknowledgments:** The authors are grateful to Yue Y. and Xiumei G. for their help and suggestions in the editing and revision of figures in this article.

**Conflicts of Interest:** The authors declare no conflict of interest.

## References

1. Emmons, P.H.; Sordyl, D.J. The state of the concrete repair industry and a vision for its future. *Concr. Repair Bull.* **2006**, *6*, 7–14.
2. Van den Heede, P.; De Belie, N.; Pittau, F.; Habert, G.; Mignon, A. Life cycle assessment of Self-Healing Engineered Cementitious Composite (SH-ECC) used for the rehabilitation of bridges. In *Life-Cycle Analysis and Assessment in Civil Engineering: Towards an Integrated Vision, Proceedings of the Sixth International Symposium on Life-Cycle Civil Engineering (IALCCE 2018), Ghent, Belgium, 28–31 October 2018*; Caspeele, R., Taerwe, T., Frangopol, D.M., Eds.; Taylor and Francis: Abingdon, UK, 2019; pp. 2269–2275.
3. Guo, X.Y.Y.; Wang, L. Research status of durability repair of concrete structure in industry environment. *Industr. Constr.* **2019**, *49*, 156–162.
4. Maalej, M.; Li, V.C. Flexural tensile-Strength ratio in engineered cementitious composites. *J. Mater. Civ. Eng.* **1994**, *6*, 513–528. [CrossRef]
5. Lepech, M.D.; Li, V.C. Application of ECC for bridge deck link slabs. *Mater. Struct.* **2009**, *42*, 1185–1195. [CrossRef]
6. Lepech, M.D.; Keoleian, G.A.; Qian, S.; Li, V.C. Design of green engineered cementitious composites for pavement overlay applications. In *Life-Cycle Civil Engineering–Biondini and Frangopol (eds), Proceedings of the First International Symposium of the International Association for Life Cycle Civil Engineering, Varenna, Lake Como, Italy, 11–14 June 2008*; Biondini, F., Frangopol, D.M., Eds.; Taylor and Francis: London, UK, 2008; pp. 837–842.
7. Zhang, Z.G.; Hu, J.; Ma, H. Feasibility study of ECC with self-Healing capacity applied on the long-Span steel bridge deck overlay. *Int. J. Pavement Eng.* **2019**, *20*, 884–893. [CrossRef]
8. Keoleian, G.A.; Kendall, A.; Dettling, J.E.; Smith, V.M.; Chandler, R.F.; Lepech, M.D.; Li, V.C. Life cycle modeling of concrete bridge design: comparison of engineered cementitious composite link slabs and conventional steel expansion joints. *J. Infrastruct. Syst.* **2005**, *11*, 51–60. [CrossRef]

9. Kim, D.; Kim, C.Y.; Urgessa, G.; Choi, J.H.; Park, C.; Yeon, J.H. Durability and rheological characteristics of high-volume ground-Granulated blast-Furnace slag concrete containing CaCO3/anhydrate-Based alkali activator. *Constr. Build. Mater.* **2019**, *204*, 10–19. [CrossRef]
10. Zhang, J.Y.; Gao, X.J.; Su, Y. Influence of poker vibration on aggregate settlement in fresh concrete with variable rheological properties. *J. Mater. Civ. Eng.* **2019**, *31*, 04019128. [CrossRef]
11. Ioannidou, K.; Kanduc, M.; Li, L.N.; Frenkel, D.; Dobnikar, J.; Del Gado, E. The crucial effect of early-stage gelation on the mechanical properties of cement hydrates. *Nat. Commun.* **2016**, *7*, 12106. [CrossRef]
12. Hou, D.S.; Li, H.B.; Zhang, L.N.; Zhang, J.R. Nano-Scale mechanical properties investigation of C-S-H from hydrated tri-Calcium silicate by nano-Indentation and molecular dynamics simulation. *Constr. Build. Mater.* **2018**, *189*, 265–275. [CrossRef]
13. Chen, X.T.; Davy, C.A.; Skoczylas, F.; Shao, J.F. Effect of heat-Treatment and hydrostatic loading upon the poro-Elastic properties of a mortar. *Cem. Concr. Res.* **2009**, *39*, 195–205. [CrossRef]
14. Honorio, T.; Benboudjema, F.; Bore, T.; Ferhat, M.; Vourc'h, E. The pore solution of cement-Based materials: Structure and dynamics of water and ions from molecular simulations. *Phys. Chem. Chem. Phys.* **2019**, *21*, 11111–11121. [CrossRef] [PubMed]
15. Pang, X.Y.; Meyer, C. Modeling cement hydration by connecting a nucleation and growth mechanism with a diffusion mechanism. Part II: Portland cement paste hydration. *SECM* **2016**, *23*, 605–615. [CrossRef]
16. Rahimi-Aghdam, S.; Bazant, Z.P.; Qomi, M.J.A. Cement hydration from hours to centuries controlled by diffusion through barrier shells of C-S-H. *J. Mech. Phys. Solids* **2017**, *99*, 211–224. [CrossRef]
17. Manning, D.G. *Effects of Traffic-Induced Vibrations on Bridge Deck Repairs*; NCHRB Synthesis: No. 86; Transportation Research Board: Washington, WA, USA, 1981.
18. Harsh, S.; Darwin, D. Effects of traffic induced vibrations on bridge deck repairs. *Concr. Internat.* **1986**, *8*, 36–44.
19. Weatherer, P.J.; Fargier-Galbadon, L.B.; Hedegaard, B.D.; Parra-Montesinos, G.J. Behavior of longitudinal joints in staged concrete bridge decks subject to displacements during curing. *J. Bridge Eng.* **2019**, *24*, 04019067. [CrossRef]
20. Wang, W.; Liu, S.; Wang, Q.Z.; Yuan, W.; Chen, M.Z.; Hao, X.T.; Ma, S.; Liang, X.Y. The impact of traffic-Induced bridge vibration on rapid repairing high-Performance concrete for bridge deck pavement repairs. *Adv. Mater. Sci. Eng.* **2014**, *2014*. [CrossRef]
21. Hong, S.; Park, S.K. Effect of vehicle-Induced vibrations on early-Age concrete during bridge widening. *Constr. Build. Mater.* **2015**, *77*, 179–186. [CrossRef]
22. Zhang, X.; Zhao, M.; Zhang, Y.J.; Lei, Z.; Zhang, Y.R. Effect of vehicle-Bridge interaction vibration on young concrete. *KSCE J. Civ. Eng.* **2015**, *19*, 151–157. [CrossRef]
23. Kwan, A.K.H.; Ng, P.L. Effects of traffic vibration on curing concrete stitch: Part I—test method and control program. *Eng. Struct.* **2007**, *29*, 2871–2880. [CrossRef]
24. Ng, P.L.; Kwan, A.K.H. Effects of traffic vibration on curing concrete stitch: Part II—cracking, debonding and strength reduction. *Eng. Struct.* **2007**, *29*, 2881–2892. [CrossRef]
25. Fabbrocino, F.; Farina, I.; Modano, M. Loading noise effects on the system identification of composite structures by dynamic tests with vibrodyne. *Compos. Part B-Eng.* **2017**, *115*, 376–383. [CrossRef]
26. On the forced vibration test by vibrodyne. Available online: https://pdfs.semanticscholar.org/badf/82156401c17ab92040a6a54bf8699ee25272.pdf (accessed on 21 September 2019).
27. Amalia, N.; Yuliza, E.; Rokhmat, M.; Wibowo, E.; Viridi, S.; Abdullah, M. The effect of hydrophilic coating on concrete pile surface in pile driving: field test. *J. Phys. Conf. Ser.* **2019**, *1204*, 012053. [CrossRef]
28. Biglari, M.; Ashayeri, I.; Bahirai, M. Modeling, vulnerability assessment and retrofitting of a generic seismically designed concrete bridge subjected to blast loading. *Int. J. Civ. Eng.* **2016**, *14*, 379–409. [CrossRef]
29. Zhang, X.; Liu, S.; Yan, C.; Wang, X.; Wang, H. Effects of vehicle-Induced vibrations on the tensile performance of early-Age PVA-ECC. *Materials* **2019**, *12*, 2652. [CrossRef]
30. Standards, C.N. *Ordinary Portland Cement GB175-2007*; Standardization Administration of China: Beijing, China, 2007.
31. Standards, C.N. *Fly Ash Used in Cement and Concrete GB/T1596-2017*; Standardization Administration of China: Beijing, China, 2017.
32. Standards, C.N. *Test Methods of Cement and Concrete for Highway Engineering JTG E30-2005*; Ministry of Transport of the People's Republic of China: Beijing, China, 2005.

33. Jiang, Z.; Xiao, X.; Li, W. Effects of vehicle-Bridge coupling vibration on the performance of early-Age concrete. *China Concre.* **2015**, *1*, 88–95.
34. Cantieni, R. *Dynamic Behavior of Highway Bridges under the Passage of Heavy Vehicles;* EMPA, Swiss Federal Laboratories for Materials Testing and Research Report No. 220: Dubendorf, Switzerland, 1992.
35. Yang, E.H.; Wang, S.; Yang, Y.; Li, V.C. Fiber-Bridging constitutive law of engineered cementitious composites. *J. Adv. Concr. Technol.* **2008**, *6*, 181–193. [CrossRef]
36. Li, V.C. Integrated structures and materials design. *Mater. Struct.* **2007**, *40*, 387–396. [CrossRef]
37. Tosun-Felekoglu, K.; Felekoglu, B. Effects of fiber-Matrix interaction on multiple cracking performance of polymeric fiber reinforced cementitious composites. *Composit. Part B-Eng.* **2013**, *52*, 62–71. [CrossRef]
38. Wu, Y.; Sun, Q.Y.; Li, W. Improved bending strength and early crack-Resistance performance of engineered cementitious composites reinforced by hybrid-Fiber. *Appl. Mech. Mate.* **2012**, *174–177*, 1047–1050. [CrossRef]

Article

# Flexural Fatigue Performance of Steel Fiber Reinforced Expanded-Shales Lightweight Concrete Superposed Beams with Initial Static-Load Cracks

**Fulai Qu [1,2], Changyong Li [1,2,*], Chao Peng [2], Xinxin Ding [1,2], Xiaowu Hu [1] and Liyun Pan [1,3,*]**

1. School of Civil Engineering and Communications, North China University of Water Resources and Electric Power, Zhengzhou 450045, China; qfl@ncwu.edu.cn (F.Q.); dingxinxin@ncwu.edu.cn (X.D.); Xiaowu-hu@stu.ncwu.edu.cn (X.H.)
2. International Joint Research Lab for Eco-Building Materials and Engineering of Henan, North China University of Water Resources and Electric Power, Zhengzhou 450045, China; zmdpengchao@163.com
3. Henan Provincial Collaborative Innovation Center for Water Resources High-efficient Utilization and Support Engineering, Zhengzhou 450046, China

* Correspondence: lichang@ncwu.edu.cn (C.L.); ply67@ncwu.edu.cn (L.P.); Tel.: +86-371-69127373 (L.P.)

Received: 15 September 2019; Accepted: 3 October 2019; Published: 6 October 2019

**Abstract:** Concerning the structural applications of steel fiber reinforced expanded-shales lightweight concrete (SFRELC), the present study focuses on the flexural fatigue performance of SFRELC superposed beams with initial static-load cracks. Nine SFRELC superposed beams were fabricated with the SFRELC depth varying from 50% to 70% of the whole sectional depth, and the volume fraction of steel fiber ranged from 0.8% to 1.6%. The fatigue load exerted on the beams was a constant amplitude sinusoid with a frequency of 10 Hz and a fatigue characteristic value of 0.10; the upper limit was taken as the load corresponded to the maximum crack width of 0.20 mm at the barycenter of the longitudinal rebars. The results showed that with the increase of SFRELC depth and the volume fraction of steel fiber, the fatigue life of the test beams was prolonged with three altered failure modes due to the crush of conventional concrete in the compression zone and/or the fracture of the tensile rebar; the failure pattern could be more ductile by the prevention of fatigue fracture by the longitudinal tensile rebar when the volume fraction of steel fiber was 1.6% and the reduction of crack growth and concrete strain in the compression zone; the fatigue life of test beams was sensitive to the upper-limit of the fatigue load, a short fatigue life appeared from the higher stress level and larger stress amplitude of the longitudinal rebar due to the higher upper-limit of the fatigue load. The methods for predicting the stress level, the stress amplitude of the longitudinal tensile rebar, and the degenerated flexural stiffness of SFRELC superposed beams with fatigue life are proposed. With the optimal composites of the SFRELC depth ratio and the volume fraction of steel fiber, the controllable failure of reinforced SFRELC superposed beams could be a good prospect with the trend curves of fatigue flexural stiffness.

**Keywords:** superposed beam; steel fiber reinforced expanded-shales lightweight concrete (SFRELC); flexural fatigue; stress level; crack width; flexural stiffness; fatigue life

## 1. Introduction

With the development of high-rise buildings and the increasing span and space of concrete structures, the disadvantage of conventional concrete with great self-weight becomes more acute compared to the imposed loads. This promotes the research and development of structural lightweight concrete with different kinds of lightweight aggregates categorized as natural lightweight aggregates and artificial lightweight aggregates [1–4]. Except for the savings in dead load for structure and on foundation due to the self-lightweight, structural lightweight concrete presents many advantages, including the rising of strength/weight ratio, the reduced risk of earthquake damage to a structure, superior thermal and sound insulation, and better durability [3–5]. However, some problems, including greater brittleness than conventional concrete for the same compressive strength, prevent the wide application of structural lightweight concrete [6,7]. One way to resolve these problems is the use of steel fibers. Based on previous research, steel fiber reinforced expanded-shales lightweight concrete (SFRELC), with market-supply sintered expanded-shales for the fine and coarse aggregates, has excellent mechanical properties, especially those related to tensile performances [8–13]. Due to lower shrinkage and reliable bond behavior with the rebar, the SFRELC can be applied to concrete structures [14–17]. To highlight the peculiarities of the tensile performance of SFRELC and the compressive property of conventional concrete, the SFRELC superposed beams and slabs were innovatively developed. The sectional characterization of this kind of superposed flexural member is composite with tensile SFRELC and compressive conventional concrete [18,19], while the fabrication is the bottom-layer of SFRELC successively followed by up-layer conventional concrete [20–22]. Based on the experimental studies and numerical analyses, the design methods for the flexural performances, including cracking resistance, crack width, flexural stiffness, and bearing capacity, of the reinforced SFRELC superposed beams under static loading were built up [23–26].

Civil engineering concrete structures are always subject to the actions of repeated loads. For example, industry plant structures suffer from the vibration from machines, bridges bear the vibration of rolling vehicles, and ocean structures are subject to repeated waves. This raises an important topic of research on the normal serviceability and reliability of engineering concrete structures under fatigue actions. Normally, the key points are concerned with the fatigue behaviors of structural materials, including the fatigue compressive performance of concrete and the fatigue fracture of tensile rebars [27–31]. At the level of structural members under fatigue load, geometric shapes, sectional composites of concrete and reinforcement, and the sectional stress distribution with different fatigue characteristics are also main factors [32–40]. For the reinforced SFRELC superposed beams, the shear fatigue behaviors were confirmed [34]. Results show that the maximum fatigue load controls the initial diagonal crack width and initial stress amplitude of stirrups, which has a great influence on the fatigue life; with an increase of maximum fatigue load, the diagonal crack width grows quickly and the fatigue failure of the test beams takes place with a large possibility of the fracture of stirrups; overload during fatigue is one of the main reasons for fatigue failure, which results in the sudden increase of diagonal crack width and stirrup' stress amplitude. Except for shear fatigue performance, a study of the flexural fatigue performance is also essential for the engineering application of reinforced SFRELC superposed beams subjected to fatigue loads. As per the previous studies on the flexural fatigue behavior of steel fiber reinforced concrete (SFRC) beams [35–40], the presence of steel fibers could promote fatigue resistance to crack growth, decrease the deflection and increase the energy dissipation at failure, and prolong the fatigue life of reinforced SFRC beams. This is due to the beneficial effect of SFRC at the tensile zone in reducing the stress level of tensile rebars. Therefore, with the integrity of the horizontal interface between SFRELC and conventional concrete, good flexural fatigue behavior of the SFRELC superposed beams could be a good prospect.

Due to a lack of investigation on the flexural fatigue of SFRELC superposed beams, the experimental study was carried out in this paper. Nine beam specimens with rectangular sections were fabricated and tested by a four-point flexural test under a constant amplitude sinusoid at a frequency of 10 Hz with a fatigue characteristic value $\rho_f = 0.10$. The main influence factors were the SFRELC depth change as 0.5, 0.6, and 0.7 times of the whole sectional depth, and the volume fraction of the steel fiber varied as 0.8%, 1.2%, and 1.6%. The upper limit of the fatigue load was taken as the load corresponding to the maximum crack width of 0.20 mm at the barycenter of the longitudinal rebar. This crack width is the limit for reinforced concrete structure cracks within the life of normal serviceability in normal environmental conditions [41]. Based on the experimental results, the crack distribution, crack width, mid-span deflection, failure patterns, and compressive strain of conventional concrete are discussed. Methods for prediction of the stress level, stress amplitude of the longitudinal tensile rebar, and flexural stiffness degeneration with fatigue life are proposed.

## 2. Experimental Work

### 2.1. Raw Materials and Basic Properties of Concrete

Common Portland cement with a strength grade of P.O. 52.5 and class-II degree fly ash were used as binders. Their properties met the requirements of the China codes GB 175 and GB/T1596 [42,43]. The content of fly ash was 20% of the total mass of the binders. The compressive and tensile strengths of the cement were 51.7 MPa and 9.2 MPa at 28 days. The water demand of the fly ash was 95% with a fineness modulus of 3.9. Sintering expanded shale, ceramisite sand, and steel fiber were used for SFRELC. Based on the maximum density principle, sintering expanded shale was sieved in continuous grading with a maximum particle size of 20 mm. Bulk and particle densities were 827 kg/m$^3$ and 1262 kg/m$^3$, cylinder compressive strength was 7.4 MPa, and 1 h water absorption was 9.1%. The ceramisite sand was made from the byproduct of sintering expanded shale with a fineness modulus of 3.56 in continuous grading with a particle size of 1.6–5 mm. Bulk and particle densities were 850 kg/m$^3$ and 1350 kg/m$^3$, and 1 h water absorption was 9.0%. The steel fiber was of a milling type with length $l_f$ = 36 mm, equivalent diameter $d_f$ = 1.35 mm, and the aspect ratio $l_f/d_f$ = 26.7. Crushed limestone with continuous grading of a 5–20 mm particle size and river sand with a 2.84 fineness modulus were used for conventional concrete. Other raw materials were the high-performance water reducer with 20% water reduction and tap-water. The mix proportions of SFRELC and conventional concrete were designed initially based on the absolute volume method at a saturated surface-drying status of raw materials [44,45] and adjusted according to previous studies [16,22,27]. The water/binder ratios of SFRELC and conventional concrete were 0.384 and 0.389.

As listed in Table 1, the slump of the concrete mixture was measured by the slump cone method in accordance with China code GB50080 [46]. The workability for all fresh mixtures was good for casting. The basic mechanical properties, including the cubic compressive strength $f_{fcu}$ ($f_{cu}$), axial compressive strength $f_{fc}$ ($f_c$), splitting tensile strength $f_{ft}$ ($f_t$), and modulus of elasticity $E_c$ of SFRELC and conventional concrete were measured by using standard cubes with a dimension of 150 mm and prisms with a dimension of 150 mm × 150 mm × 300 mm. These standard specimens were cast and cured in the same conditions accompanied by the test beams in accordance with the China codes GB50081 and GB50152 [47,48].

**Table 1.** Basic properties of steel fiber reinforced expanded-shales lightweight concrete (SFRFLC) and conventional concrete.

| Concrete | $v_f$ (%) | Slump (mm) | $f_{fcu}$ (MPa) | $f_{fc}$ (MPa) | $f_{ft}$ (MPa) | $E_c$ (GPa) |
|---|---|---|---|---|---|---|
| FRFLC | 0.8 | 165 | 44.9 | 41.9 | 3.25 | 22.8 |
| | 1.2 | 145 | 43.4 | 37.1 | 3.35 | 22.0 |
| | 1.6 | 140 | 44.2 | 41.8 | 3.73 | 23.7 |
| Conventional concrete | | 160 | 69.1 | 64.7 | 2.71 | 37.0 |

### 2.2. Preparation of Test Beams

Nine test beams with rectangular sections were designed with the prospect of flexural failure [23,41]. The dimensions were, width $b$ = 150 mm, depth $h$ = 300 mm, length $L$ = 3.0 m, and span $l_0$ = 2.7 m. As presented in Figure 1, each beam had 2 Φ18 mm hot-rolled deformed HRB400 longitudinal tensile rebars, 2 Φ8 mm hot-rolled plain HPB300 longitudinal construction rebars, and Φ6 mm hot-rolled plain HPB300 stirrups with spacing of 120 mm. The yield strength and ultimate strength of the Φ18 mm rebars were 451 MPa and 564 MPa. The thickness of the concrete cover for the longitudinal tensile rebars was $c$ = 25 mm. The sectional depth ($h_1$) of SFRELC changed from 150 mm to 210 mm, corresponding to the ratio of the whole sectional depth $\alpha_h$ = 0.5, 0.6, and 0.7, respectively. The volume fraction of the steel fiber of SFRELC was $v_f$ = 0.8%, 1.2%, and 1.6%. The details of the tested beams are presented in Table 2.

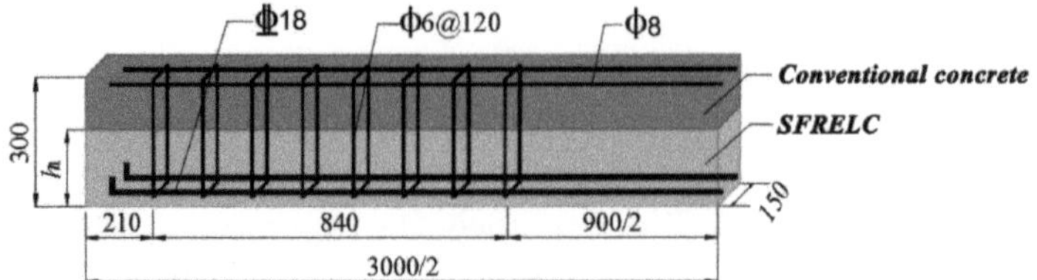


**Figure 1.** 3D view of reinforced SFRELC superposed beam (unit: mm).

**Table 2.** Characteristics of test beams.

| Beam No. | $b$ (mm) | $h$ (mm) | $v_f$ (%) | $\alpha_h$ |
|---|---|---|---|---|
| BF0.8-0.5a | 147 | 403 | 0.8 | 0.5 |
| BF0.8-0.6a | 160 | 403 | 0.8 | 0.6 |
| BF0.8-0.6b | 156 | 406 | 0.8 | 0.6 |
| BF0.8-0.7a | 162 | 409 | 0.8 | 0.7 |
| BF0.8-0.7b | 158 | 406 | 0.8 | 0.7 |
| BF1.2-0.6a | 153 | 403 | 1.2 | 0.6 |
| BF1.2-0.6b | 157 | 404 | 1.2 | 0.6 |
| BF1.6-0.6a | 156 | 402 | 1.6 | 0.6 |
| BF1.6-0.6b | 154 | 404 | 1.6 | 0.6 |

The procedure for the fabrication of the test beams was the same as previous studies [23–27]. The steel reinforcement framework was made of steel bars with designed dimensions and geometric shapes, and the formwork was assembled from formed steel plates. The mixture of concrete was mixed by a 500 L horizontal-shaft forced mixer cast into the formwork and compacted by the vibrators that hung on the outside surface of the formwork. After that, the top surface of the SFRELC was covered by a plastic film for 48 h. Then, the formwork was demolded, and the test beams were cured for 7 d by spraying water.

### 2.3. Test Method

The four-point bending tests were carried out on the test beams by an MTS fatigue testing machine. The concentrated loads were measured and controlled by the MTS loading system on the top surface of the beam, as exhibited in Figure 2. Test beams were firstly loaded under static load to make the initial cracks. To accelerate the test procedure and get failure under fatigue load, the upper limit of the fatigue load, $P_{max}$, on each beam was taken as the load corresponding to the maximum crack width of 0.20 mm at the side surface barycenter of the longitudinal tensile rebars. In accordance with China code GB50152 [48], the fatigue load was taken as a constant amplitude sinusoid at a frequency of 10 Hz with the fatigue characteristic value $\rho_f$ = 0.10. The test procedure was as follows: (1) pre-load within 20% $P_{max}$ was exerted for 2–3 times to ensure the loading and data collection system worked well; (2) the static repeat load tests within $P_{max}$ were conducted twice, and the fatigue test was started at the initial load of 0.1$P_{max}$; (3) test data were read while the fatigue number $N$ reached 5000, 10,000, 50,000, 100,000, 200,000, and 500,000; after that, data were collected at fatigue numbers with an increment of 500,000; (4) the test was conducted until the beam was damaged.

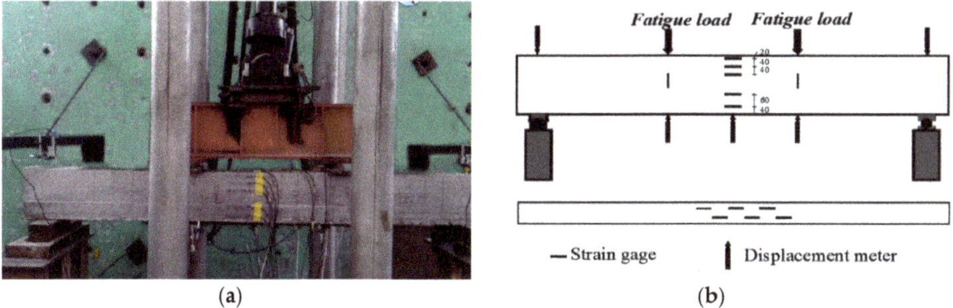

(a) (b)

**Figure 2.** Fatigue testing devices of test beams: (**a**) live-action photo; (**b**) gauge installation.

As displayed in Figure 2, the concrete strains at the mid-span section of the test beam were measured by the six strain gauges bonded on the side surface, and the strain gauges were continuously bonded on the top and bottom surfaces. The deflection of the test beam was measured by the electrical displacement meters installed on the supports and mid-span section. Crack width at the side surface barycenter of the longitudinal tensile rebars was read by using a crack visualizer with a precision of 0.02 mm. To simplify the discussion on the general variation of the compressive concrete strain, the concrete strains at the top surface of the test beams are computed as an average for each group. The width of the main cracks and the mid-span deflection of the test beams in each group are also computed as an average in this study.

## 3. Results

### 3.1. Crack Distribution and Growth

Figure 3 presents the crack distribution of test beams before fatigue failure. The main cracks at pure bending sections almost appeared during the procedure of static loading before fatigue. Combined with the changes of the maximum and average crack width presented in Figure 4, the cracks elongated and opened rapidly under the fatigue within 5000 numbers. After that, the development of cracks became steady with some root cracks at the bottom of the test beams.

With the increase of the SFRELC depth ratio $\alpha_h$, more cracks appeared on the test beams while the maximum and average main crack width decreased. In the steady development period, the maximum crack width decreased by about 11% when $\alpha_h$ increased from 0.5 to 0.6, and continuously decreased by about 15% when $\alpha_h$ increased from 0.6 to 0.7. This is due to the increasing sectional depth of SFRELC which provided a restraining effect on the crack growth and strengthened the bearing capacity of SFRELC in the tension zone [24,25]. Meanwhile, the tensile stress of the longitudinal rebars was reduced as presented in relative studies on reinforced SFRC beams [38–40].

With the volume fraction of steel fiber $v_f$ increasing from 0.8% to 1.6%, the elongation of cracks reduced and the crack width became uniform. In the steady development period, the maximum crack width decreased by about 8% when $v_f$ increased from 0.8% to 1.2% and continuously decreased by about 4% when $v_f$ increased from 1.2% to 1.6%. This exhibits that the presence of steel fibers in the tension zone of the test beams improved the internal condition of the mid-span section in flexure [35–40,49,50].

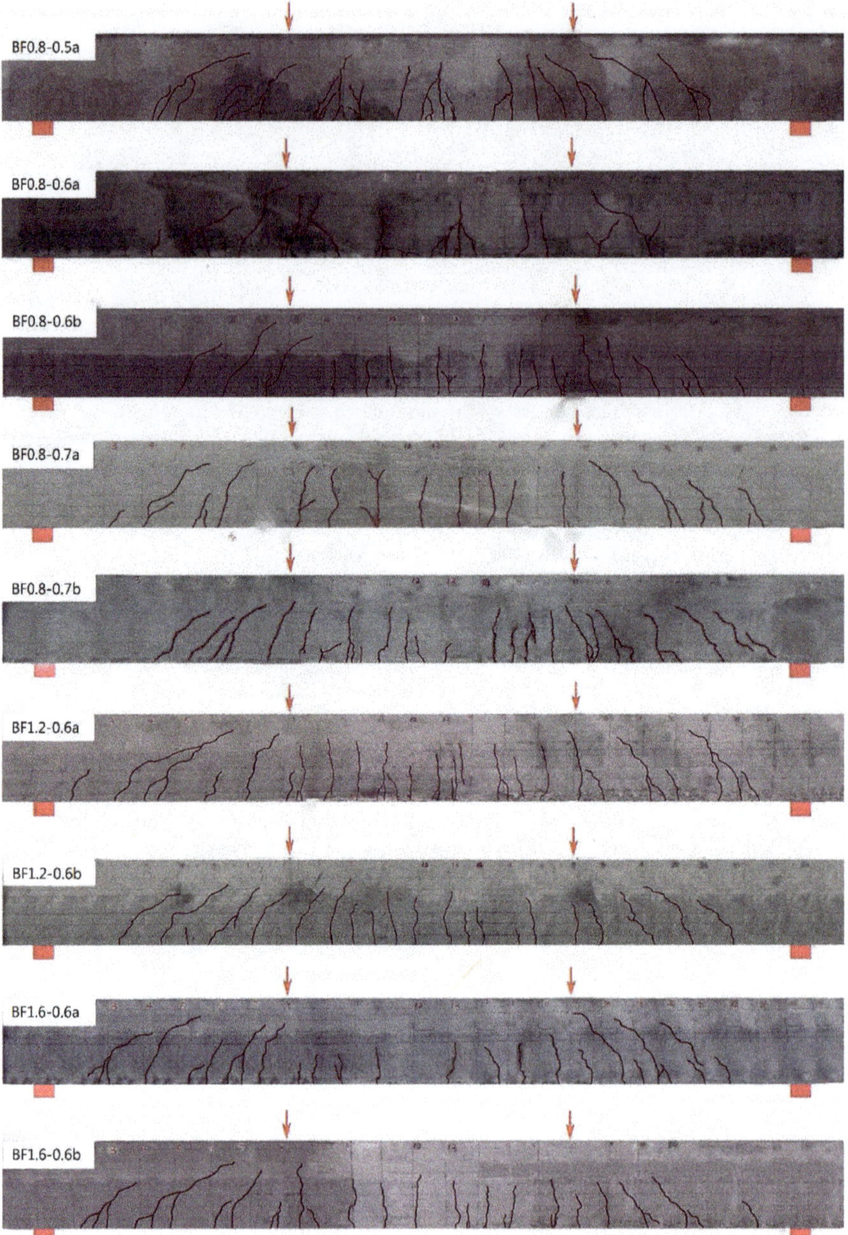


**Figure 3.** Crack distribution of test beams with different $\alpha_h$ or $v_f$.

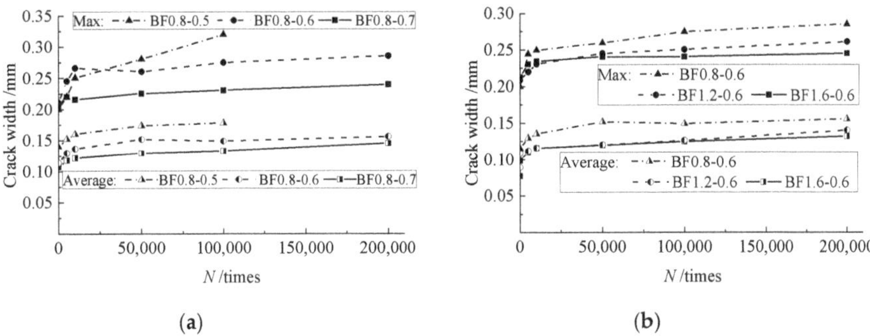


**Figure 4.** Changes of crack width affected by: (**a**) $\alpha_h$; and (**b**) $v_f$.

### 3.2. Failure Pattern

As exhibited in Figure 5 showing the interesting graphical locations of the fatigue failure patterns of the test beams, when the fatigue was close to test-beam failure, one of the main cracks grew upward quickly accompanied by the pull-out of steel fibers. In the case of the same $v_f$ = 0.8%, sudden failure took place on test beam BF0.8-0.5a with $\alpha_h$ = 0.5 along the main crack with a sudden fracture of the rebar. While in the case of the test beams BF0.8-0.6a/b and BF0.8-0.7a/b with $\alpha_h$ = 0.6 and 0.7, the conventional concrete crushed in the compression zone almost with the simultaneous fracture of the rebar. This indicates the beneficial effect of the higher SFRELC depth on the improvement of failure brittleness. Meanwhile, the stagger deformation on two sides of the failure section led to a greater splitting force on the horizontal interface between SFRELC and superposed conventional concrete, which resulted in peeling along the horizontal interface on test beams BF0.8-0.7a/b with $\alpha_h$ = 0.7.

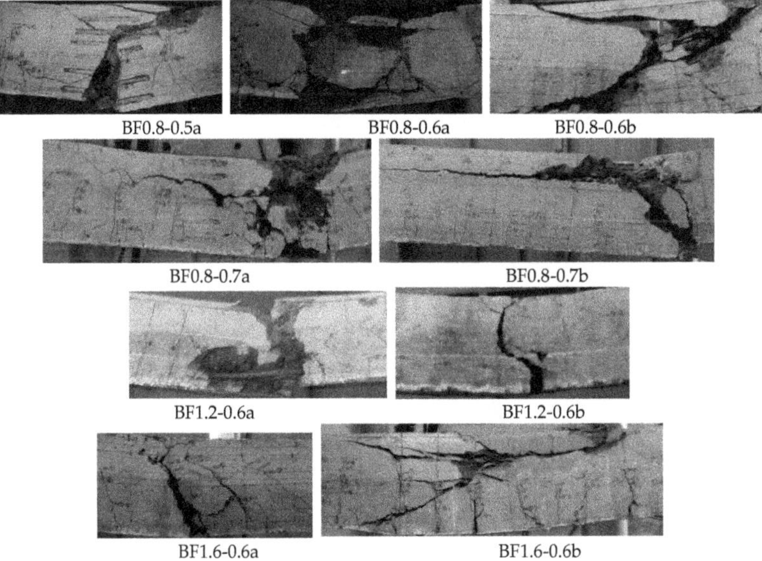


**Figure 5.** Interesting graphical locations of the fatigue failure patterns of test beams.

In case of the same SFRELC depth ratio $\alpha_h$ = 0.6, the crush of conventional concrete almost with the simultaneous fracture of the rebar occurred on test beams BF0.8-0.6a/b with $v_f$ = 0.8% and BF1.2 0.6a/b with $v_f$ − 1.2%, while the crush of conventional concrete only took place on test beams

BF1.6-0.6a/b with $v_f$ = 1.6%. This confirms that the presence of steel fibers in the tension zone of the test beams would not only improve the toughness of failure but also change the failure pattern [16,22]. An obvious phenomenon of more main cracks with shorter extension was observed in BF1.6-0.6b. This is beneficial to the redistribution of tensile force along the span into several main crack sections and the assistance of steel fibers bearing tensile force at the cross-section, and finally contributes to the increase of fatigue lift due to the reduction of tensile stress of the longitudinal tensile rebars.

Table 3 summarizes the fatigue lift and the characteristics of the failure pattern of the test beams. To get the same maximum crack width under the initial static-load, the actual upper-limit of the fatigue load $P_{max}$ was different on the two beams of each group due to the variance of the appearance and growth of cracks. The shorter fatigue life of a test beam existed with a higher $P_{max}$, although the same failure pattern appeared in the test beams of a group. This displays that the fatigue life was sensitively affected by the upper-limit of fatigue load.

**Table 3.** Fatigue life and failure mode of test beams.

| Beam No. | $P_{max}$ (kN) | N (number) | Failure Pattern |
|---|---|---|---|
| BF0.8-0.5a | 95 | 183,654 | Abrupt along a section of the main crack with sudden fracture of a rebar. |
| BF0.8-0.6a | 95 | 196,891 | Crush of concrete in the compression zone and fracture of a rebar. |
| BF0.8-0.6b | 90 | 274,562 | |
| BF0.8-0.7a | 90 | 325,302 | Crush of concrete in the compression zone and fracture of a rebar, peeling along the horizontal interface. |
| BF0.8-0.7b | 100 | 178,997 | |
| BF1.2-0.6a | 95 | 208,963 | Crush of concrete in the compression zone and fracture of a rebar. |
| BF1.2-0.6b | 95 | 213,255 | |
| BF1.6-0.6a | 100 | 245,968 | Crush of concrete in the compression zone. |
| BF1.6-0.6b | 95 | 301,256 | |

### 3.3. Compressive Concrete Strain

Under the fatigue load, the total strain of the material increases with the fatigue life in three stages: the first stage is about 5%–10% fatigue life, the strain grows rapidly due to the inherent flaw initiation, such as the presence of air voids and weak regions between aggregates and binder paste; the second stage is about 80% fatigue life, the strain increases slowly and steadily due to the steady development of internal damage; the third stage is about 5%–10% fatigue life, the strain increases fast due to the unstable development of internal damage over a critical limit [37,40,49,50]. In this study, as presented in Figure 6, the increase of concrete strain in the top surface of the compressive zone of the test beams had similar trends as mentioned above in the first two stages with cyclic numbers, while the third stage was not measured due to the defect of measures. During these states, the strain development in the compression zone of the superposed beam did not change with the presence of steel fibers in the tension zone.

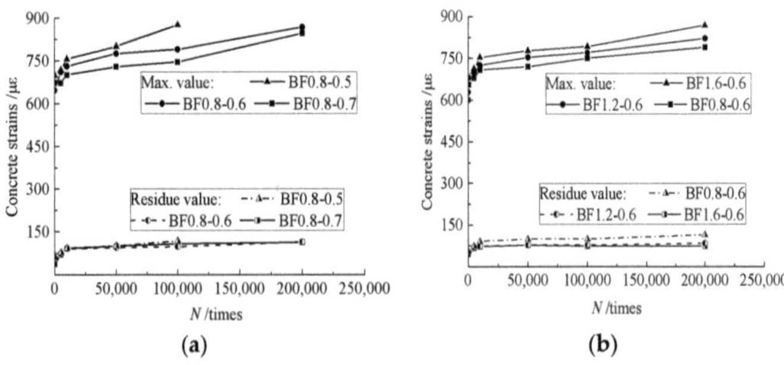


**Figure 6.** Changes of compressive strain at the top-surface of the mid-span section affected by: (**a**) $\alpha_h$; and (**b**) $v_f$.

In the case of the same $v_f = 0.8\%$, the initial compressive strain at the top surface of the test beams was approximately equal due to the almost average value of $P_{max}$. However, at the same cyclic numbers of fatigue load, the maximum strain gradually reduced with the increase of $\alpha_h$. This is due to the gradual fracture of steel fibers in the tensile zone with the increase of cyclic numbers of fatigue load. As a result, the neutral axis moved up. The superposed beam with larger $\alpha_h$ had more steel fibers in the tensile zone subjected to the tensile force to restrain the upward movement of the neutral axis, which gives a relative larger compression zone with a slowly increased strain compared to the beams with smaller $\alpha_h$.

In the case of the same $\alpha_h = 0.6$, the maximum strain increased with the increase of $v_f$. This is due to the increased upper-load exerted on the beams. The root is in the higher restraining of the steel fibers on cracks rather than the improvement on the ultimate bearing capacity of the reinforced SFRELC superposed beams [23–26]. As a result, the steel fibers subjected to higher tensile stress lost the loading function rapidly and the neutral axis moved up quickly, and the compressive strain increased quickly with the reduced compression zone.

### 3.4. Changes of Mid-Span Deflection

The mid-span deflection is a macro index reflecting the sectional deformation of test beams. Similar to the changes in the compressive strain of concrete, as presented in Figure 7, the mid-span deflection of test beams passed through the same stages under the fatigue load. Due to the decrease of the flexural stiffness of the test beams with the increase of $\alpha_h$ [24,25], the mid-span deflection of the test beams increased with $\alpha_h$ in the case of the same $v_f$. Meanwhile, in the case of the same $\alpha_h$, the confinement of steel fibers to cracks enhanced the integrality of the test beam and benefited the flexural stiffness. This led to a reduction in the mid-span deflection of the test beams with the increasing volume fraction of steel fiber $v_f$.

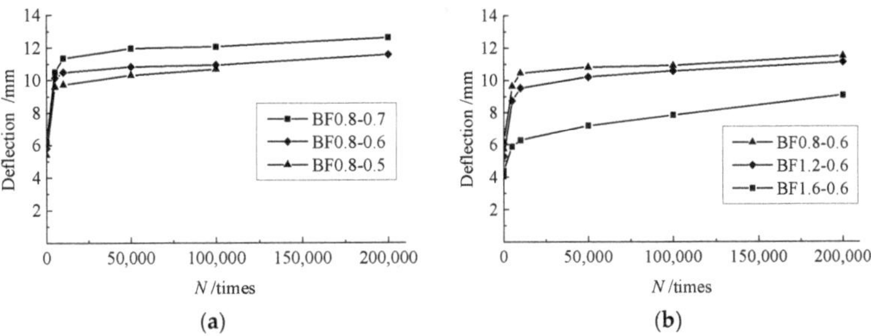


**Figure 7.** Changes of mid-span deflection affected by: (**a**) $\alpha_h$; and (**b**) $v_f$.

## 4. Discussion

The flexural fatigue performance of the reinforced SFRELC superposed beam has the same basic fatigue mechanisms in flexure as the reinforced SFRC beams and the reinforced conventional concrete flexural members [37–40,49,50]. This should be the compound response of the fatigue compressive strength of conventional concrete, the flexural fatigue behavior of SFRELC, and the fatigue property of the tensile steel bars based on the sectional composition. On the design premise of the tested reinforced SFRELC superposed beams that failed with certain flexural ductility under static loading [23,41], the typical fatigue response with three stages divided into the fast first and rapid third stages accompanied with a second steady stage for the compressive strain of conventional concrete provides the foundation for the flexural fatigue performance of the reinforced SFRELC superposed beams [30,41]. In fact, a similar fatigue response with three stages are presented for the SFRC and steel fiber reinforced lightweight concrete in compression [27–29], the SFRC in flexure [35,36], and the steel bars in tension [31]. Therefore, the good properties of materials are the basis for good performances of structural members. On this premise, how to give a full display of the advantages of the materials needs to be given attention, as it is an important issue for the optimization of the composite section within different materials.

For the reinforced SFRELC superposed beams under fatigue loads, the design object should be the utilization of the compression capacity of conventional concrete and the tensile peculiarity of SFRELC to a maximum degree. This has a broader guiding significance for the implication of these kinds of superposed flexural members in civil engineering structures. Meanwhile, the additional phenomena of fatigue failure of the reinforced SFRELC superposed beams may be presented due to the laniation of the horizontal interface between SFRELC and conventional concrete. This can be controlled by the optimization of the SFRELC depth and the volume fraction of steel fibers. If larger shear-tensile stress exists along the horizontal interface due to the higher SFRELC depth, a direct method is to insert vertical reinforcement for the horizontal interface to enhance the entirety of the composite sections [51].

## 5. Prediction of Fatigue Life

### 5.1. Stress Level of Superposed Beam

As the upper-load was determined by the 0.2 mm maximum crack width at the barycenter of the longitudinal rebar, the concrete, steel fiber, and longitudinal rebar were under high stress levels during the fatigue load. The ultimate bending capacity of the reinforced SFRELC superposed beam can be computed with equations as follows [24,41],

$$\alpha_1 f_c b x = f_y A_s + k_t \alpha_t \lambda_f f_t b x_t, \tag{1}$$

$$M_u = f_y A_s (h_0 - x/2) + k_t \alpha_t \lambda_f f_t b x_t (h - x/2 - x_t/2), \tag{2}$$

$$x_t = h - x/k_1 \beta_1, \tag{3}$$

where, $\alpha_1$ is the coefficient of the simplified stress figure for compressive concrete, $f_c$ is the axial compressive strength of conventional concrete, $b$ is the sectional width, $x$ is the equivalent depth of the compressive zone, $f_y$ is the yield strength of the longitudinal tensile rebar, $A_s$ is the sectional area of the longitudinal tensile rebar, $k_t$ is the adjustive coefficient for the equivalent force of tensile stress, here, $k_t$ = 1.25; $\alpha_t$ is the strengthening coefficient of steel fiber to SFRELC, in this study $\alpha_t$ = 0.31; $\lambda_f$ is the steel fiber factor, here, $\lambda_f = 26.7v_f$; $f_t$ is the tensile strength of concrete with the same grade of SFRELC, $x_t$ is the equivalent depth of the tensile zone, $k_1$ is the adjustive coefficient for the equivalent depth of the tensile zone, here $k_1$ = 0.85; $\beta_1$ is the coefficient of the simplified stress figure for compressive concrete, here $\beta_1$ = 0.8.

Table 4 presents the computed stress level of fatigue $S = P_{max}/P_u$. The values are quite high from 0.735 to 0.867. This led to a short fatigue life of test beams due to the unrecoverable plasticity of concrete [49,50].

**Table 4.** Stress level of test beams and stress amplitude of longitudinal tensile rebars.

| Test Beam | $\alpha_h$ | $P_{max}$ (kN) | $\rho_f$ | $N$ (number) | $P_u$ (kN) | $S$ | Average $S$ | $\sigma^f_{s,max}$ (MPa) | $\sigma^f_{s,min}$ (MPa) | $\Delta\sigma^f_s$ (MPa) |
|---|---|---|---|---|---|---|---|---|---|---|
| BF0.8-0.5a | 0.5 | 95 | 0.1 | 183,654 | 113.3 | 0.838 | 0.838 | 262.6 | 26.3 | 236.4 |
| BF0.8-0.6a | 0.6 | 95 | 0.1 | 196,891 | 114.5 | 0.830 | 0.808 | 262.8 | 26.3 | 236.5 |
| BF0.8-0.6b | 0.6 | 90 | 0.1 | 274,562 | | 0.786 | | 248.9 | 24.8 | 224.1 |
| BF0.8-0.7a | 0.7 | 90 | 0.1 | 325,302 | 115.4 | 0.780 | 0.823 | 248.4 | 24.8 | 223.5 |
| BF0.8-0.7b | 0.7 | 100 | 0.1 | 178,997 | | 0.867 | | 275.9 | 27.6 | 248.4 |
| BF1.2-0.6a | 0.6 | 95 | 0.1 | 208,963 | 120.6 | 0.788 | 0.788 | 269.4 | 26.9 | 242.4 |
| BF1.2-0.6b | 0.6 | 95 | 0.1 | 213,255 | | 0.788 | | 269.4 | 26.9 | 242.5 |
| BF1.6-0.6a | 0.6 | 100 | 0.1 | 245,968 | 129.3 | 0.773 | 0.754 | 269.3 | 26.9 | 242.3 |
| BF1.6-0.6b | 0.6 | 95 | 0.1 | 301,256 | | 0.735 | | 255.8 | 25.6 | 230.2 |

Normally, the relationship of stress level and fatigue life is represented by the *S-N* curve as follows [28],

$$S = a + b \log N, \tag{4}$$

where $a$ and $b$ are the coefficients dependent on test data.

Based on test data of this study, $a = 2.304$, $b = -0.278$. The comparison between test data and the computed results is exhibited in Figure 8. A good fit can be given with Pearson's $R = 0.826$, between the line and test data.

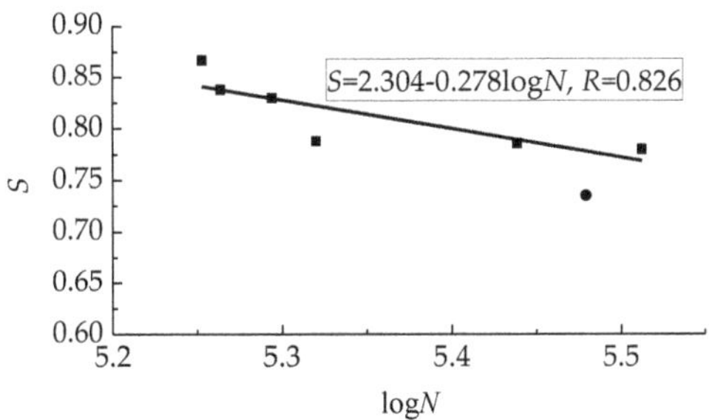


**Figure 8.** Fitness of *S-N* curve with test data.

### 5.2. Stress Amplitude of Longitudinal Rebar

According to the China code [41], the stress amplitude of the longitudinal rebar can be computed as follows,

$$\Delta\sigma_{s}^{f} = \sigma_{s,max}^{f} - \sigma_{s,min}^{f}, \tag{5}$$

$$\sigma_{s,max}^{f} = \alpha_{E}^{f} M_{max}^{f} (h_0 - x_0)/I_0, \tag{6}$$

$$\sigma_{s,min}^{f} = \alpha_{E}^{f} M_{min}^{f} (h_0 - x_0)/I_0, \tag{7}$$

where, $M_{max}^{f}$ and $M_{min}^{f}$ are the maximum and minimum moment under the upper and lower-limit of fatigue loads, $\sigma_{s,max}^{f}$ and $\sigma_{s,min}^{f}$ are the maximum and minimum stress of the longitudinal rebar, $\alpha_{E}^{f}$ is the ratio of the modulus of elasticity of the rebar to the modulus of fatigue deformation of conventional concrete, $I_0$ is the inertia moment of the transformed section, $x_0$ is the depth of the compression zone of the transformed section, $h_0$ is the effective depth of the transformed section.

The sectional characteristic parameters of the transformed section in the condition of unchanged initial centroid can be computed as follows [24],

$$W_0 = I_0/(h - y_0), \tag{8}$$

$$y_0 = \left( \frac{b(h-h_1)^2}{2} + \alpha_{Ec} b h_1 (h - \frac{h_1}{2}) + (\alpha_{Ec} - 1) A_s (h - a_s) \right) / (b(h-h_1) + \alpha_{Ec} b h_1 + (\alpha_{Ec} - 1) A_s), \tag{9}$$

when $y_0 \geq h - h_1$,

$$I_0 = \frac{b(h-h_1)^3}{12} + b(h-h_1)(y_0 - \frac{h-h_1}{2})^2 + \frac{\alpha_{Ec} b \left( (y_0 - h + h_1)^3 + (h - y_0)^3 \right)}{3} + (\alpha_{Es} - 1) A_s (h - a_s - y_0)^2, \tag{10}$$

when $y_0 < h - h_1$,

$$I_0 = \frac{b(h-h_1-y_0)^3}{3} + \frac{b y_0^3}{3} + \alpha_{Ec} b h_1 (h - y_0 - \frac{h_1}{2})^2 + \frac{\alpha_{Ec} b h_1{}^3}{12} + (\alpha_{Es} - 1) A_s (h - a_s - y_0)^2, \tag{11}$$

where, $\alpha_{Ec}$ is the ratio of the modulus of elasticity between SFRELC and conventional concrete, $y_0$ is the distance of the centroid to the compressive edge of the transformed section.

The computation results are presented in Table 4. As per China code GB50010 [41], when $\rho_f = 0.1$, $\Delta\sigma_s^f = 162$ MPa for the HRB400 rebar. This means that the test stress amplitude was over 38.3%–53.1% of the limit. Therefore, fracture of the longitudinal tensile rebar was commonly inevitable. However, with the assistant of steel fiber in the tension zone, the tensile stress of the rebar should decrease with an increasing volume fraction of the steel fiber. The test beams with $v_f = 1.6\%$ failed without fracture of the longitudinal rebar.

### 5.3. Flexural Stiffness of Normal Section

Flexural stiffness is an important index reflecting the safety of reinforced concrete beams. Based on the principle of minimum flexural stiffness for the computation of the deflection of reinforced flexural members, the flexural stiffness of test beams can be calculated as follows [16,41],

$$B_s = 0.1065Ml_0^2/a_f \tag{12}$$

where, $M$ is the mid-span moment, $l_0$ is the span, $a_f$ is the mid-span deflection.

Based on a previous study, the flexural stiffness $B_s$ can be computed as follows [16,26],

$$B_s = \frac{0.85E_cI_0}{1+(1.16-M_{cr}/M)/(6\alpha_{Es}\rho)}, \tag{13}$$

$$M_{cr} = \gamma f_{ft}W_0, \tag{14}$$

$$\gamma = 1.55(0.73+60/h), \tag{15}$$

where, $E_c$ is the modulus of elasticity of conventional concrete, $W_0$ is the elastic resistance moment of the transformed section to tensile edge, $f_{ft}$ is the tensile strength of SFRELC, $\alpha_{Es}$ is the ratio of the modulus of elasticity between the rebar and SFRELC, $\rho$ is the reinforcement ratio of longitudinal tensile rebars, $M_{cr}$ is the cracking moment of the mid-span section.

Under the upper-load, the tested and calculated values of initial flexural stiffness are presented in Table 5. A good agreement between them is indicated by the average ratio of tested to calculated values of 1.10 with a variation coefficient of 0.056. Meanwhile, the initial flexural stiffness reduced with increasing $\alpha_h$, but increased 6.1% and 17.4% in beams with $v_f = 1.2\%$ and 1.6% compared to those with $v_f = 0.8\%$.

**Table 5.** The measured stiffness of beams under static load and calculated stiffness contrast.

| Test Beam | $a_f$ (mm) | $B_s$ ($\times 10^{12}$) | | Tested/Calculated |
|---|---|---|---|---|
| | | Tested | Calculated | |
| BF0.8-0.5a | 8.9 | 3.532 | 3.302 | 1.070 |
| BF0.8-0.6a | 9.4 | 3.530 | 3.274 | 1.078 |
| BF0.8-0.6b | 9.2 | 3.417 | 3.301 | 1.035 |
| BF0.8-0.7a | 9.4 | 3.345 | 3.276 | 1.021 |
| BF0.8-0.7b | 9.6 | 3.639 | 3.225 | 1.128 |
| BF1.2-0.6a | 9.0 | 3.687 | 3.233 | 1.140 |
| BF1.2-0.6b | 9.0 | 3.687 | 3.233 | 1.140 |
| BF1.6-0.6a | 8.5 | 4.110 | 3.382 | 1.215 |
| BF1.6-0.6b | 8.2 | 4.047 | 3.412 | 1.186 |

With the increasing cyclic numbers of fatigue load, the compressive strain of conventional concrete and the tensile strains of SFRELC and longitudinal rebar increased. The depth of the compression zone was minimized with the gradual upward extension of cracks. This causes the reduction of the inertia moment of the transformed section and the reduction of the flexural stiffness. Therefore, a revised coefficient $\zeta$ should be multiplied as follows to compute the flexural stiffness of fatigue $B_s^f$,

$$B_s^f = \zeta B_s, \tag{16}$$

$$\zeta = 1 - 0.04\lg N. \tag{17}$$

Table 6 presents the tested and computed values of the flexural stiffness of test beams. The ratios are compared in Figure 9. It gives 1.056 average, with a variation coefficient of 0.099.

**Table 6.** Comparison between measured and calculated values of fatigue stiffness.

| Test Beam | Item | Cyclic Numbers of Fatigue Load | | | | | |
|---|---|---|---|---|---|---|---|
| | | Initial | 5000 | 10,000 | 50,000 | 100,000 | 200,000 |
| BF0.8-0.5 | tested | 3.532 | 2.886 | 2.861 | 2.789 | 2.743 | —— |
| | calculated | 3.302 | 2.790 | 2.750 | 2.656 | 2.615 | —— |
| BF0.8-0.6 | tested | 3.474 | 2.738 | 2.693 | 2.649 | 2.638 | 2.564 |
| | calculated | 3.287 | 2.800 | 2.760 | 2.666 | 2.625 | 2.584 |
| BF0.8-0.7 | tested | 3.492 | 2.620 | 2.515 | 2.447 | 2.437 | 2.382 |
| | calculated | 3.251 | 2.788 | 2.747 | 2.654 | 2.613 | 2.572 |
| BF1.2-0.6 | tested | 3.687 | 3.096 | 2.975 | 2.874 | 2.826 | 2.758 |
| | calculated | 3.233 | 2.740 | 2.700 | 2.609 | 2.569 | 2.528 |
| BF1.6-0.6 | tested | 4.078 | 3.643 | 3.548 | 3.372 | 3.244 | 3.041 |
| | calculated | 3.397 | 2.890 | 2.848 | 2.751 | 2.708 | 2.666 |

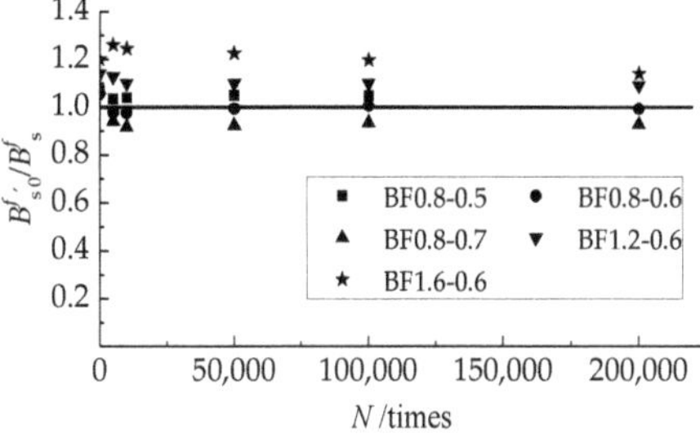


**Figure 9.** Comparison of tested to computed flexural stiffness at different cyclic numbers of fatigue load.

Figure 10 exhibits the trend curves of degenerated flexural stiffness of test beams, where the relative stiffness is the ratio of the stiffness at a cyclic number to the initial stiffness, and the relative life is the ratio $N/N_f$ of a cyclic number to fatigue life.

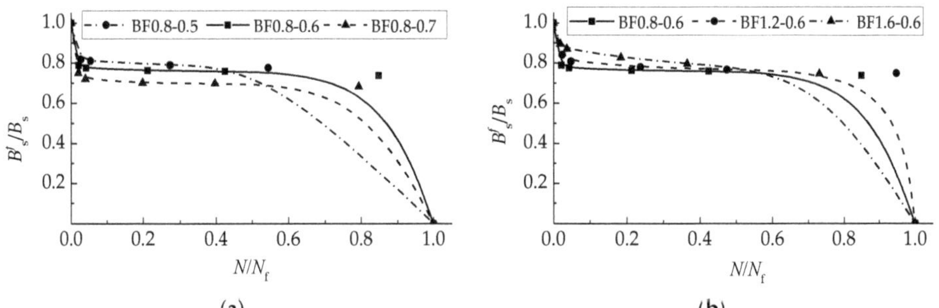


**Figure 10.** Relative stiffness degradation curve of beam affected by: (**a**) $\alpha_h$; and (**b**) $v_f$.

The curves display three shapes due to the different failure patterns of the test beams. For the test beams in groups BF0.8-0.6, BF0.8-0.7, and BF1.2-0.6, the curves composited by three stages with the first rapidly, the second steady, and the third rapidly. This is similar to that of reinforced concrete flexural members which failed as compressive concrete accompanied by a fracture of the tensile rebars [40,49,50]. For the test beam BF0.8-0.5, the first two stages of the curve are normal, the third drops linearly which represents the sectional fracture without obvious omen. For the test beams in group BF1.6-0.6, the declines in the three stages of the curve are continuous. This exhibits a sustained degeneration of flexural stiffness due to the accumulation of defects in SFRELC and conventional concrete as well as rebars. In this case, the plasticity of the longitudinal tensile rebar has the opportunity to function. Therefore, the controllable failure of the reinforced SFRELC superposed beams can be prospected by the optimal composites of $\alpha_h$ and $v_f$.

## 6. Conclusions

Based on the experimental study, the main conclusion can be drawn as follows:

(1) In the case of the same volume fraction of steel fiber, three failure patterns of test beams under fatigue load took place with the increase of SFRELC depth ratio from 0.5 to 0.7. The first is abrupt along a section of the main crack with the sudden fracture of a tensile rebar, the second is the crush of conventional concrete in the compression zone accompanied with the fracture of the tensile rebar, and the third is only the crush of conventional concrete in the compression zone. It should be noted that the horizontal interface between SFRELC and conventional concrete was laniated during the failure of test beams with an SFRELC depth ratio of 0.7. In this case, the entirety of the composite section should be enhanced.

(2) In the case of the same SFRELC depth ratio, two failure patterns of test beams under fatigue load took place with the increasing volume fraction of steel fiber ranging from 0.8% to 1.6%. Test beams with the steel fiber volume fractions of 0.8% and 1.2% failed with the crush of conventional concrete in the compression zone accompanied with the fracture of the tensile rebar, while the test beams with the steel fiber volume fraction of 1.6% failed with the crush of conventional concrete in the compression zone. This indicated the presence of steel fiber with a larger volume fraction could improve the failure pattern to be more ductile for reinforced SFRELC superposed beams.

(3) The fatigue life was sensitive to the upper limit of the fatigue load. The higher upper limit of the fatigue load led to the higher stress level and larger stress amplitude of the longitudinal rebar. This shortened the fatigue life of the test beams in this study. Formulas are proposed to evaluate the stress level of test beams, the stress amplitude of the longitudinal tensile rebar, and the degenerated flexural stiffness.

(4) The trend curves of fatigue flexural stiffness exhibit the different mechanisms of fatigue failure of test beams. This provides the prospect of controllable failure of reinforced SFRELC superposed beams by the optimal composites of the SFRELC depth ratio and the volume fraction of steel fiber.

**Author Contributions:** Methodology, F.Q. and C.L.; tests, data interpretation and writing—original draft preparation, C.P., X.D. and X.H.; writing—review and editing, C.L. and F.Q.; funding acquisition, C.L. and L.P.

**Funding:** This study was funded by [State Key Research and Development Plan, China] grant number [2017YFC0703904], [Key Scientific and Technological Research Project of University in Henan, China] grant number [14B560004 and 12A560008], and [Innovative Sci-Tech Team of Eco-building Material and Structural Engineering of Henan Province, China] grant number [YKRZ-6-066].

**Conflicts of Interest:** The authors declare no conflict of interest.

## References


1. Ministry of Housing and Urban-Rural Construction of the People's Republic of China. *Technical Specification of Lightweight Aggregate Concrete*; JGJ 51-2002; China Building Industry Press: Beijing, China, 2002. (In Chinese)
2. Standardization Administration of the People's Republic of China. *Lightweight Aggregates and Its Test Methods—Part 1: Lightweight Aggregates*; GB/T 17431.1-2010; Standards Press of China: Beijing, China, 2010. (In Chinese)
3. Zhao, M.S.; Ma, Y.Y.; Pan, L.Y.; Li, C.Y. *An Overview of Study on Natural Lightweight Aggregate for Concrete in China*; DEStech Publications, Inc.: Lancaster, PA, USA, 2015; pp. 125–133.
4. Ma, Y.Y.; Fan, Y.J.; Pan, L.Y.; Li, C.Y. *Study and Application of Cinder as Lightweight Aggregate of Concrete in China*; DEStech Publications, Inc.: Lancaster, PA, USA, 2015; pp. 134–140.
5. Zhao, M.S.; Zhang, X.Y.; Song, W.H.; Li, C.Y.; Zhao, S.B. Development of steel fiber reinforced expanded-shale lightweight concrete with high freeze-thaw resistance. *Adv. Mater. Sci. Eng.* **2018**. [CrossRef]
6. Ministry of Housing and Urban-Rural Construction of the People's Republic of China. *Technical Specification for Lightweight Aggregate Concrete Structures*; JGJ12-2006; China Building Industry Press: Beijing, China, 2006. (In Chinese)
7. Hassanpour, M.; Shafigh, P.; Mahmud, H.B. Lightweight aggregate concrete fiber reinforcement—A review. *Constr. Build. Mater.* **2012**, *37*, 452–461. [CrossRef]
8. Zhao, S.B.; Li, C.Y.; Qian, X.J. Experimental study on mechanical properties of steel fiber reinforced full lightweight concrete. *Geotech. Special Pub.* **2011**, *212*, 233–239.
9. Pan, L.Y.; Yuan, H.; Zhao, S.B. Experimental study on mechanical properties of hybrid fiber reinforced full lightweight aggregate concrete. *Adv. Mater. Res.* **2011**, *197*, 911–914. [CrossRef]
10. Li, C.Y.; Chen, H.; Zhao, S.B. Mechanical properties of steel fiber reinforced light-aggregate concrete. *Adv. Mater. Res.* **2012**, *366*, 12–15. [CrossRef]
11. Zhao, M.L.; Zhao, M.S.; Chen, M.H.; Li, J.; Law, D. An experimental study on strength and toughness of steel fiber reinforced expanded-shale lightweight concrete. *Constr. Build. Mater.* **2018**, *183*, 493–501. [CrossRef]
12. Zhao, S.B.; Zhao, M.S.; Zhang, X.Y.; Peng, Z.J.; Huang, T.H. Study on complete stress-strain curves of steel fiber reinforced lightweight-aggregate concrete under uniaxial compression. *J. Build. Struct.* **2019**, *5*, 1–10. (In Chinese)
13. Zhao, M.S.; Zhang, B.X.; Shang, P.R.; Fu, Y.; Zhang, X.Y.; Zhao, S.B. Complete stress-strain curve of self-compacting steel fiber-reinforced expanded-shale lightweight concrete under uniaxial compression. *Materials* **2019**, *18*, 2979. [CrossRef]
14. Zhao, S.B.; Li, C.Y.; Zhao, M.S.; Zhang, X.Y. Experimental study on autogenous and drying shrinkage of steel fiber reinforced lightweight-aggregate concrete. *Adv. Mater. Sci. Eng.* **2016**. [CrossRef]
15. Zhao, M.S.; Zhang, X.Y.; Yan, K.; Fei, T.; Zhao, S.B. Bond performance of deformed rebar in steel fiber reinforced lightweight-aggregate concrete affected by multi-factors. *Civ. Eng. J.* **2018**, *3*, 276–290. [CrossRef]
16. Zhao, M.S.; Su, J.Z.; Shang, P.R.; Zhao, S.B. Experimental study and theoretical prediction of flexural behaviors of reinforced SFRELC beams. *Constr. Build. Mater.* **2019**, *208*, 454–463. [CrossRef]
17. Li, X.K.; Li, C.Y.; Zhao, M.L.; Yang, H.; Zhou, S.Y. Testing and prediction of shear performance for steel fiber reinforced expanded-shale lightweight concrete beams without web reinforcements. *Materials* **2019**, *10*, 1594. [CrossRef] [PubMed]
18. Li, C.Y.; Yang, H.; Liu, Y.; Gao, K. Flexural behavior of reinforced concrete beams superposing with partial steel fiber reinforced full-lightweight concrete. *Appl. Mechan. Mater.* **2013**, *438*, 800–803. [CrossRef]
19. Pei, S.W.; Du, Z.H.; Li, C.Y.; Shi, F.J. Research of punching performance of reinforced SFRLAC superposed two-way slabs. *Appl. Mechan. Mater.* **2013**, *438*, 667–672. [CrossRef]
20. Nes, L.G.; Øverli, J.A. Structural behaviour of layered beams with fibre-reinforced LWAC and normal density concrete. *Mater. Struct.* **2015**. [CrossRef]
21. Iskhakov, I.; Ribakov, Y. A design method for two-layer beams consisting of normal and fibered high strength concrete. *Mater. Des.* **2007**, *5*, 1672–1677. [CrossRef]
22. Holschemacher, K.; Iskhakov, I.; Ribakov, Y.; Mueller, T. Laboratory tests of two-layer beams consisting of normal and fibered high strength concrete: Ductility and technological aspects. *Mechan. Adv. Mater. Struct.* **2012**, *19*, 513–522. [CrossRef]

23. Li, C.Y.; Zhao, S.B.; Chen, H.; Gao, D.Y. Experimental study on flexural capacity of reinforced SFRFLC superposed beams. *J. Build. Struct.* **2015**, *2*, 257–264. (In Chinese)
24. Li, C.Y.; Ding, X.X.; Zhao, S.B.; Zhang, X.Y.; Li, X.K. Cracking resistance of reinforced SFRFLC superposed beams with partial ordinary concrete in compression zone. *Open Civ. Eng. J.* **2016**, *10*, 727–737. [CrossRef]
25. Li, C.Y.; Zhao, Z.F.; Ding, X.X.; Zhao, S.B. Experimental study on crack width of reinforced HSC-on-SFRFLC superposed beams. *Adv. Eng. Res.* **2016**, *44*, 16–22.
26. Pan, L.Y.; Shang, Y.Q.; Kang, X.X.; Li, C.Y. Experimental research on flexural stiffness of reinforced steel fiber reinforced full-lightweight concrete superposed beams. *J. North China Univ. Water Resour. Electric Power (Nat. Sci. Ed.)* **2015**, *1*, 43–46.
27. Goel, S.; Singh, S.P. Fatigue performance of plain and steel fibre reinforced self compacting concrete using S-N relationship. *Eng. Struct.* **2014**, *74*, 65–73. [CrossRef]
28. Zhang, C.; Gao, D.Y.; Gu, Z.Q. Fatigue behavior of steel fiber reinforced high-strength concrete under different stress levels. *IOP Conf. Series: Mater. Sci. Eng.* **2017**, *274*, 012067. [CrossRef]
29. Choi, S.J.; Mun, J.S.; Yang, K.H.; Kim, S.J. Compressive fatigue performance of fiber-reinforced lightweight concrete with high-volume supplementary cementitious materials. *Cem. Concr. Compos.* **2016**, *73*, 89–97. [CrossRef]
30. Lantsoght, E.O.L.; van der Veen, C.; de Boer, A. Proposal for the fatigue strength of concrete under cycles of compression. *Constr. Build. Mater.* **2016**, *107*, 138–156. [CrossRef]
31. Hawileh, R.; Rahman, A.; Tabatabai, H. Evaluation of the low-cycle fatigue life in ASTM A706 and A615 grade 60 steel reinforcing bars. *J. Mater. Civ. Eng.* **2010**, *22*, 65–76. [CrossRef]
32. De Corte, W.; Boel, V.; Helincks, P.; De Schutter, G. Fatigue assessment of a lightweight steel-concrete bridge deck concept. *Bridge Struct.* **2016**, *12*, 11–19. [CrossRef]
33. Chen, H.J.; Liu, T.H.; Tang, C.W.; Tsai, W.P. Influence of high-cycle fatigue on the tension stiffening behavior of flexural reinforced lightweight aggregate concrete beams. *Struct. Eng. Mechan.* **2011**, *40*, 847–866. [CrossRef]
34. Li, C.Y.; Nie, X.; Kang, X.X.; Peng, C.; Zhang, X.Y.; Zhao, S.B. Experimental study on shear fatigue behavior of reinforced SFRFLC superposed beams. *J. North China Univ. Water Resour. Electric Power* **2017**, *6*, 16–24.
35. Singh, S.P.; Kaushik, S.K. Fatigue strength of steel fibrous concrete in flexure. *Cem. Concr. Compos.* **2003**, *25*, 779–786. [CrossRef]
36. Germano, F.; Tiberti, G.; Plizzari, G. Post-peak fatigue performance of steel fiber reinforced concrete under flexure. *Mater. Struct.* **2016**, *49*, 4229–4245. [CrossRef]
37. Batson, G. Flexural fatigue strength of steel fiber reinforced concrete beams. *ACI J.* **1972**, *11*, 673–677.
38. Goel, S.; Singh, S.P.; Singh, P. Felxural fatigue strength and failure probability of self compacting fibre reinforced concrete beams. *Eng. Struct.* **2012**, *40*, 131–140. [CrossRef]
39. Banjara, N.K.; Ramanjaneyulu, K. Experimental investigations and numerical simulations on the flexural fatigue behaviors of plain and fiber-reinforced concrete beams. *J. Mater. Civ. Eng.* **2018**, *8*, 04018151. [CrossRef]
40. Parvez, A.; Foster, S.J. Fatigue behavior of steel-fiber-reinforced concrete beams. *J. Struct. Eng.* **2015**, *4*, 04014117. [CrossRef]
41. Ministry of Housing and Urban-Rural Construction of the People's Republic of China. *Design Code for Concrete Structures*; GB 50010-2010; China Building Industry Press: Beijing, China, 2011. (In Chinese)
42. General Administration of Quality Supervision, Inspection and Quarantine of the People's Republic of China. *Common Portland Cement*; GB 175-2007; China Standard Press: Beijing, China, 2007. (In Chinese)
43. General Administration of Quality Supervision, Inspection and Quarantine of the People's Republic of China. *Fly Ash Used for Cement and Concrete*; GB/T 1596-2005; China Standard Press: Beijing, China, 2005. (In Chinese)
44. Ministry of Housing and Urban-Rural Construction of the People's Republic of China. *Specification for Mix Proportion Design of Ordinary Concrete*; JGJ 55-2011; China Building Industry Press: Beijing, China, 2011. (In Chinese)
45. Ministry of Housing and Urban-Rural Construction of the People's Republic of China. *Steel Fiber Reinforced Concrete*; JG/T472-2015; China Building Industry Press: Beijing, China, 2015. (In Chinese)
46. Ministry of Housing and Urban-Rural Construction of the People's Republic of China. *Standard for Test Method of Performance on Ordinary Fresh Concrete*; GB/T 50080-2002; China Building Industry Press: Beijing, China, 2002. (In Chinese)

47. Ministry of Housing and Urban-Rural Construction of the People's Republic of China. *Standard for Test Method of Mechanical Properties on Ordinary Concrete*; GB/T 50081-2002; China Building Industry Press: Beijing, China, 2002. (In Chinese)
48. Ministry of Housing and Urban-Rural Construction of the People's Republic of China. *Standard for Testing Method of Concrete Structures*; GB 50152-2012; China Building Industry Press: Beijing, China, 2012. (In Chinese)
49. Song, Y.P.; Zhao, S.B.; Wang, R.M.; Li, S.Y. Full-range nonlinear analysis of fatigue behaviors of reinforced concrete structures by finite element method. *Acta Oceanol. Sin.* **1994**, *1*, 143–154.
50. Zhao, S.B. Experimental study on fatigue behaviors of reinforced concrete plates. *J. Basic Sci Eng.* **1999**, *3*, 289–297. (In Chinese)
51. Zhao, S.B. Experimental research on shear behavior of reinforced concrete composite beams under uniformly distributed load. *China Ocean Eng.* **1998**, *4*, 467–476.

*Article*

# Structural Safety Evaluation of Precast, Prestressed Concrete Deck Slabs Cast Using 120-MPa High-Performance Concrete with a Reinforced Joint

**Jae-Hyun Bae [1], Hoon-Hee Hwang [1,*] and Sung-Yong Park [2]**

1 R&D Center, Korea Road Association, Seongnam-si 13647, Korea; bjh@kroad.or.kr
2 Department of Infrastructure Safety Research, Korea Institute of Civil Engineering and Building Technology, Goyang-si 10223, Korea; sypark@kict.re.kr
* Correspondence: poonhee@kroad.or.kr; Tel.: +82-10-2208-5324

Received: 9 August 2019; Accepted: 17 September 2019; Published: 19 September 2019

**Abstract:** Prestressed concrete structures are used in various fields as they can reduce the cross-sectional area of members compared with reinforced concrete structures. In addition, the use of high-performance and strength concrete can help reduce weight and achieve excellent durability. Recently, structures have increasingly been constructed using high-performance and strength concrete, and therefore, structural verification is required. Thus, this study experimentally evaluated the structural performance of a long-span bridge deck slab joint, regarded as the weak point of structures. The specimens were designed with a 4 m span for application to cable-stayed bridges. To ensure the required load resistance and serviceability, the specimens comprised of 120 MPa high-performance fiber-reinforced concrete and were prestressed. The deck slabs satisfied all static and fatigue performance as well as serviceability requirements, although they were thinner than typical concrete bridge deck slabs. The study also verified whether the deck slabs were suitable to help implement precast segmental construction methods. Finally, the results confirmed that the structural performance of the developed prestressed concrete (PSC) deck slab was sufficient for the intended bridge application as it achieved a sufficiently large safety factor in the static and fatigue performance tests, relative to the design requirement.

**Keywords:** ultra-high-performance concrete; deck slab; prestressed concrete; static test; fatigue test; deck slab joint

---

## 1. Introduction

High-performance concrete not only improves the durability of structures by ensuring high quality, but also enables weight reduction owing to its excellent structural performance. Recently, high-performance materials such as ultra-high-performance concrete, whose strength has been significantly increased compared with normal concrete, have been attracting attention as they help realize innovative performance improvements and light-weight structures [1,2].

Bridge deck slabs are frequently damaged members because they directly support live loads through contact with the wheel load of vehicles. Therefore, continuous efforts have been invested to improve the quality and structural performance of these slabs by utilizing high-performance materials. In particular, many research and construction examples that utilize a precast method for accelerated bridge construction can be found in the literature [3]. The precast deck should be assembled with several prefabricated panels to complete the deck members; thus, there is a risk of lack of continuity. Therefore, joints between panels that can guarantee continuity are necessary; further, their safety must be verified as they could potentially become weak points of the structure. Several studies have been conducted on the joints of precast decks, and various types of joint details such as female–female joints, match–cast joints, male–to–female joints, spiral confinement methods, lap-spliced joints, and post-tensioning methods have been proposed [4]. All these joints have been experimentally verified to provide adequate structural performance; however, their details need to be improved considering the characteristics of high-performance materials. The bridge deck slab to which the high-performance materials are applied can exhibit high load-resistance performance, irrespective of their thickness and help reduce the weight of structures; therefore, joints with suitable joint details are required. The performance of ultra-high-performance concrete (UHPC) joints applied to the precast deck was evaluated using several joint details that could utilize the high adhesion performance of the UHPC [5]. Experiments on UHPC lap-spliced joints reported that the lap-spliced length of the steel rebar required to achieve the same level of flexural performance as without the joint was 200 mm [6]. The experimental research on the ultra-high-performance fiber reinforced concrete (UHPFRC) joint between precast concrete panels reinforced with glass-fiber-reinforced polymer (GFRP) bar concluded that using a lap-spliced length of 200 mm can maintain continuity of joints until failure occurs in the adjacent precast body [7]. Another study demonstrated that polymer methyl methacrylate-polymer concrete (PMMA-PC) could be an alternative to economic joint filler materials owing to their extremely high adhesion performance. In particular, the required lap-spliced length is smaller than UHPC and 40% of normal concrete [8].

These previous studies were focused on RC–type–joints because of the high adhesion performance of UHPC. However, research on prestressed concrete (PSC)–type–joints suitable for a long-span structure is relatively insufficient.

In this study, the behavior of joints in thin long-span bridge deck slabs was evaluated experimentally. The specimens were fabricated with a thickness of 150 mm and a span of 4 m. The high-performance concrete of 120 MPa was applied to resist the flexural moment caused by this relatively thin long span. A method to apply prestress using post-tension was adopted for this type of joint. This was because it could address problems such as deflection and cracks that could be disadvantageous owing to the relatively thin thickness of the slabs.

## 2. Description of Specimens

### 2.1. Specimens

To verify the structural performance of the 120 MPa prestressed concrete (PSC) deck slab with reinforced steel fibers, various specimens were designed and fabricated, as summarized in Table 1. To verify the performance of the deck slabs with joints, a specimen without a joint (SC120f-PSC-NJ) was fabricated as a reference. In addition, two specimens for static performance tests (SC120f-PSC-J1, J2) and a specimen for fatigue performance testing (SC120f-PSC-JF) were fabricated.

**Table 1.** Summary of specimens used in this study.

| Specimen | Joint | Test Type |
|---|---|---|
| SC120f-PSC-NJ | No | Static |
| SC120f-PSC-J1 | Yes | Static |
| SC120f-PSC-J2 | Yes | Static |
| SC120f-PSC-JF | Yes | Fatigue |

The cross-section of the deck slab specimens was designed in accordance with the Korean Highway Bridge Design Code (Limit State Design) (2015) [9] and the Structural Design Guideline for Fiber-Reinforced SUPER Concrete [10]. The dimensions of the specimens were 2 m (W) × 4.2 or 5.4 m (L) × 0.15 m (H). Those specimens with a length of 4.2 m were used for static testing. The fatigue test specimen was 5.4 m long, longer than the static test specimen, because additional space was required to prevent the specimen from being dislodged from its supporting points as a load was repeatedly applied. In the experiments, a relatively long span of 4.0 m was adopted, given that the deck slabs were primarily investigated for application to cable-stayed bridges. For those specimens with joints, the interfaces where the segments met were fitted with shear keys to improve the performance by mechanical interlocking. The types and cross-sectional joint details of the test specimens based on the purpose of the experiment are summarized in Table 1 and shown in Figure 1.

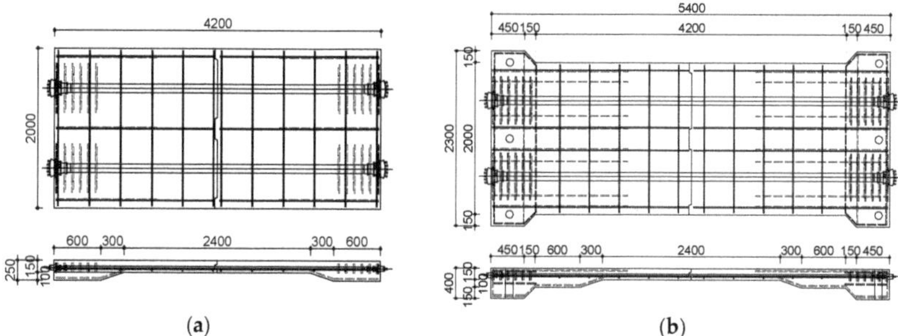


**Figure 1.** Dimensions of test specimens (unit: mm): (**a**) static specimens (SC120f-PSC-NJ, J1, J2); (**b**) fatigue specimens (SC120f-PSC-JF).

The tendon material was SWPC 7BL, which is a low-relaxation steel with seven strands and a 15.2 mm nominal diameter. The yield and tensile strengths of the strands were 1600 and 1881 MPa, respectively. Full prestressing was applied using the post-tension method to prevent tensile stress being applied to the overall section, while the tendons were placed in a straight line in groups of five at 1.0 m intervals. Moreover, the eccentricity of the tendons was 20 mm, and flat-type anchors, typically used for prestressed deck slabs, were used.

### 2.2. Material Properties of SC120f

High-strength concrete with a compressive strength of about 120 MPa, sustained tensile strength through internal fiber reinforcement, and exceptional durability as compared with conventional concretes was applied to the test specimen. The properties of the materials used in SC120f are listed in Table 2. Ordinary Portland cement was adopted as cement, and silica fume was used as reactive powder. Fine aggregates comprising quartz sand, with diameters smaller than 0.5 mm, and $SiO_2$ content larger than 90%, were applied to SC120f. Coarse aggregates were not used. The steel fiber used exhibited tensile strength higher than 2000 MPa and a diameter of 0.2 mm. Further, polycarboxylate superplasticizer (density 1.01 g/cm$^3$, dark brown liquid, solid content 30%) was used. A combination of calcium sulfoaluminate (CSA) type expansion admixture and glycol-based shrinkage reducing agent was used as the shrinkage reducing agent.

**Table 2.** Basic mix composition of SC120f.

| W/B | Mass Unit (kg/m$^3$) | | | | | | |
|---|---|---|---|---|---|---|---|
| | Water | Premix Binder | Fine Aggregates | Steel Fiber | | Super Plasticizer | Air-Entraining Agent |
| | | | | 16 mm | 20 mm | | |
| 0.23 | 210 | 1180 | 847 | - | 78 | 7 | 17 |

Premix binder: cement, Zr, BS, filler, expansion agent, shrinkage reducing agent premix; Zr:BS = 30:70.

### 2.3. Fabrication

All the specimens, except for SC120f-PSC-NJ with no joint, were fabricated by match casting. After installing a formwork, the rebar and tendons were placed, and then the concrete was first poured into one side of the joint. The poured concrete was allowed to cure for approximately 2–3 days, and then the concrete was poured into the other side using match casting. Subsequently, high-temperature steam curing (at about 90 °C) was conducted. After curing, epoxy was applied to the interface of the joint where the segments abutted, and prestressing was applied using a hydraulic jack. At this time, the jacking force was determined to maintain the effective prestress considering the immediate losses such as elastic shortening of the concrete, anchorage losses, and frictional losses. After jacking, grouting was conducted to complete the specimens (Figure 2).

**Figure 2.** Fabrication of the specimens: (**a**) assembly of mold and reinforcement placing, (**b**) concrete casting (one side), (**c**) concrete casting (other side, match casting), (**d**) epoxy spreading, (**e**) jacking, and (**f**) grouting.

## 3. Static Performance Evaluation

### *3.1. Static Loading Test (SC120f-PSC-NJ, J1, and J2 Specimens)*

In the static loading tests, the deflection, strain, and joint widening of each deck slab specimen were measured, and the behavior of the specimens was examined using a 3500-kN static actuator (MTS, Eden Prairie, MN, USA), with the load being applied until failure. The load was applied using a displacement-controlled method at a rate of 3 mm/min. Furthermore, the load was applied using a third-point loading method by ASTM C78/C78M [11] such that the largest bending moment would be generated at the center of the specimen (where the joint was located). Under actual boundary conditions, the bridge deck slab would be combined with the girder or crossbar; however, simple support was applied to induce conservative experimental results (Figure 3).

**Figure 3.** Static testing of specimens SC120f-PSC-J2.

## 3.2. *Result of Static Loading Test*

### 3.2.1. Failure Mode

The results of the static load test revealed that the SC120f-PSC-NJ specimen without the joint exhibited typical flexural behavior similar to general beam members. As the load increased, many tensile cracks appeared and grew at the bottom of the deck, while the concrete was crushed at the top of the deck slab between the load points during the final failure. At the moment of failure, the maximum load was 435.65 kN, and the deflection of the center of the deck slab under the maximum load was 88.06 mm.

For the SC120f-PSC-J1 and J2 specimens with joints, joint widening occurred from the start of loading, and then cracking occurred in the concrete adjacent to the joint. The joint widened as the load increase, and failure occurred as the concrete at the top of the joint was crushed. The SC120f-PSC-J1 and J2 specimens with joints had relatively few cracks compared with the SC120f-PSC-NJ specimen without a joint, given that the cracks form around the initial cracks initiated by the separation of the joints. The measured maximum loads were 322.07 and 289.04 kN for the SC120f-PSC-J1 and J2 specimens, respectively, while the deflections of the center of the deck slab under the maximum load were 58.6 and 60.18 mm, respectively (Figure 4).

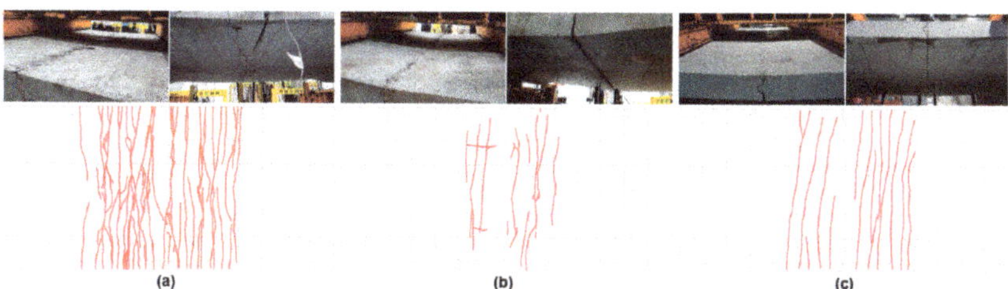


**Figure 4.** Photographs and maps of failure modes in static loading tests. (**a**) SC120f-PSC-NJ, (**b**) SC120f-PSC-J1, and (**c**) SC120f-PSC-J2.

### 3.2.2. Evaluation of Flexural Strength

To evaluate the level of safety in terms of strength, the experimental results were compared with the ultimate limit state load required by the design code. The ultimate limit load was calculated by considering the third-point load test conditions from the ultimate moments induced by the ultimate limit state load combination specified in the design code. As a result of verifying the structural performance of the PSC deck slabs with joints, the maximum loads of the SC120f-PSC-J1 and J2 specimens were 74% and 66% of that of SC120f-PSC-NJ, respectively. Although the presence of the joints caused the maximum load to decrease, the values were 3.67 and 3.29 times higher than the ultimate limit load specified in the design code, as shown in Figures 5 and 6. Therefore, the fabricated

PSC deck slabs had a sufficient safety factor compared with the design requirement despite the presence of a joint (Table 3).

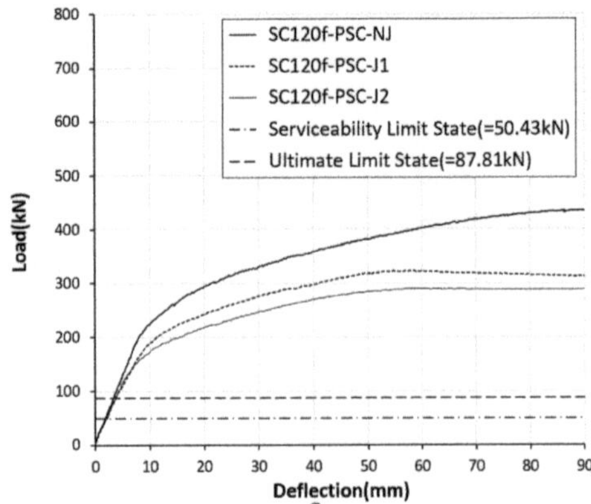


**Figure 5.** Load–deflection curves for static loading test specimens.

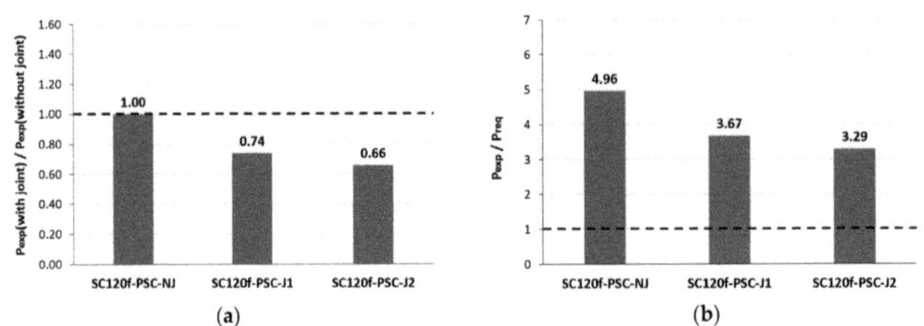


**Figure 6.** Comparison of maximum load for static loading test specimens; (**a**) $P_{exp}$ (with joint)/$P_{exp}$ (without joint), and (**b**) $P_{exp}/P_{req}$.

**Table 3.** Comparison of maximum load for static loading test specimens.

| Specimen | Deflection at Maximum Load | Maximum Load | Ultimate Limit State Load | $P_{exp}$ (with Joint)/ $P_{exp}$ (without Joint) | $P_{exp}/P_{req}$ |
|---|---|---|---|---|---|
| SC120f-PSC-NJ | 88.06 mm | 435.65 kN | | 1.00 | 4.96 |
| SC120f-PSC-J1 | 58.6 mm | 322.07 kN | 87.81 kN | 0.74 | 3.67 |
| SC120f-PSC-J2 | 60.18 mm | 289.04 kN | | 0.66 | 3.29 |

Note: $P_{exp}$ (with joint) = experimental maximum load of SC120f-PSC-J1 or J2; $P_{exp}$ (without joint) = experimental maximum load of SC120f-PSC-NJ; $P_{req}$ = ultimate limit state load.

### 3.3. Evaluation of Serviceability

The Korean Highway Bridge Design Code (Limit State Design) specifies that deflection of a bridge concrete deck slab caused by live loads and their dynamic effects must not exceed the following limits, where L is the distance between the centers of the deck slab supports: deck slab with no passage of people, L/800; deck slab with passage of a limited number of people, L/1000; deck slab with passage of many people, L/1200. Furthermore, the design code specifies a maximum crack width of 0.2 mm for PSC members and 0.3 mm for reinforced concrete members. For the SC120f-PSC-NJ specimen, both the deflection and crack width at the center of the deck slab, generated at the serviceability limit state, satisfied the values specified in the design code (Figure 7). For the SC120f-PSC-J1 and J2 specimens, both the deflection and joint widening at the center of each specimen generated at the serviceability limit state also satisfied the specified limits (Figure 8).

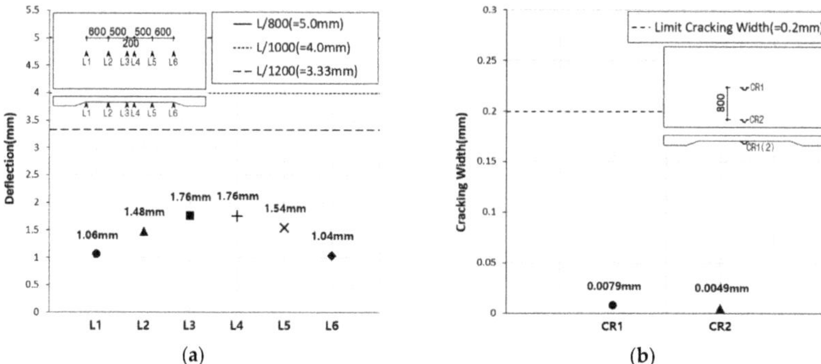


(a) (b)

**Figure 7.** Deflection and crack width of SC120f-PSC-NJ (static loading test). (**a**) Deflection, and (**b**) crack width.

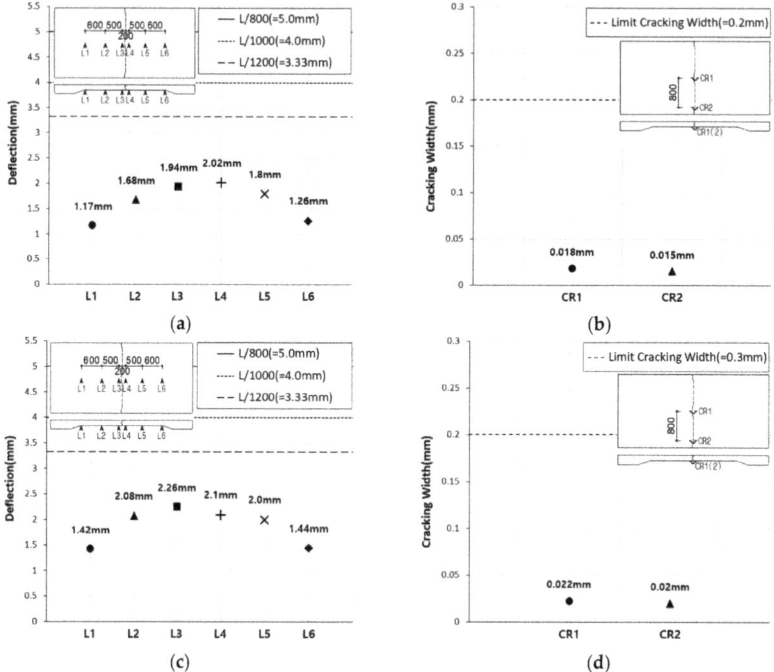


(a) (b)

(c) (d)

**Figure 8.** Deflection and crack width of SC120f-PSC-J1, J2 (static loading test). (**a**) Deflection (SC120f-PSC-J1), (**b**) crack width (SC120f-PSC-J1), (**c**) deflection (SC120f-PSC-J2), and (**d**) crack width (SC120f-PSC-J2).

## 4. Fatigue Performance Evaluation

### *4.1. Fatigue Loading Test and Residual Strength Test (SC120f-PSC-JF Specimens)*

In the fatigue loading test (SC120f-PSC-JF), a total of two million fatigue load cycles were applied to the specimen using a 1000 kN dynamic actuator (MTS, Eden Prairie, MN, USA). When simple supports are applied as the boundary conditions for the fatigue loading test, a load imbalance can arise due to the impact of the load or the movement of the specimen during fatigue loading. Therefore, fatigue loads were applied after the specimen was fixed to the structural test frame with steel rods through the holes in the fixed part of the specimen (Figures 1 and 9). In addition, to examine the effects of the accumulated fatigue, data were measured at the first cycle and then after $10^2$, $10^3$, $10^4$, $10^5$, $10^6$, and $2 \times 10^6$ cycles. The fatigue loads were applied at a rate of 2–3 times/s using a fixed-point fatigue–loading method. The fatigue load (110 kN), obtained by multiplying the standard truck (KL-510) rear wheel load (96 kN) of the Korean Highway Bridge Design Code (Limit State Design) (2015) by the impact load factor of 15%, was applied to verify the performance in a more severe situation than the fatigue level required by design (80% of the standard truck rear wheel load was considered for fatigue examination). Moreover, the performance of the joint was verified in the harshest environment by applying the fatigue load to the joint interface of the specimen over a wheel–ground contact area of 231 mm × 577 mm. After $2 \times 10^6$ fatigue–loading cycles, the residual strength test was conducted in the same manner as the static loading test to enable a comparison under the same conditions. Therefore, the boundary conditions were adjusted again in the same way as in the static loading test.

**Figure 9.** Fatigue testing of SC120f-PSC-JF specimen.

### 4.2. *Result of Fatigue Loading Test and Residual Strength Test*

#### 4.2.1. Failure Mode

As a result of applying fatigue loads to the SC120f-PSC-JF specimen, the joint slightly widened up to $2 \times 10^6$ fatigue loading cycles, although no cracking was observed in the concrete deck slab body. In the residual strength test, a failure mode similar to those used for static loading tests (SC120f-PSC-J1 and J2) was observed, as shown in Figure 10.

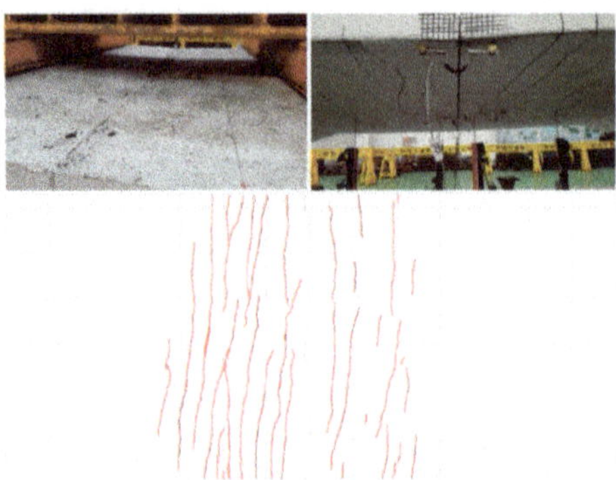

**Figure 10.** Failure mode of SC120f-PSC-JF.

4.2.2. Evaluation of Flexural Strength

The load–deflection curve of the fatigue loading test specimen (SC120f-PSC-JF) obtained from the residual strength test is compared with those of the static loading test specimens (SC120f-PSC-J1 and J2) in Figures 11 and 12. The maximum load applied to the SC120f-PSC-JF (304.53 kN) was similar to those of the static loading test specimens (322.07 and 289.04 kN). Moreover, the maximum load was approximately 3.47 times higher than the ultimate limit state load specified in the design code, indicating that the structure had a sufficient safety factor despite the accumulation of fatigue loads (Table 4).

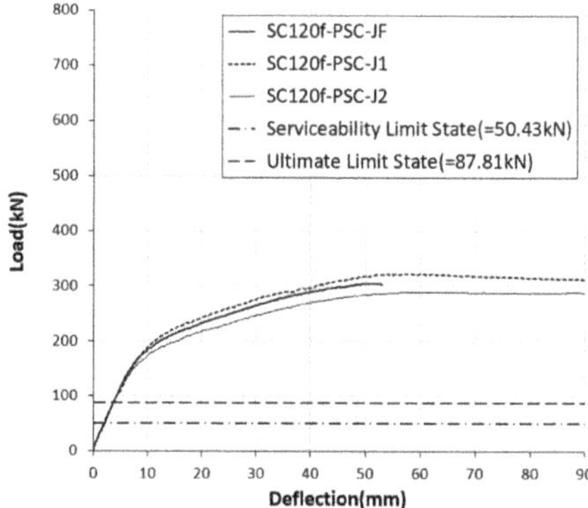


**Figure 11.** Load–deflection curves of test specimens with joints.

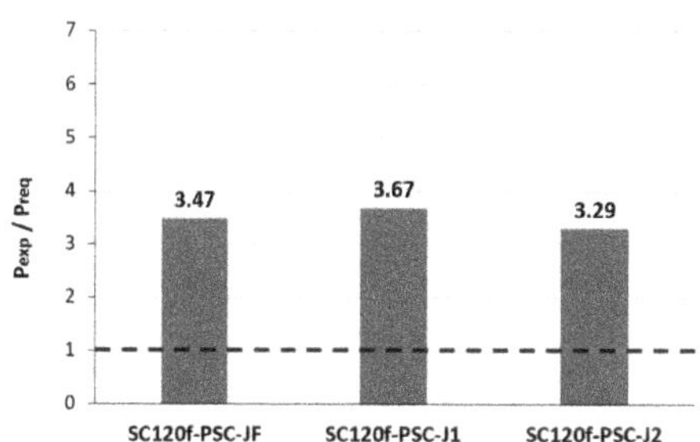


**Figure 12.** Comparison of maximum load for test specimens with joints.

**Table 4.** Comparison of maximum load for test specimens with joints.

| Specimen | Deflection at Maximum Load | Maximum Load | Ultimate Limit State Load | $P_{exp}/P_{req}$ |
|---|---|---|---|---|
| SC120f-PSC-JF | 50.48 mm | 304.53 kN | | 3.47 |
| SC120f-PSC-J1 | 58.6 mm | 322.07 kN | 87.81 kN | 3.67 |
| SC120f-PSC-J2 | 60.18 mm | 289.04 kN | | 3.29 |

Note: $P_{exp}$ = experimental maximum load; $P_{req}$ = ultimate limit state load.

### 4.3. *Evaluation of Serviceability*

The deflection of the SC120f-PSC-JF fatigue loading test specimen shown in Figure 13a, which shows the deflection at the serviceability limit according to the number of loads, was about 28.2–37.2% of the allowable deflection specified in the design code. Figure 13b shows that the joint widening of the specimen is 19.2–24.8% of the limit of 0.2 mm for a PSC member. These results indicate that the deck slabs tested in this study would exhibit no serviceability problems until $2 \times 10^6$ fatigue cycles have occurred. Similarly, in the residual strength measurement test conducted after the occurrence of $2 \times 10^6$ fatigue cycles, the deflection and cracking were within the limits specified in the design code (Figure 14).

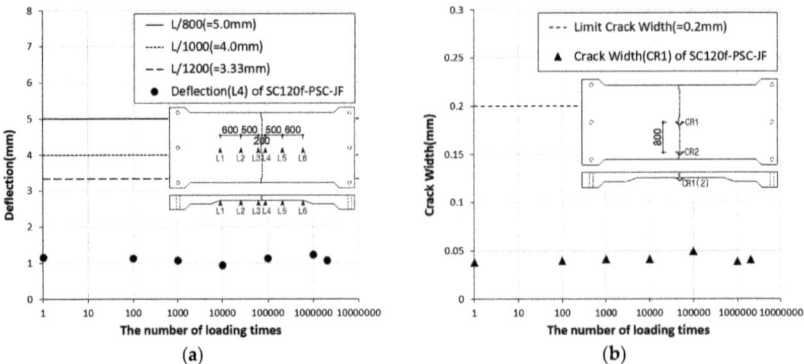


**Figure 13.** (**a**) Deflection and (**b**) crack width of SC120f-PSC-JF (fatigue test).

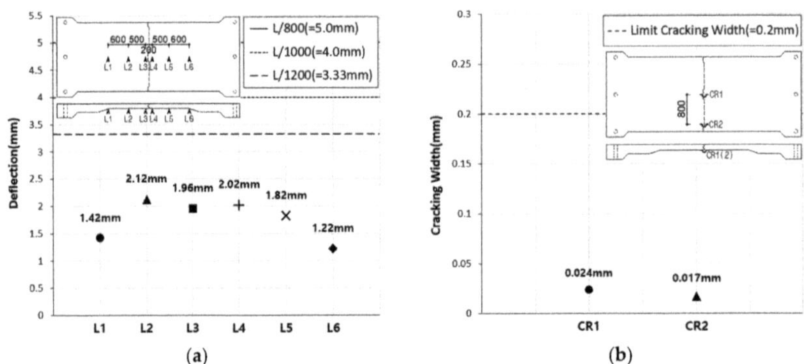


**Figure 14.** (**a**) Deflection and (**b**) crack width of SC120f-PSC-JF (residual strength test).

## 5. Conclusions

This study verifies the structural performance of joints in thin long-span bridge deck slabs that used 120 MPa UHPC reinforced with steel fibers. The effect of the PSC–type–joint by post-tensioning methods on static and fatigue performance were verified experimentally. The following conclusions were derived:

(1) In the static loading test, the flexural strengths of the specimens with joints were 74% and 66% of that of the specimen without a joint. However, they were about 3.7 and 3.3 times higher than the levels required by the design code.
(2) In addition, the deflections and crack widths of the specimens with joints satisfied the limit of the design code at the serviceability limit state.
(3) After two million fatigue load cycles were applied, the residual strength of the specimen with a joint was similar (95%, 105%) to the flexural strengths of the specimens with joints in the static loading test.
(4) In the fatigue loading test, the deflection at the serviceability limit according to the number of loads, was about 28.2–37.2% of the allowable deflection specified in the design code, and the joint widening of the specimen was 19.2–24.8% of the limit.

**Author Contributions:** Conceptualization, J.-H.B. and H.-H.H.; writing and original draft preparation, J.-H.B.; writing, review, and editing, H.-H.H. and S.-Y.P.

**Funding:** This research was funded by a grant (13SCIPA02) from the Smart Civil Infrastructure Research Program funded by the Ministry of Land, Infrastructure and Transport (MOLIT) of the Korean Government and the Korea Agency for Infrastructure Technology Advancement (KAIA). The funding agency had no role in the study design, the collection, analysis, or interpretation of data, the writing of the report, or the decision to submit the article for publication.

**Conflicts of Interest:** The authors declare no conflicts of interest.

## References

1. Graybeal, B.A. UHPC in the U.S. Highway Transportation System. In Proceedings of the Second International Symposium on Ultra High Performance Concrete, Kassel, Germany, 5–7 March 2008; pp. 11–17.
2. Rebentrost, M.; Wight, G. Experience and Applications of Ultra-High Performance Concrete. In Proceedings of the Second International Symposium on Ultra High Performance Concrete, Kassel, Germany, 5–7 March 2008; pp. 19–30.
3. Graybeal, B.A. *Design and Construction of Field-Cast UHPC Connections*; Report No. FHWA-HRT-14-084; USDOT FHWA: Washington, DC, USA, 2014.
4. Badie, S.S.; Tadros, M.K. *Full-Depth Precast Concrete Bridge Deck Panel Systems*; NCHRP Report 584; Transportation Research Board: Washington, DC, USA, 2008.
5. Graybeal, B.A.; Swenty, M. UHPFRC for Prefabricated Bridge Component Connections. In Proceedings of the Third International Symposium on Ultra High Performance Concrete, Kassel, Germany, 7–9 March 2012; Volume 1223.
6. Hwang, H.H.; Park, S.Y. A study on the flexural behavior of lap-spliced cast-in-place joints under static loading in ultra-high performance concrete bridge deck slabs. *Can. J. Civ. Eng.* **2014**, *41*, 615–623. [CrossRef]
7. Arafa, A.; Farghaly, A.; Ahmed, E.; Benmokrane, B. Laboratory testing of GFRP-RC panels with UHPFRC joints of the Nipigon River Cable-Stayed Bridge in Northwest Ontario, Canada. *J. Bridge Eng.* **2016**, *21*, 05016006. [CrossRef]
8. Mantawy, I.; Chennareddy, R.; Genedy, M.; Taha, M.R. Polymer Concrete for Bridge Deck Closure Joints in Accelerated Bridge Construction. *Infrastructures* **2019**, *4*, 31. [CrossRef]
9. MOLIT (Ministry of Land, Infrastructure and Transport). *Korean Highway Bridges Design Code (Limit State Design Method)*; MOLIT: Sejong-si, Korea, 2015.
10. KCI (Korea Concrete Institute). *The Structural Design Guideline of the Fiber Reinforced SUPER Concrete*; KCI: Seoul, Korea, 2019.
11. *ASTM C78/C78M, Standard Test Method for Flexural Strength of Concrete (Using Simple Beam with Third-Point Loading)*; ASTM International: West Conshohocken, PA, USA, 2018.

Article

# Experimental and Numerical Study on Interface Bond Strength and Anchorage Performance of Steel Bars within Prefabricated Concrete

Zhijian Hu [1], Yasir Ibrahim Shah [1,*] and Pengfei Yao [2]

[1] Department of Road and Bridge Engineering, School of Transportation, Wuhan University of Technology, Wuhan 430070, China; hzj@whut.edu.cn
[2] Central and Southern China Municipal Engineering Design & Research Institute Co., Ltd., Wuhan 430010, China; yaopf@citic.com
* Correspondence: yasiribrahimshah@whut.edu.cn

**Abstract:** This study investigates the interface bond strength and anchorage performance of steel bars within prefabricated concrete. Twenty-two specimens were designed and manufactured to study the interface bond behavior of deformed and plain steel bars under a larger cover thickness. Diameter of steel bars, strength grade of concrete, and anchorage length were considered influential factors. The finite element method (ABAQUS) was used for the validation of experimental results. The interface bond's failure mechanism and the anchorage length in the prefabricated concrete under different concrete strength levels were explored and compared to national and international codes. A suitable value of the basic anchoring length for the prefabricated structure was recommended. The results show that the interface bond strength of prefabricated bridge members is directly proportional to the strength grade of the concrete, inversely proportional to the reinforcement diameter, and less related to anchorage length. The effect of the cover thickness of the surrounding concrete is negligible. Conversely, the bearing capacity of prefabricated bridge members depends on the strength of the concrete, the diameter of the steel bar, and the anchorage length. Furthermore, it is concluded that the mechanical bond strength accounts for 88% of the bond strength within prefabricated concrete.

**Keywords:** bond strength; prefabricated concrete structure; anchorage performance; mechanical bond strength

**Citation:** Hu, Z.; Shah, Y.I.; Yao, P. Experimental and Numerical Study on Interface Bond Strength and Anchorage Performance of Steel Bars within Prefabricated Concrete. *Materials* **2021**, *14*, 3713. https://doi.org/10.3390/ma14133713

Academic Editor: Eva O. L. Lantsoght

Received: 23 May 2021
Accepted: 14 June 2021
Published: 2 July 2021



## 1. Introduction

In recent years, prefabricated reinforced concrete structures are widely used to construct commercial buildings, temporary safety protection structures, and large and medium-size bridges. The prefabricated structure has the advantages of a short construction periods, industrialized production, high dimensional accuracy, and less environmental pollution. The sufficient bond strength and anchorage performance of steel bars within prefabricated concrete are the key to ensure the service performance of the structure [1–3].

At present, the design of the anchorage length of the connecting reinforcement within the prefabricated concrete structure is usually considered according to the relevant provisions of the cast-in-place structure. However, due to the different characteristics of the prefabricated connection, the thickness of the protective layer of the connecting reinforcement is generally more than 50 mm larger than that of the cast-in-place structure and so the influence of the thickness of the protective layer can be ignored in the calculation of the anchorage length and interface bond strength. Presently, the existing relevant specifications of the assembled concrete structure still utilizes the relevant provisions of the cast-in-place structure for the anchorage length of the connecting steel bars, which is not suitable for calculating the bond strength and anchorage performance of the prefabricated assembled bridge with large protective layer thickness.

There are only a few publications for the case of prefabricated concrete that have been published; this is the motivation for conducting the presented research. For the general case of a bond between reinforcement and concrete, however, much work has been conducted and reported in publications; the most relevant is mentioned here and shortly discussed. Steel bars have a significant effect on the mechanical properties and bond strength of concrete [4]. Li et al. established the formula of the ultimate bond strength by an experimental study on bond anchorage performance of 1860-grade high-strength prestressed steel strands and lightweight aggregate concrete [5].

The experimental study analyzes the compressive bond anchorage properties of 500 MPa steel bars in concrete. Five influence factors, including concrete strength, the steel bar's diameter, concrete cover, embedment length, and transverse reinforcement, were considered. The result shows that the influence of the surrounded concrete cover thickness on compressive bond strength is more than the steel bar's diameter [6]. Saeed et al. concluded from an experimental study that the anchor strength and stiffnesses are directly proportional to the bond length; the cross-sectional area ratio of Carbon Fibre-Reinforced Polymer (CFRP) rods to anchor borehole affects the stiffness and bonding capacity of the anchor [7]. Dang et al. proposed the standard test to investigate the bond performance of 18 mm prestressing strands used in precast/prestressed concrete applications. The pull-out resistance of steel can be improved by controlling the crack growth inside the concrete [8]. Hayashi et al. used the three-dimensional discrete model to analyze reinforced concrete (RC) anchorage performance. Results indicate that concrete strength is reduced if reinforcement spacing between the column and the embedded is very close because of a non-homogeneous behavior of concrete anchorage performance in a multidirectional arrangement of reinforcement bars [9].

Bond performance between concrete and steel bars was examined under corrosion level and temperature and it was found, from the study, that bond strength was influenced by temperature and corrosion [10]. John et al. revised how the anchorage's contribution was calculated and recognized the contribution of end bearing to laps and anchorages of compression bars. Bond influences the width and spacing of transverse cracks, tension stiffening, and flexural curvature. At the ultimate limit state, the bond is responsible for the strength of end anchorages and lapped joints of reinforcement and influences the rotation capacity of plastic hinge regions [11]. The hysteretic behavior of the anchorage slip is examined in reinforced concrete structures. Reinforced concrete columns subjected to axial compression and inelastic lateral deformation reversals develop significant rotations due to anchorage slip [12].

Anchorage and pull-out behavior depend on the geometry of steel fibers and is also related to the characteristics of the matrix [13]. The form of wet connection is to weld, lap, or mechanically connect the reserved connecting steel bar or connecting rod at the connection part, anchor the steel bar through post cast concrete or other grouting materials, and to connect the different prefabricated components. The dry connection is to embed the steel connecting parts in the prefabricated concrete components and then to connect them into one through bolt connection or welding a holistic approach [14]. The failure mode of reinforced concrete central pull-out specimen and the whole failure process of the reinforced concrete bond interface is divided into two stages: the elastic stage without crack and the working stage with crack. Based on the experimental results, the corresponding calculation methods of bond interface energy in different stages are obtained. The process of bond failure of reinforced concrete was analyzed from the perspective of energy [15].

Much numerical research has been carried out on the stochastic character of concrete [16–21]. Concrete is a primary construction material composed of cement and aggregates and the geometry and distribution of aggregates significantly affect the interface bond performance of the concrete structure. Jonak et al. used a contact interface between concrete and undercut anchor by using the finite element method (ABAQUS) and studied the cone failure occurring in the pull-out test. The result shows that the break-out angle of the undercut anchor was considerably less than the concrete capacity design method [18,19]. Ombres et al. conducted direct single-lap shear tests on 20 specimens in order to study the bond behavior of steel reinforce grout to concrete joints. Experimental results were compared with finite element simulation and spectated to be in good agreement [21]. Funari et al. proposed a moving mesh numerical model using interface elements to calculate debonding mechanisms, crack opening, and cracks propagation in fiber-reinforced polymer (FRP) concrete beams [22].

It is concluded from the literature review that research on the bond strength and anchorage performance of reinforcement-concrete used in prefabricated bridges is still limited in number. There are still some differences in the design standards of the assembly type concrete and anchoring. However, due to the differences in test conditions and test design set by different scholars, the current test results are relatively discrete and the conclusions are quite different. Moreover, the influence of the failure mode on the interface bond strength calculation has not been clearly distinguished in the existing experimental studies. The complete interface failure method was used to calculate the bond strength, leading to the smaller calculated value of bond strength between steel and prefabricated concrete. A high value of the anchorage length is not suitable for the design and construction of prefabricated bridges. Existing codes for prefabricated structures stipulate the anchorage length of a post-cast straight anchor connecting steel bars of precast concrete members (JGJ 145-2004) (GB 50010-20118) [23,24]. The mechanical characteristics of reinforced concrete bonding interface under sufficient cover thickness are an essential issue in the design and construction of prefabricated bridges.

This study investigates the influential factors that affect the interface bond strength and the anchorage performance of steel bars within prefabricated concrete. Twenty-two specimens were manufactured for the pull-out test by utilizing a larger cover thickness. Based on experimental results and finite element simulation, the failure mode, ultimate load, the load-displacement curves, and the effect of the different influential factors on interface bond strength and anchorage performance of steel bars within prefabricated concrete were analyzed.

Furthermore, the bond strength calculation formulas and anchorage length for the steel bars within prefabricated concrete were fitted and derived. The national and international codes for the anchorage length and interface bond strength of steel bars within prefabricated concrete under different concrete strength levels were compared and analyzed. Recommended values of bond strength and the anchorage length of steel bars in the design of prefabricated bridges are given.

## 2. Experimental Program and Analysis of Test Results

### 2.1. Specimen Design and Fabrication

Twenty-two reinforced concrete specimens were designed and manufactured for pull-out tests using larger cover thicknesses are shown in Figure 1. The section size of each specimen is 300 mm × 300 mm. The concrete strength grades were C30 and C50, the anchorage lengths of reinforcement were 200 mm and 300 mm, the diameters of the ribbed steel bar were 16 mm and 20 mm, and the diameters of plain steel bars were 8 mm and 20 mm.

### 2.2. Materials Test and Properties

Hot-rolled Ribbed Bar (HRB400) [25] with the diameter of 12 mm and 16 mm was used in this study. Elastic modulus and compressive strength of C50 and C30 were calculated according to relevant specifications [26]. Three groups of standard cubes 150 mm × 150 mm × 150 mm were made for the compression test of C50 and C30 (each group has three specimens). An elastic modulus test was carried out on a prism block of the size 150 mm × 150 mm × 300 mm; three groups of specimens were designed with three test specimens in each group. The mix proportions of C30 and C50 are given in Table 1.

**Figure 1.** Reinforced concrete test specimens.

**Table 1.** Mix proportion of C30 and C50.

| C30 | | Mix Ratio | C50 | | Mix Ratio |
|---|---|---|---|---|---|
| Cement | 440 | 1 | Cement | 450 | 1 |
| Sand | 532 | 1.209 | Sand | 682 | 1.515 |
| Aggregate | 1243 | 2.82 | Aggregate | 1113 | 2.47 |
| Water | 185 | 0.420 | Water | 155 | 0.344 |
| - | - | - | Fly ash | 50 | 0.111 |

The tensile test was carried out according to the Chinese code [27,28] out on HRB400 with a diameter of 12 mm and 16 mm. The average value of material performance test results of concrete and steel bars are shown in Table 2.

**Table 2.** Material test results of concrete and steel bar.

| Specimen | Average of Elastic Modulus (MPa) | Compressive Strength (MPa) |
|---|---|---|
| C50 | 34,429 | 53.8 |
| C30 | 32,521 | 31.2 |
| Steel bar | Yield strength (MPa) | Ultimate strength (MPa) |
| HRB400Φ12 | 525 | 645 |
| HRB400Φ16 | 605 | 705 |

*2.3. Test Instruments*

A bolt comprehensive parameter tester was used as a loading device. An intelligent digital pressure gauge indicates the load's value and the loading end displacement was measured by electronic displacement, as shown in Figure 2. At the first stage of loading, the load increment was 2 kN–5 kN and, at the second stage of loading, the load increment was 5 kN–10 kN. Electronic displacement meters were simultaneously set at the loading end and free end, respectively, with the spacing of 70 mm at the end and 100 mm in the middle.

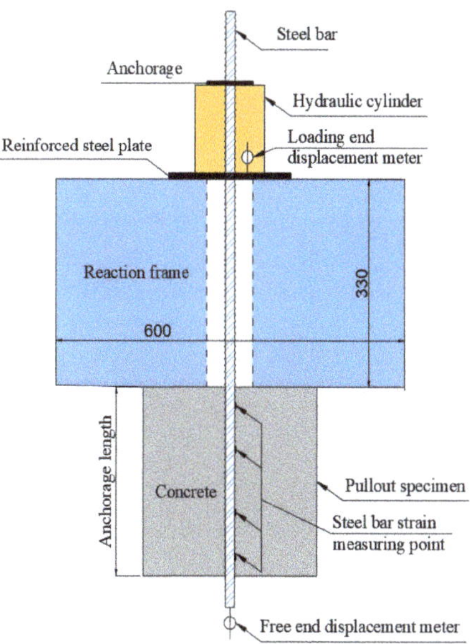


**Figure 2.** Schematic diagram of the loading device.

### 2.4. Pull-Out Test Results

The failure mode and ultimate load of each specimen in the pull-out test are shown in Table 3. The specimens in Table 3 are numbered according to the concrete strength grade-reinforcement diameter-anchorage length, such as C30-12-300.

**Table 3.** Failure modes of reinforced concrete pull-out specimens.

| Specimen Number | Failure Mode | Ultimate Load (kN) | Specimen Number | Failure Mode | Ultimate Load (kN) |
|---|---|---|---|---|---|
| C30-12-200 | Reinforcement failure | 59.5 | C50-12-200 | Reinforcement failure | 72.5 |
| C30-12-300 | Reinforcement failure | 64.75 | C50-12-300 | Reinforcement failure | 68.9 |
| C30-16-200 | Reinforcement failure | 60 | C50-16-200 | Reinforcement failure | 125.1 |
| C30-16-300 | Reinforcement failure | 106.82 | C50-16-300 | Reinforcement failure | 128 |
| C50-ø8-300 | Pull-out of a plain bar | 11.48 | C50-16-400 | Reinforcement failure | 119.7 |
| C50-ø20-300 | Pull-out of a plain bar | 18.1 | C50-20-300 | Reinforcement failure | 153.77 |

Test results of all the specimens show the tensile failure of reinforcement. The ultimate load difference of C50 and C30 indicates that, for prefabricated concrete members with larger cover thicknesses, the concrete strength grade and reinforcement diameter significantly influences the ultimate load of the bond interface between reinforcement and concrete. The ultimate load of the ribbed steel bar specimen C50-20-300 is 8.5 times higher than that of the plain steel bar specimen C50-ø20-300.

It can be estimated that the mechanical interlocking force accounts for about 88% of the bond strength between deformed steel bars and concrete. In addition, by comparing the failure modes of different specimens, it is concluded that, with the decrease in concrete strength and the increase in steel bar diameter, the failure mode of specimens gradually changes from the tensile failure of reinforcement to pull-out failure.

### 2.5. Load-Displacement Curves

The load-displacement curve from the experimental results is drawn and demonstrates the tensile failure of the specimen, as shown in Figure 3. Specimen C50-12-300 is taken as an example.

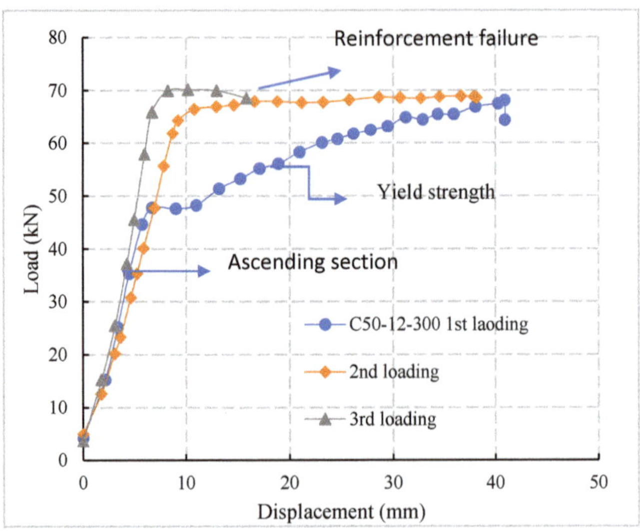


**Figure 3.** Load-displacement curve-tensile failure of reinforcement.

Test specimens were loaded three times. At the first stage of loading, the specimen's ductility was large and there is a yield strengthening stage before the steel bars yield. In the second stage of loading, when the load value reaches its maximum value of the first load the steel bars begin to yield and there is a longer yield deformation stage. At third stage of loading, when the load reaches the yield value the steel bars break and there is no obvious post-yield strengthening stage. The diameter of the reinforcement determines the ultimate bearing capacity of the specimen and it possesses a positive linear correlation with the cross-sectional area of the reinforcement.

### 2.6. Failure Mode

There were two main failure modes of the steel bar pull-out specimens in prefabricated concrete structures: reinforcement pull-out failure and tensile failure of the steel bar.

(1) Pull-out failure of reinforcement.

When the anchorage is insufficient, the pull-out force was greater than the bond interface bearing capacity of reinforced concrete and the failure of the bond interface occurs in the specimens. It can be seen from Figure 4a when the ribbed bar is pulled out, the concrete at the loading end was damaged in a cone shaped manner and there was no obvious necking of the bar. It can be seen from Figure 4b that the transverse rib of the reinforcement at the bond between the pulled-out reinforcement and concrete was intact Under the pull-out load, the reinforcement was pulled out together with some intercostal concrete. The residual intercostal concrete accounts for about 50% of the spacing between the transverse ribs of the reinforcement.

(2) Tensile failure of the steel bar.

As shown in Figure 4c, when the reinforcement was sufficiently anchored, the bearing capacity of the reinforced concrete interface was more significant compared to the tensile bearing capacity of the reinforcement. The tensile failure of the reinforcement occurs in the specimens with a strength grade of C50.

### 2.7. Analysis of Plain and Ribbed Steel Bar Diameter

Figure 5 shows the load-displacement curve of plain and ribbed reinforcement with different diameters. It can be observed from the figure that the pull-out failure of steel bars occurs in both groups of the test specimens. With the increase in steel bar diameter, the ultimate load of the specimens increases gradually. The ultimate load of C50-G8-300 and C50-G20-300 is 11.48 kN and 18.1 kN, respectively. The analysis shows that, due to the Poisson's ratio effect of steel, the reinforcement will produce radial deformations under the pull-out force. If the diameters of steel bars were large, there would be more obvious shrinkage.

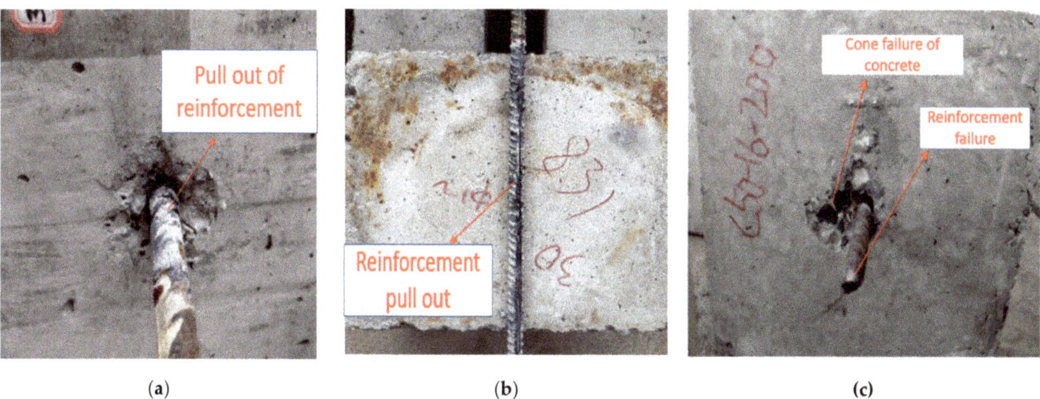


**Figure 4.** (**a**) Pull-out failure of reinforcement at loading end. (**b**) Cone-shaped failure after pulling out. (**c**) Tensile failure of reinforcement.

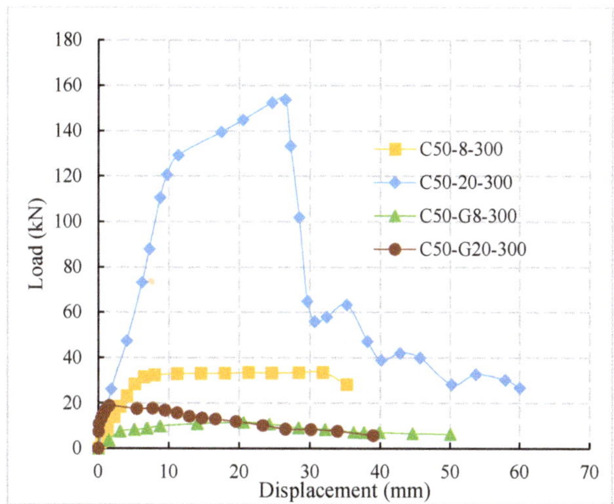


**Figure 5.** Load-displacement curve of plain and ribbed reinforcement.

The bond interface between reinforcement-concrete was to be debonding and bond stress decreases between plain reinforcement and concrete with the increase in reinforcement diameter. Comparing the ultimate load of C50-8-300 and C50-G8-300 specimens with C50-20-300 and C50-G20-300, it is found that the interface bond strengths of plain steel bar are mainly composed of chemical bond strength and friction force between reinforced concrete, the mechanical interlocking force of bond interface is small or negligible, the bearing capacity of the bond interface between the plain steel bar and concrete is weak, and the ultimate load is far less than that of the deformed steel bar.

## 3. Modelling

### 3.1. Specimen Design

The finite element model (ABAQUS) of reinforced concrete pull-out specimens was established to simulate the interface bond and anchorage characteristics of steel bars within prefabricated concrete. In order to ensure the accuracy of the model and improve the calculation efficiency, the bilinear axisymmetric quadrilateral reduced integral element CGAX4R was used for both reinforcement and concrete. The mesh density of the bonding interface, reinforcement axis, and concrete edges were 0.5 mm, 0.5 mm, and 2 mm, respectively. By establishing the surface size of the transverse rib, the influences of the artificial definition of interface elements on the analysis results were reduced. The model diagram and mesh generation are shown in Figure 6. The minimum mesh size of the model was 0.4 mm.

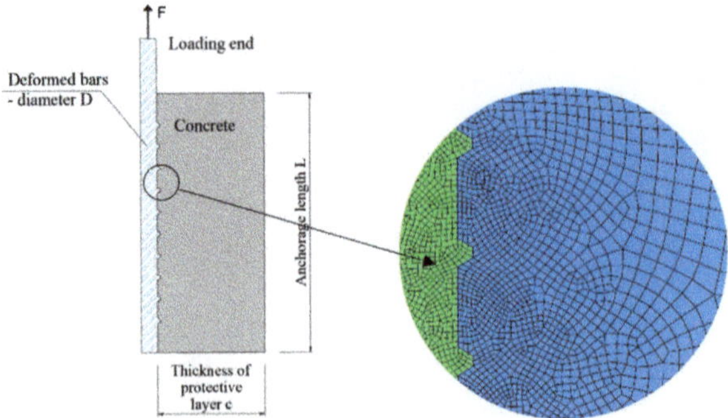


**Figure 6.** Mesh generation of the finite element model.

### 3.2. Contact Problem Simulation

The detailed characteristics of the bond interface between steel and concrete were considered. The normal behavior of the contact surface was simulated by "Hard Contact" and implemented by the Classical Lagrange Multiplier Method.

The transmitted compressive stress between the contact surfaces was unlimited. When the pressure on the contact surface becomes negative or zero, the two contact surfaces were separated. The tangential behavior was simulated by "Penalty Friction." In the Classical Coulomb Friction Model, the critical friction stress depends on the contact pressure and the elastic slip was allowed on the contact surface. Assuming that the friction coefficient $\mu$ between the contact surfaces was the same 0.1 [29], axial symmetry was adopted for boundary conditions.

### 3.3. Parameter Selection

In order to study the interface bond strength and anchorage performance between the reinforcement and the concrete of the prefabricated bridge, the larger thickness of the protective layer was adopted according to the structural characteristics. The cross-sectional dimensions of the specimens were 300 mm $\times$ 300 mm according to the prefabricated structure specification, as per experimental design. The test parameters include the diameter of reinforcement, concrete strength, and anchorage length. The concrete strength grade was C30 and C50. The reinforcement diameter was 12 mm, 16 mm, and 20 mm and the anchorage length was 150 mm, 200 mm, and 300 mm, respectively.

### 3.4. Material Constitutive Model

In order to simulate the failure and crack development of the bond interface of reinforced concrete, the elastic-plastic damage constitutive model was adopted for concrete, which can be used to observe and analyze the development law of cracks by using the cloud diagram of concrete tensile damage (DAMAGET). The pull-out phenomenon of reinforced concrete's bonding interface was observed through concrete compression damage (DAMAGEC) [30]. The constitutive model of concrete compression damage and tension damage is shown in Figure 7.

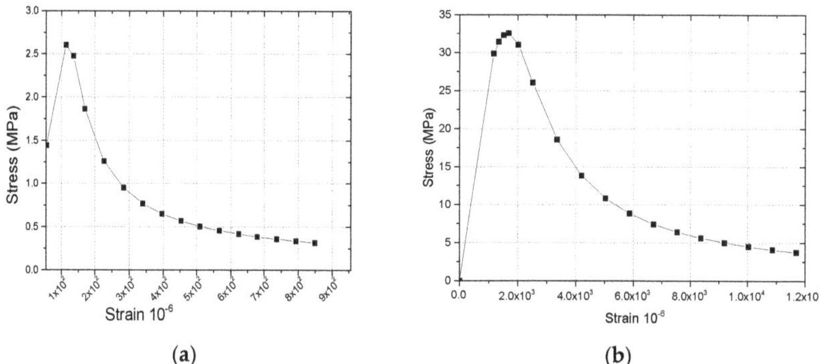


**Figure 7.** Plastic damage constitutive curve of concrete. (**a**) Compression damage constitutive curve of concrete. (**b**)Tensile damage constitutive curve of concrete.

The constitutive model can be expressed in the following:

$$\sigma_{t,c} = (1 - d_{t,c})E_0(\varepsilon_{t,c} - \tilde{\varepsilon}_{t,c}^{pl}) \tag{1}$$

where $d_{t,c}$ is the damage factor of concrete, $E_0$ is the elastic modulus of concrete, $\varepsilon_{t,c}$ is the concrete strain, and $\tilde{\varepsilon}_{t,c}^{pl}$ is the equivalent plastic strain of concrete.

In the simulation of reinforced concrete structure, the interface effect between reinforcement and concrete (such as bond-slip and locking behavior) was simulated by introducing "Tensile Hardening" into the concrete model. The tensile hardening data were defined according to the cracking strain $\tilde{\varepsilon}_t^{pl}$. The relationship between the equivalent plastic strain $\tilde{\varepsilon}_t^{ck}$ and the cracking strain $\tilde{\varepsilon}_t^{pl}$ in the model is described as follows $\tilde{\varepsilon}_t^{ck}$.

$$\tilde{\varepsilon}_t^{pl} = \tilde{\varepsilon}_t^{ck} - \frac{d_t}{(1 - d_t)}\frac{\sigma_t}{E_0} \tag{2}$$

According to the definition of compression hardening, the hardening data are defined according to the inelastic strain $\tilde{\varepsilon}_c^{inl}$. The relationship between the equivalent plastic strain $\tilde{\varepsilon}_c^{pl}$ and the inelastic strain $\tilde{\varepsilon}_c^{inl}$ in the model is as follows:

$$\tilde{\varepsilon}_c^{pl} = \tilde{\varepsilon}_c^{inl} - \frac{d_c}{(1 - d_c)}\frac{\sigma_c}{E_0} \tag{3}$$

$$\sigma_s = \begin{cases} E_s\varepsilon_s & \varepsilon_s \leq \varepsilon_y \\ f_y & \varepsilon_s > \varepsilon_y \end{cases} \tag{4}$$

where $\sigma_s$ is the stress of reinforcement, $E_s$ is the elastic modulus of reinforcement, $f_y$ is the yield strength of reinforcement, $\varepsilon_s$ is the strain of reinforcement, and $\varepsilon_y$ is the yield strain of reinforcement. Properties of concrete and reinforcement are indicated in Table 4.

### 3.5. Finite Element Model Validation

Taking C50-12-300 as an example, the finite element model was established. According to the results of finite element analysis, the load-displacement curve of the specimen was extracted and compared with the experimental result curve, as shown in Figure 8. It can be observed that the rising stage and ultimate load of the finite element model are consistent with that of the test specimen. The errors between the limit load of the finite element model and the experimental value were 2.4% and 0.8%, respectively.

**Table 4.** Material properties of concrete and reinforcement.

| Specimen | Poisson's Ratio | Young's Modulus (MPa) | Compressive Strength (MPa) |
|---|---|---|---|
| C50 | 0.2 | 34,429 | 53.8 |
| C30 | 0.2 | 32,521 | 31.2 |
| Steel bar | / | / | Ultimate strength (MPa) |
| HRB400Φ12 | 0.3 | 206,000 | 645 |
| HRB400Φ16 | 0.3 | 206,000 | 705 |

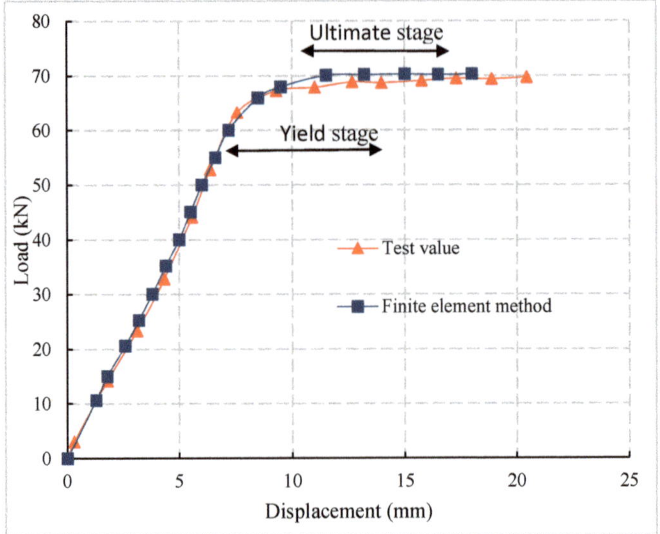


**Figure 8.** Test finite element comparison.

### 3.6. *Analysis of Interface Bond Failure Process*

Under the pull-out load, the failure mode of the bond interface of reinforced concrete was the same as that of the experimental test results. The development of cracks cannot be directly observed in the test, but the development law of cracks can be observed and analyzed by using the cloud picture of concrete tensile damage (DAMAGET) in the finite element results. The failure appearance of the bonding interface of reinforced concrete can be observed through concrete compression damage (DAMAGEC) and the main failure mode is shown in Figure 9.

The bond strength between reinforcement and concrete was determined by the properties of the interface between them, mainly including three factors: (1) the chemical bond force between the concrete substrate and the surface coating of reinforcement; (2) the relative sliding friction resistance between reinforcement and concrete along with the interface; (3) the mechanical interlocking force caused by the unevenness of the interface between reinforcement and concrete [30–33]. The test results show that the failure modes of reinforced concrete interface bond specimens under pull-out load are mainly divided into the yield fracture of reinforcement and interface concrete failure. The bond strength of ribbed steel bar and concrete was composed of chemical bond force, friction force, and mechanical interlocking force.

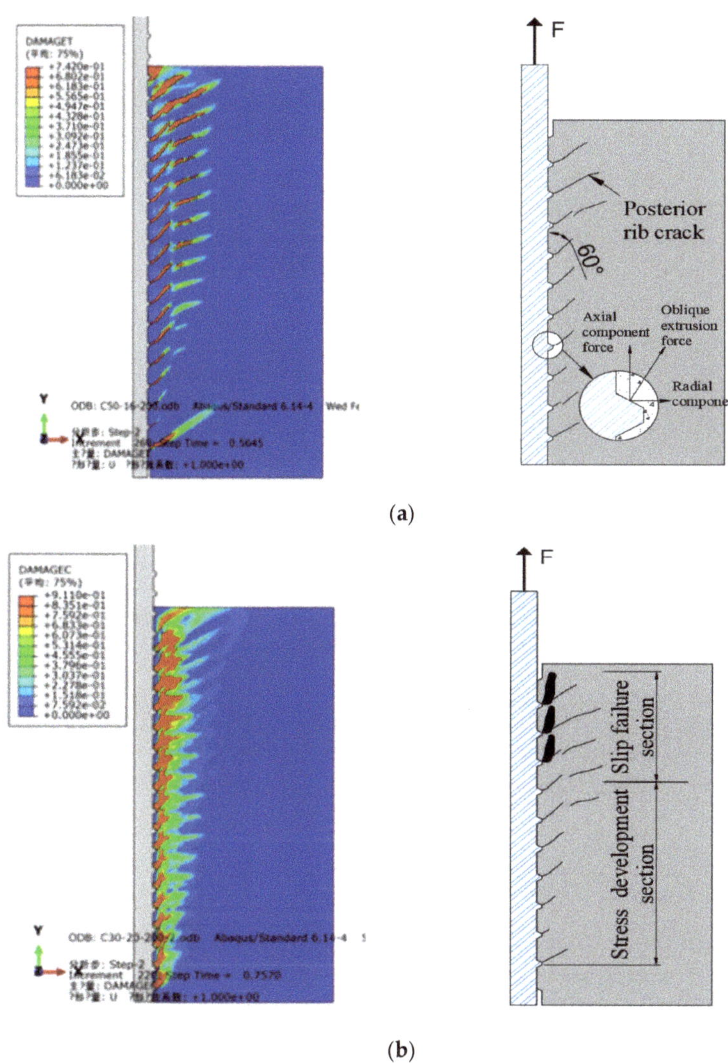


**Figure 9.** Failure process of the bonding interface. (**a**) Concrete crack development. (**b**) Interface concrete failure.

At the initial loading stage, the bonding interface's slip resistance was assumed by the chemical bonding force. In contrast, the mechanical interlocking force and friction force do not play a role temporarily. With the increase in the load, the chemical bond force fails, the bond interface slips relatively, the mechanical interlocking force and friction force begin to play a role, and the interface slip resistance is then provided by the oblique extrusion force between the transverse rib and the concrete. The axial component of the oblique extrusion force renders the concrete between the ribs subject to bending and shearing as a cantilever beam. The radial component of the oblique extrusion force results in the concrete around the reinforcement producing circumferential tensile stress. At this time, the concrete around the reinforcement was in a three-phase stress state. As shown in Figure 9a, the concrete behind the transverse rib of the steel bar was pulled by the oblique extrusion force, while the concrete in front of the rib was pressed. With the increase in the load, the radial cracks first appear behind the rib and develop along the direction of 60°, with the axial direction of the steel bar (the inclination angle of the transverse rib of the reinforcing bar). The radial crack depths of the cracks were approximately equal to the spacing between the transverse ribs of the reinforcing bar. As shown in Figure 9b, the radial failure depth was about two times the transverse rib height of the reinforcement. This failure process is called the shear bond failure of ribbed bars.

### 3.7. Load-Displacement Curve

According to the results of the finite element analysis, the load-displacement curve of the specimen was extracted, the stress characteristics and failure mechanism of different stages were analyzed, and the influences of concrete strength grade, reinforcement diameter, and anchorage length on the load-displacement curve were compared.

As shown in Figure 10, the constitutive interface model can be divided into an initial linear elastic stage and failure stage. The failure process can be divided into the micro-slip stage, internal crack slip-stage, and pull-out fluctuation decline stage. In the micro-slip stage, the slip at the loading end is minimal and there is no slip at the free end, which shows a linear phase on the load slip curve. It can be considered that the bond force gradually transfers from the near end to the far end and the adhesive force was complete during the internal crack sliding stage. When the load continues to increase, the bond force was transferred to the far end, a small amount of slip begins to appear at the far end, and the adhesive force disappears. The interface bond was mainly maintained by the friction force and mechanical interlocking forces between the concrete and reinforcement ribs. In the pull-out stage, the interface concrete was damaged, the load suddenly decreases, and the reinforcement is gradually pulled-out.

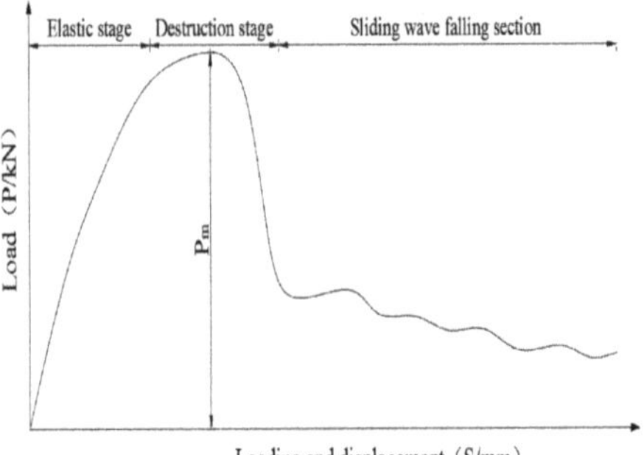


**Figure 10.** Load-displacement curve.

## 4. Parametric Study

### 4.1. Scope of Investigation

Parametric analysis was carried out with concrete strength, reinforcement diameter, and anchorage length as a variable. The parameters, failure mode, ultimate load, and bond strength of each specimen are shown in Table 5. The specimens in the Table are numbered according to the concrete strength grade—reinforcement diameter—anchorage length, such as C30-12-300. The calculation formula of bond stress in the Table is as follows:

$$\tau = \frac{F}{\pi d l} \tag{5}$$

where $\tau$ is the bond strength of reinforced concrete, $F$ is the pull-out load, $d$ is the diameter of reinforcement, and $l$ is the effective bond length.

**Table 5.** Analysis results.

| Scheme | Failure Mode | Ultimate Load [kN] | Bond Strength /MPa | Specimen Number | Failure Mode | Ultimate Load [kN] | Bond Strength /MPa |
|---|---|---|---|---|---|---|---|
| C30-12-200 | Reinforcement failure | 42.71 | 14.78 | C50-16-300 | Reinforcement failure | 76.3 | 14.52 |
| C30-16-200 | Reinforcement failure | 75.96 | 11.53 | C50-12-300 | Reinforcement failure | 44.71 | 15.82 |
| C30-20-200 | Reinforcement pull-out | 99.97 | 10.83 | C50-20-200 | Reinforcement failure | 124.27 | 13.86 |
| C30-16-300 | Reinforcement failure | 74.4 | 11.81 | C50-16-200 | Reinforcement failure | 75.74 | 14.57 |

*4.2. Comparison of Load-Displacement Curves of FEM and Test Result*

The load-displacement curves of the experiment and FEM are compared, which are in good agreement and shown in Figure 11. Load-displacement comparison shows that with the increase in the concrete strength, the steel bar's pull-out failure changes into the tensile failure of the steel bar. The ultimate load-bearing capacity of the specimen increases significantly. It is concluded that in engineering practice, high-strength concrete can improve the bond strength of steel bars and reduce the requirements for anchorage length of steel bars to a certain extent.

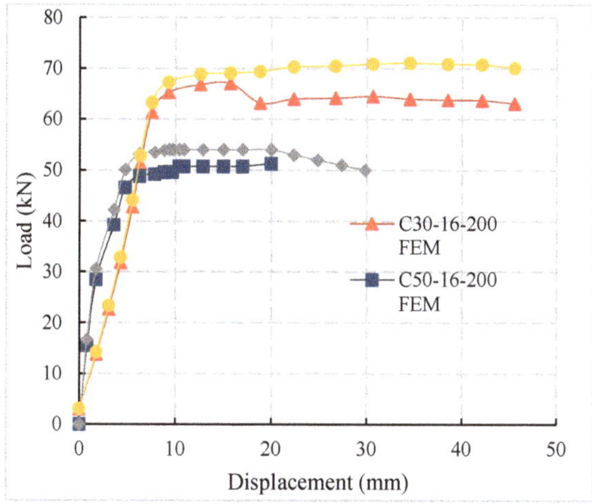


**Figure 11.** Comparison of load-displacement curves.

*4.3. Influence of the Concrete Strength Grade*

The bond strength and ultimate load of specimens with different concrete strengths were compared in Figure 12. The d16-100 series in Figure 12 represent five specimens with different concrete strengths with a reinforcement diameter of 16 mm and anchorage length of 100 mm.

According to Figure 12a, the bond strength between reinforcement and concrete increases with concrete strength. Compared with C30, C40, C50, and C60 specimens, the bond strength of C80 specimens increases by 44.2%, 34.7%, 16.4%, and 5.8%, respectively.

At the same time, it can be observed from Figure 12a that the correlation between the bond strength and the anchorage length of the specimens with different anchorage lengths was small for the connection reinforcement of prefabricated assembled concrete structure with larger cover thickness. The influence of anchorage length can be ignored in the calculation of bond strength. It can be observed from Figure 12b that with the increase in concrete strength, the failure mode of the specimen changes from the pull-out failure to the tensile failure of the reinforcement and the ultimate bearing capacity was significantly increased. The results show that when the pull-out failure occurs, the anchorage length was insufficient and the concrete strength possesses a significant impact on the ultimate bearing capacity of the specimens. However, when the steel bar was failing, the anchorage length was sufficient, such as the d16-200 series and d16-300 series, and the concrete strength had little influence.

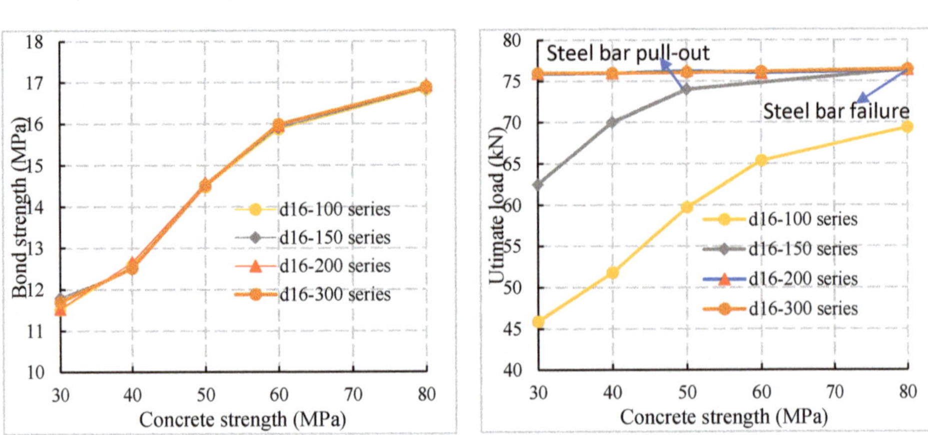


**Figure 12.** Influence of concrete strength grade. (**a**) Comparison of bond strength and concrete strength. (**b**) Ultimate load concrete strength comparison.

## *4.4. Influence of the Reinforcement Diameter*

The load-displacement curves of specimens with different reinforcement diameters were compared, as shown in Figure 13a. When the anchorage length and concrete strength grades were the same, the specimens' bearing capacity gradually increased with the increase in the reinforcement diameter. However, the failure mode of the specimens changed from the tensile failure (C30-12-200 and C30-16-200) to the pull-out failure (C30-20-200 and C30-25-200), indicating that the specimens did not pull out and the anchorage length required for failure was significantly increased; that is, the anchorage length closely related to the diameter of reinforcement.

From the previous analysis, it can be seen that the interfacial bond strength between the reinforcement and the concrete was mainly composed of the mechanical interlocking force. Shape parameters determine the mechanical interlocking force, the transverse rib's height, and the spacing between the transverse ribs. The relative rib area (the ratio of the projected area of the transverse rib on the surface of the reinforcement to the reinforcement's surface area) is taken as the index to evaluate the bond performance.

When the relative rib area was larger, the bond performance should be improved According to (GB 1499.2-2018) [25], the relative rib area of reinforcement decreases with the increase in the diameter of steel bars and the bond performance between the connecting reinforcement and the concrete also decreases, as shown in Figure 13b Compared with the diameter of the specimen of 8 mm, 12 mm, 16 mm, and 20 mm, the bond strength of the 25 mm steel bar decreases by 33.9%, 28.1%, 15.6%, and 11.1%. It can be seen from Figure 13c that the ultimate load is linearly related to the cross-sectional area of the reinforcement As the reinforcement's diameter increases, the specimen's failure mode changes from tensile failure to pull-out failure. The diameter of the reinforcement is no longer the main influencing factor of the ultimate load (such as the C30-200 series).

When the diameter of reinforcement increases, the bond area and mechanical interlocking depth of reinforcement concrete increases. The increment of interface bearing capacity caused by the increase in bond area was more significant than the decrease caused by the decrease in bond strength under the same conditions. It was concluded that the ultimate load still increases with the increase in reinforcement diameter.

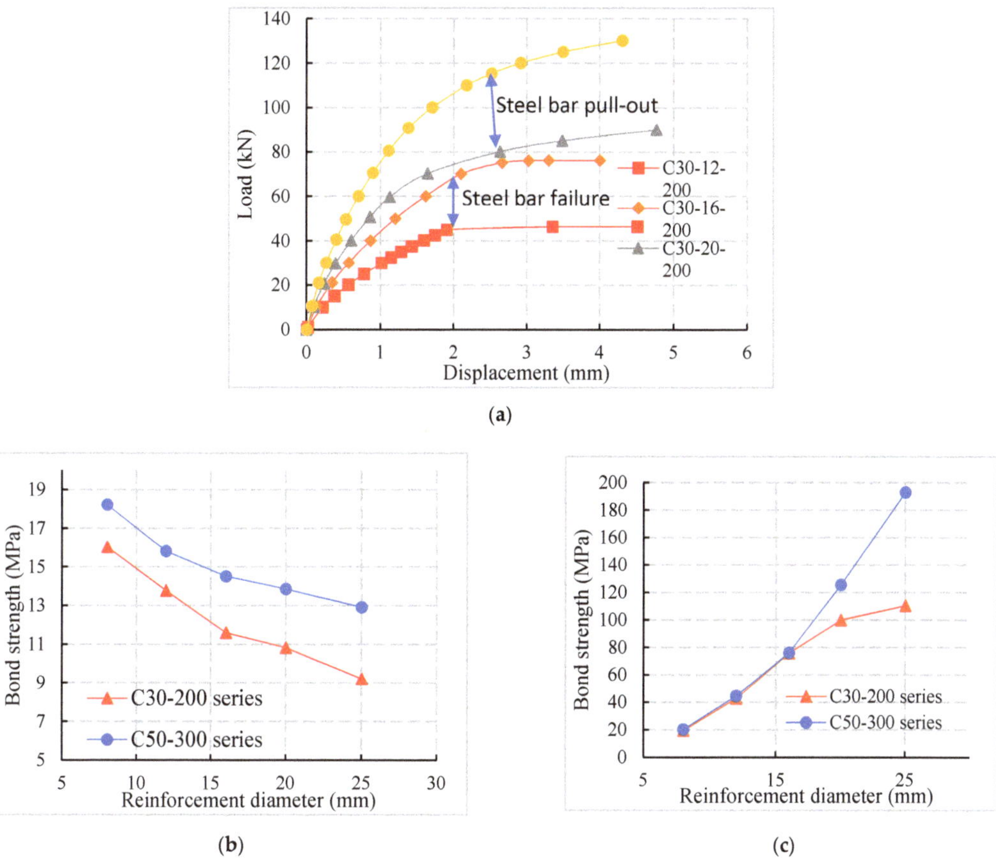


**Figure 13.** (**a**) Comparison of load-displacement curves of specimens with different reinforcement diameters. Influence of steel bar diameter: (**b**) Bond strength and steel bars diameter; (**c**) Ultimate load and steel bars diameter.

### *4.5. Influence of Anchorage Length*

It can be observed from the C50-16-100 specimen in Figure 14a that when the steel bars were not sufficiently anchored, the pull-out failure of the specimen occurs. Before the steel bars were pulled out, the interface stiffness decreases rapidly with the load increase and the bond interface between reinforcement and concrete fails. When the reinforcement is fully anchored (C50-16-200 and C50-16-300 specimens in Figure 14a and the interface stiffness decreases with the increase in load before the reinforcement yields, the change amplitude was small. It can be observed from Figure 14b that the anchorage length has little effect on the bond strength between the connecting steel bar and the concrete.

By comparing the ultimate load of different specimens in Figure 14c, it was found that the ultimate load of specimens with an anchorage length of 200 mm in the C30-16 series increases by 21.5% and 64.2%, respectively, compared with specimens with anchorage lengths of 150 mm and 100 mm. It was concluded that when the anchorage length was not sufficient, the ultimate load increases with the increase in anchorage length and the failure mode of the specimen will change from pull-out failure to tensile failure.

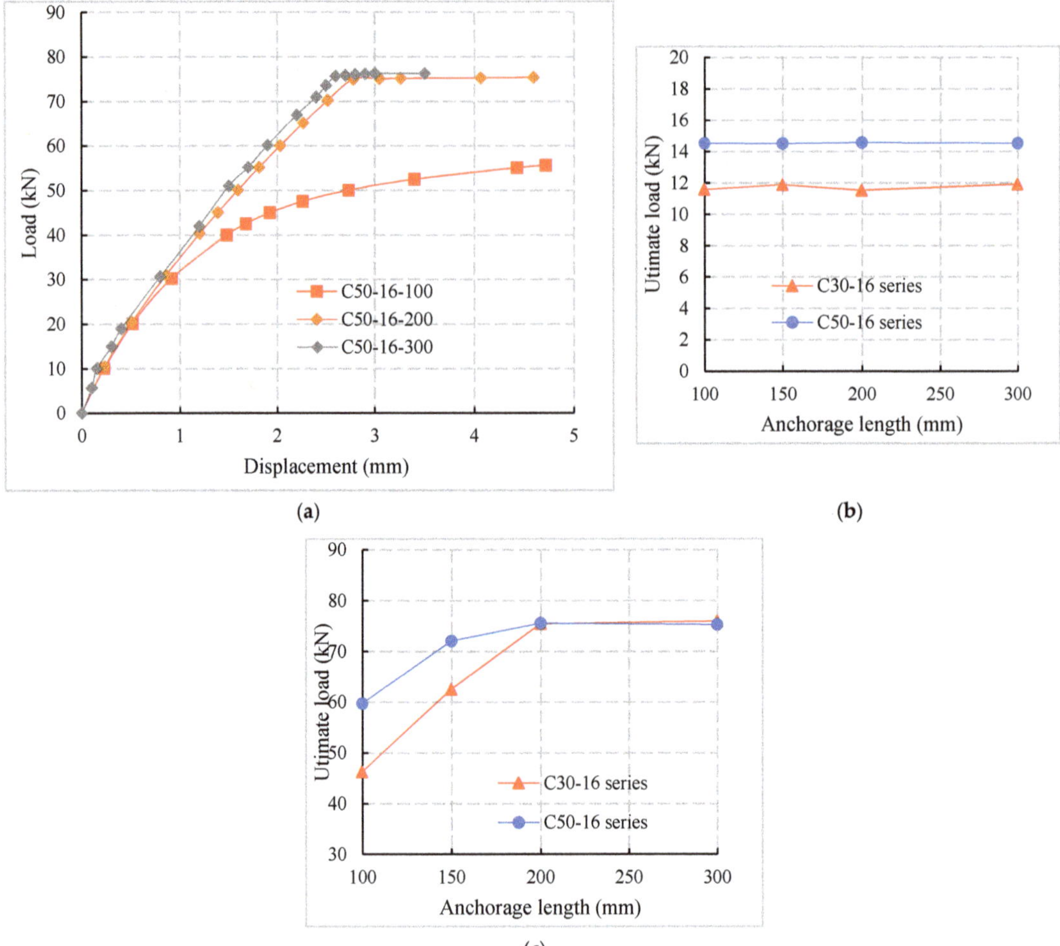


**Figure 14.** Influence of anchorage length. (**a**) Load-displacement curves for anchorage lengths. (**b**) Bond strength and anchorage length. (**c**) Relationship b/w ultimate load and anchorage length.

## 5. Simplified Calculation Formula

### *5.1. Calculation Formula for the Bond Strength*

For calculating the bond strength of steel bars within the concrete, scholars and codes have given the corresponding semi-empirical and semi-theoretical calculation formulas [23,24,33,34] by comprehensively considering different factors. The relevant formulas consider the concrete strength, protective layer thickness, anchorage length, and reinforcement diameter. The typical calculation formula of bond strength is shown in Table 6. The concrete structure design code [24] only calculates the bond strength between steel bar and concrete from the perspective of concrete tensile strength, without considering the influence factors such as steel bar type, steel bar diameter, anchorage length, and cover thickness. The Australian code [33] and the American code [34] considered the concrete strength, which is the ratio of concrete cover thickness to steel bar diameter, as the key index of bond strength calculation.

**Table 6.** Calculation formula of bond strength.

| Standard | Calculation Formula of Bond Strength |
|---|---|
| Australia AS3600 [33] | $\tau_u = 0.3 \times (0.5 + c/d)\sqrt{f_{cu}}$ |
| America ACI318-11 [34] | $\tau_u = 0.08 \times (1.2 + 3c/d + 50d/l)\sqrt{f_{cu}}$ |
| Literature [35] | $\tau_u = (1.6 + 0.7c/d + 20\rho_{sv})(0.82 + 0.9d/l)f_t$ |

Although the concrete cover in the fabricated structure is sufficient, the bond strength of the connecting steel bar will not be affected due to the too-small cover thickness. Therefore, the existing formula for calculating the bond strength of reinforcement is not suitable for the calculation of the connecting reinforcement of prefabricated bridges, which will lead to the length of the reserved connecting reinforcement of prefabricated components and will have an adverse impact on the construction difficulty and assembly accuracy control.

Therefore, the existing formula of bond strength of steel bars is not suitable for calculating the connection reinforcement of prefabricated bridges, which will lead to the long reserved connection reinforcement of prefabricated members and will have an adverse effect on the construction difficulty and accuracy control of the assembly. According to the above analysis, the main factors influencing the bond strength of the prefabricated and assembled concrete structure connecting steel bars are the diameter of the reinforcement and the concrete's strength. Therefore, the calculation formula of the bond strength of the connecting steel bar can be analyzed.

In the Table 6, $\tau_u$ is the interface bonding strength of reinforced concrete interface, $f_{cu}$ is the standard value of concrete compressive strength, $f_{t,r}$ is the characteristic value of concrete tensile strength, and $f_t$ is the splitting strength of concrete. $d$ is the diameter of reinforcement and $c$ is the thickness of the protective layer. The anchorage length is $l$ and $\rho_{sv}$ is the reinforcement ratio. Multiple linear regression analysis was used to determine the influence proportion of each factor of the above formulas. The results are shown in Figure 15. It can be seen that the bond strength of the interface between reinforcement and concrete is positively correlated with the strength of concrete and negatively correlated with the diameter of reinforcement. The formula for calculating the bond strength of connecting bars in prefabricated concrete structure can be fitted as follows:

$$\tau_u = 0.1f_{cu} - 0.35d + 14.9 \tag{6}$$

where $\tau_u$ is the bond strength of the interface between steel bar and concrete, $f_{cu}$ is the standard value of concrete compressive strength, and $d$ is the diameter of the steel bar.

The minimum error between the fitting formula and the test value in literature is 2%, the maximum error is 11%, and the overall error is 7%. The minimum error is 0.6%, the maximum error is 7%, and the overall error is 4%. The average ratio of the fitting value to the test value is 1.04, and the standard deviation and coefficient of variation of the ratio are 0.06 and 5.77%, respectively. The results show that the fitting values are in good agreement with the experimental values and the dispersion is low.

According to the principle of the complete failure of interface, the formula (6) was compared with this paper's test results and the references [35,36]. The comparison results are shown in Figure 16. It was found that the average ratio of the bond strength fitting value to the test value is 1.04, the standard deviation and coefficient of variation of the ratio are 0.06 and 5.77%, respectively, and the goodness of fit $R^2$ between the fitting value and the measured value is 0.87, which indicates that the fitting value is in good agreement with the test value and that the dispersion is low.

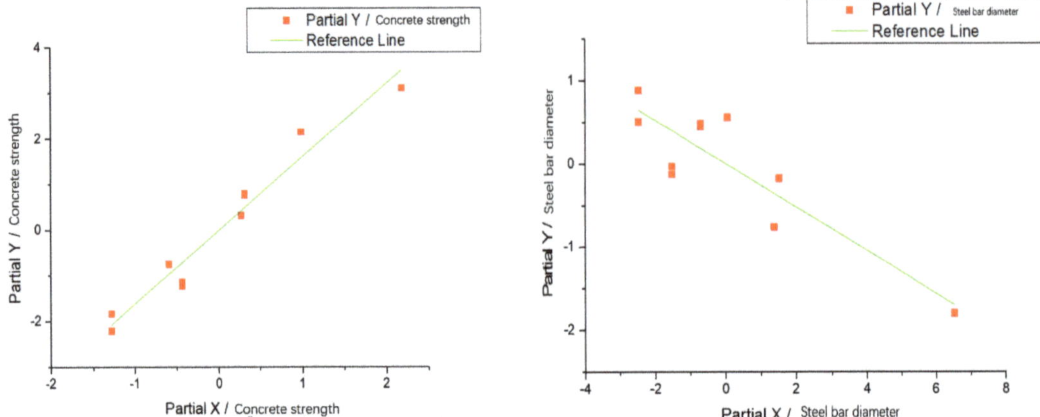


**Figure 15.** Multiple linear regression curve.

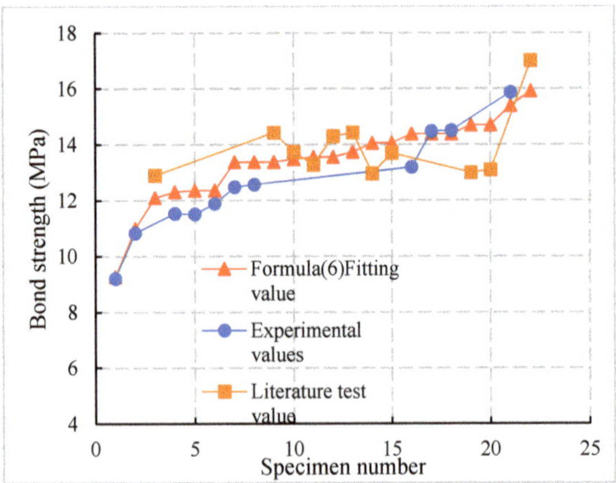


**Figure 16.** Comparison curve between fitting value and test value of bond strength.

Taking the diameter of a steel bar, 25 mm, and the thickness of the protective layer 85 mm commonly used in the connection of prefabricated concrete members as an example, the differences of bond strength calculation between domestic and foreign codes and the fitting formula in this paper are shown in Figure 17. It can be observed that the bond strength calculation value of ACI 318-11 [34] is 25% and 15.9% higher than that of AS3600 [33] and JTG 3362-2018 [27], respectively, and this indicates that there are some differences in bond strength calculation between domestic and foreign codes. Compared with AS3600 [33], JTG 3362-2018 [27], and ACI318-11 [34], the bond strength of the fitting formula (6) was increased by 62.7%, 50.9%, and 30.25%, respectively, and this indicates that the calculation results of formula (6) in China and abroad are smaller without considering the influence of cover thickness, which is not suitable for the calculation of bond strength between reinforcement and concrete of prefabricated concrete members.

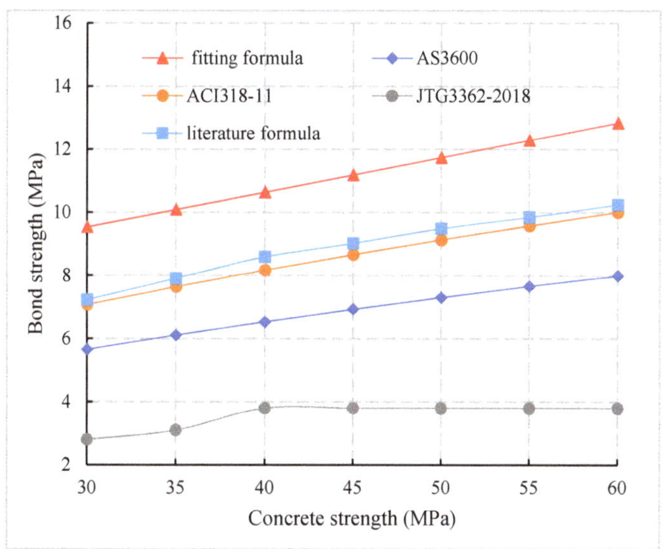


**Figure 17.** Comparison of calculation formulas of bond strength in China and abroad.

### *5.2. Calculation Formula for the Anchorage Length*

When the bond strength between reinforcement and concrete was constant, the failure mode of the specimen was determined by the anchorage length. There was a critical length between the pull-out failure and the tensile failure of the reinforcement, which is called the critical anchorage length $l_{cr}$. When the anchorage length was less than the critical anchorage length, the pull-out failure occurs.

When the anchorage length was greater than the critical anchorage length, the bearing capacity of the bonding interface was greater than the ultimate tensile load of the reinforcement and the tensile failure of the reinforcement occurs. For the calculation of the critical anchorage length of the connection reinforcement in the prefabricated concrete structure, the calculation formula of the critical anchorage length can be given according to the bond strength calculation formula as follows:

$$l_{cr} = \frac{P_u}{\pi d \tau_u} = \frac{\sigma_s d}{4\tau_u} = \frac{\sigma_s d}{4(0.10 f_{cu} - 0.35d + 14.9)} \tag{7}$$

where $P_u$ is the ultimate tensile load of reinforcement, $\sigma_s$ is the ultimate strength of reinforcement, $d$ is the diameter of reinforcement, and $\tau_u$ is the bond strength.

The basic anchorage length $L_a$ of reinforcement specified in the code is generally determined based on the critical anchorage length calculated by the corresponding bond strength and multiplied by the corresponding safety factor. The calculation formula of the basic anchorage length of reinforcement, the ratio $L_a/L_{cr}$ of the basic anchorage length, and the critical anchorage length in domestic and foreign codes are shown in Table 7.

The basic anchorage length of the tensile reinforcement is $l_a$, $n$ is the concrete strength coefficient, $f_y$ is the design value of the tensile strength of the reinforcement, $f_{cu}$ is the standard value of the concrete compressive strength, and $f_t$ is the design value of the axial tensile strength of the concrete. The diameter of the anchored reinforcement is $d$, $\alpha$ is the shape coefficient of the anchored reinforcement, and $c$ is the thickness of the protective reinforcement layer. $\psi_t$, $\psi_e$, and $\psi_s$ are the reinforcement location coefficient, coating coefficient, and variety coefficient. $\lambda$ is a concrete variety coefficient, $K_{tr}$ is the reinforcement coefficient, $k_1$ and $k_2$ are reinforcement location coefficient of AS3600 [33] specification, and $A$ is a cross-sectional area of anchorage reinforcement. By comparison of critical anchorage length between domestic and foreign codes (Figure 18) it can be observed that the critical anchorage length corresponding to formula (7) is 0.66 times, 0.61 times, and 0.79 times of the critical anchorage length of JTG3362-2018 [27], ACI318-11 [34] and AS3600 [33] respectively, which indicates that the anchorage length requirement of the precast assembled concrete structure is far less than the standard value and special consideration should be given to it in the design and calculation.

**Table 7.** Calculation formula of basic anchorage length (domestic and foreign codes).

| Standard | Calculation Formula of Basic Anchorage Length | $L_a/L_{cr}$ |
|---|---|---|
| JTG3362-2018 [27] | $l_a = nd$ | 1.0 |
| AustraliaAS3600 [33] | $l_a = \frac{k_1 k_2 f_y A}{(2c+d)\sqrt{f_{cu}}} \geq 25k_1 d$ | 1.01 |
| U.S.A ACI318-11 [34] | $l_a = \frac{\psi_t \psi_e \psi_s \lambda}{c+K_{tr}} \times \frac{f_y d^2}{1.1\sqrt{f_{cu}}} \geq 300$ mm | 1.22 |

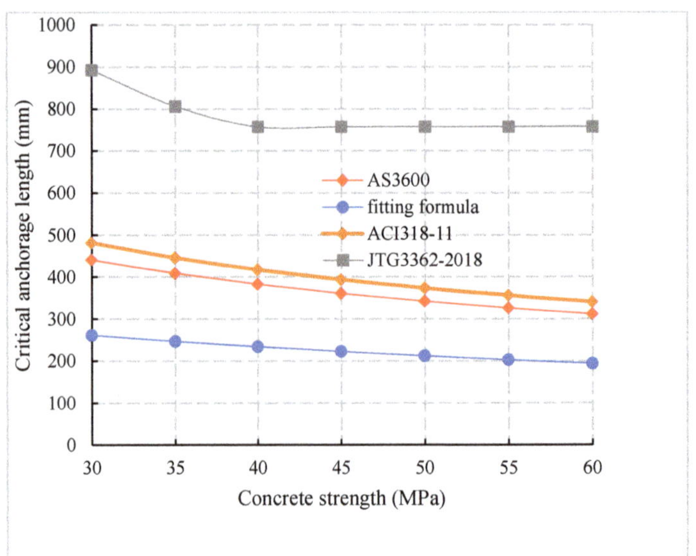


**Figure 18.** Comparison of critical anchorage length calculation between domestic and foreign codes.

Compared with the provisions of domestic and foreign codes for the basic anchorage length of reinforcement (Figure 19), it can be observed that the basic anchorage length of JTG3362-2018 [27] is 1.61 times and 2.14 times of that of ACI318-11 [34] and AS3600 [33], respectively, compared with Figure 18 and, in Figure 18, the relationship between the critical anchorage length and the basic anchorage length in Chinese and foreign codes can be obtained. The basic anchorage length La of JTG3362-2018 [27], ACI318-11 [34], and AS3600 [33] was 1.14, 1.2, and 1.0 times of the critical anchorage length Lcr, respectively, which indicates that the Chinese code is more conservative than the foreign code. By considering 1.7 times of safety factor, the basic anchorage length of connecting reinforcement of prefabricated concrete members can be obtained. Compared with the basic anchorage length of domestic and foreign codes, the basic anchorage length of the connecting reinforcement of prefabricated bridge is 0.49 times, 0.79 times, and 1.05 times of that of JTG3362-2018 [27], ACI318-11 [34] and AS3600 [33], respectively. In Table 8, the comparison of anchorage length between domestic and foreign codes and formula (7).

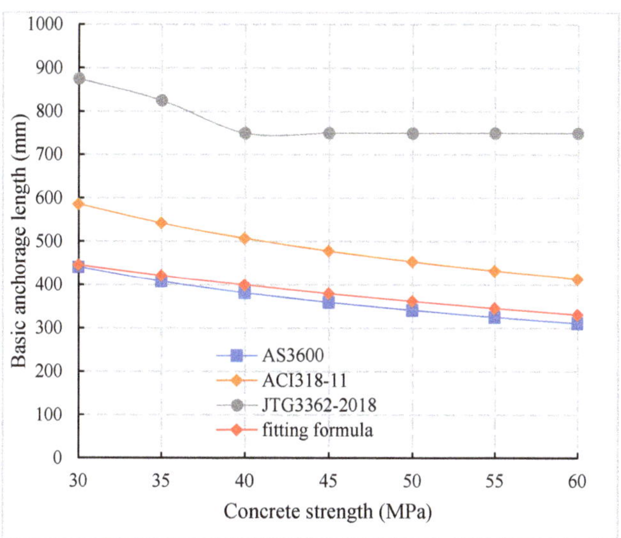


**Figure 19.** Comparison of calculation of basic anchorage length in domestic and foreign codes.

**Table 8.** Comparison of anchorage length between domestic and foreign codes.

| Standard | Lcr-Formula (11)/Lcr | La/Lcr | (La-JTG)/La | 1.7 Lcr(11)/La |
|---|---|---|---|---|
| JTG-3362-2018 [27] | 0.32 | 1.14 | 1.0 | 0.49 |
| ACI-318-11 [33] | 0.56 | 1.22 | 1.61 | 0.79 |
| AS3600 [34] | 0.79 | 1.0 | 2.14 | 1.05 |

In short, if the current domestic codes of prefabricated concrete will still be followed for the design of the anchorage length of prefabricated members, the size of prefabricated components will be too large, which will increase the construction cost and construction difficulty. Therefore, it is necessary to modify the anchorage length of the existing specifications. By considering the safety factor of 1.7 times, it is suggested to take it as 0.5 times the design value of the existing JTG3362-2018 [27].

## 6. Discussion

Calculation of the interface bond strength and anchorage length of steel bar within prefabricated concrete are usually considered according to the relevant provisions of the cast-in-place structure. However, the cover thickness is generally more than 50 mm larger than that of the cast-in-place structure. The complete interface failure method was used to calculate the bond strength, which may lead to the smaller calculated value of bond strength and high values of the anchorage length are not suitable for the design and construction of prefabricated bridges. Combined with experimental research and numerical analysis methods, this paper calculates the formula for interface bond strength and anchorage length by considering the main influencing factors and compares it with national and international codes. It is recommended that the basic anchorage length should be 18 d when the concrete strength grade is C35 or less and 15 d when the strength grade of the concrete is C40 or more.

## 7. Conclusions

From experimental research and numerical analysis methods, the following calculations are made.

1. It is concluded that the effect of cover thickness of the surrounding concrete is negligible for calculating interface bond strength within prefabricated structures.
2. Compared with C50-12-300 and C50-12-400, the ultimate load of C50-16-300 and C50-16-400 increased by 85.7% and 49.9%, respectively, and 19.8% and 20.8%, respectively when compared with C30-16-300 and C30-16-400, which indicates that for precast concrete members with larger cover thicknesses, the concrete strength grade and reinforcement diameter possess a significant influence on the ultimate load of the bond interface between reinforcement and concrete.
3. The ultimate load of the ribbed steel bar specimen C50-20-300 is 8.5 times higher than that of the plain steel bar specimen C50-G20-300. It is estimated that the mechanical interlocking force accounts for about 88% of the bond strength between deformed steel bars and concrete.
4. It is concluded that domestic codes design the anchorage length of prefabricated members, leads to a large size, and increases the construction cost and construction difficulty. It is suggested to take it as 0.5 times the design value of the existing JTG3362-2018 [27] by considering the safety factor of 1.7 times.

**Author Contributions:** Z.H., Supervision, Investigation, Funding acquisition. Y.I.S., Methodology, Visualization, Writing-review &editing. P.Y., Formal analysis, Data curation, Software. All authors have read and agreed to the published version of the manuscript.

**Funding:** This research was financially supported by the Project of National Key Research & Development (2017YFC0806000) and the "5511" Innovation Driven Project of Jiangxi Province (20165ABC28001).

**Institutional Review Board Statement:** Not applicable.

**Informed Consent Statement:** Not applicable.

**Data Availability Statement:** Some or all data, models, or codes generated or used during the study are proprietary or confidential in nature and may only be provided with restrictions.

**Conflicts of Interest:** The authors declare that they have no known conflict competing for financial interest or personal relationships that could have appeared to influence the work reported in this paper.

## References

1. Khaleghi, B.; Schultz, E.; Seguirant, S.; Marsh, L.; Haraldsson, O.; Eberhard, M.; Stanton, J. Accelerated bridge construction in Washington State: From research to practice. *PCI J.* **2012**, *57*, 34–49. [CrossRef]
2. Baghchesaraei, O.R.; Hossein, H.L.; Alireza, B. Behavior of prefabricated structures in developed and developing countries. *Bull. Soc. Sci. Liège* **2016**, *85*, 1229–1234.
3. Culmo, M.P. *Accelerated Bridge. Construction: Experience in Design, Fabrication, and Erection of Prefabricated Bridge. Elements and Systems: Final Manual. No. FHWA-HIF-12-The United States. Federal Highway Administration*; Office of Bridge Technology: Washington, DC, USA, 2011.
4. Baran, E.; Tolga, A.; Seda, Y. Pull-out behavior of prestressing strands in steel fibre reinforced concrete. *Constr. Build. Mater.* **2012**, *28*, 362–371. [CrossRef]
5. Li, S.; Song, C. Experimental research on bond anchorage performance of 1860-grade high-strength steel strands and lightweight aggregate concrete. *Constr. Build. Mater.* **2020**, *235*, 117482. [CrossRef]
6. Li, H.-T.; Deeks, A.J.; Su, X.-Z. Experimental Study on Compressive Bond Anchorage Properties of 500 MPa Steel Bars in Concrete. *J. Struct. Eng.* **2013**, *139*, 04013005. [CrossRef]
7. Saeed, Y.M.; Al-Obaidi, S.M.; Al-Hasany, E.G.; Rad, F.N. Evaluation of a new bond-type anchorage system with expansive grout for a single FRP rod. *Constr. Build. Mater.* **2020**, *261*, 120004. [CrossRef]
8. Dang, C.N.W.; Micah, H.; José, R.M.-V. Quantification of bond performance of 18-mm prestressing steel. *Construct. Build. Mater.* **2018**, *159*, 451–462. [CrossRef]
9. Hayashi, D.; Nagai, K.; Eddy, L. Mesoscale Analysis of RC Anchorage Performance in Multidirectional Reinforcement Using a Three-Dimensional Discrete Model. *J. Struct. Eng.* **2017**, *143*, 04017059. [CrossRef]
10. Zhang, B.; Zhu, H.; Chen, J.; Yang, O. Evaluation of bond performance of corroded steel bars in concrete after high temperature exposure. *Eng. Struct.* **2019**, *198*, 109–479. [CrossRef]
11. Cairns, J. Bond and anchorage of embedded steel reinforcement in fib Model Code. *Struct. Concr.* **2015**, *16*, 45–55. [CrossRef]

12. Saatcioglu, M.; Alsiwat, J.M.; Ozcebe, G. Hysteretic Behavior of Anchorage Slip in R/C Members. *J. Struct. Eng.* **1992**, *118*, 2439–2458. [CrossRef]
13. Abdallah, S.; Fan, M. Anchorage mechanisms of novel geometrical hooked-end steel fibres. *Mater. Struct.* **2017**, *50*, 139. [CrossRef]
14. Sulaiman, M.F.; Ma, C.-K.; Apandi, N.M.; Chin, S.; Awang, A.Z.; Mansur, S.A.; Omar, W. A Review on Bond and Anchorage of Confined High-strength Concrete. *Structures* **2017**, *11*, 97–109. [CrossRef]
15. Ogura, N.; Bolander, J.E.; Ichinose, T. Analysis of bond splitting failure of deformed bars within structural concrete. *Eng. Struct.* **2008**, *30*, 428–435. [CrossRef]
16. Sucharda, O.; Mateckova, P.; Bilek, V. Non-Linear Analysis of an RC Beam without Shear Reinforcement with a Sensitivity Study of the Material Properties of Concrete. *Slovak J. Civ. Eng.* **2020**, *28*, 33–43. [CrossRef]
17. Yu, Q.; Chen, Z.; Yang, J.; Rong, K. Numerical Study of Concrete Dynamic Splitting Based on 3D Realistic Aggregate Mesoscopic Model. *Materials* **2021**, *14*, 1948. [CrossRef] [PubMed]
18. Jonak, J.; Siegmund, M.; Karpiński, R.; Wójcik, A. Three-Dimensional Finite Element Analysis of the Undercut Anchor Group Effect in Rock Cone Failure. *Materials* **2020**, *13*, 1332. [CrossRef]
19. Jonak, J.; Karpiński, R.; Wójcik, A.; Siegmund, M. The Influence of the Physical-Mechanical Parameters of Rock on the Extent of the Initial Failure Zone under the Action of an Undercut Anchor. *Materials* **2021**, *14*, 1841. [CrossRef]
20. Falkowicz, K.; Ferdynus, M.; Rozylo, P. Experimental and numerical analysis of stability and failure of compressed composite plates. *Compos. Struct.* **2021**, *263*, 113657. [CrossRef]
21. Ombres, L.; Verre, S. Experimental and Numerical Investigation on the Steel Reinforced Grout (SRG) Composite-to-Concrete Bond. *J. Compos. Sci.* **2020**, *4*, 182. [CrossRef]
22. Funari, M.F.; Spadea, S.; Fabbrocino, F.; Luciano, R. A Moving Interface Finite Element Formulation to Predict Dynamic Edge Debonding in FRP-Strengthened Concrete Beams in Service Conditions. *Fibers* **2020**, *8*, 42. [CrossRef]
23. *JGJ 145-2004 Technical Specification for Post Anchorage of Concrete Structures*; Ministry of construction of the People's Republic of China: Beijing, China, 2004.
24. *Ministry of Housing and Urban.-Rural Development of the People's Republic of China. Code for Design of Concrete Structures (GB 50010-2010)*; Construction Industry: Beijing, China, 2010.
25. *National Standard of the People's Republic of China, Steel for Reinforced Concrete Part. 2: Hot Rolled Ribbed Steel Bar: GB1499.2-Beijing: China*; Quality Inspection Press: Beijing, China, 2018.
26. *State Administration of Market. Supervision and Administration. GB / T50082-Standard for Test. Methods of Long-Term Performance and Durability of Ordinary Concrete*; Construction Industry Press: Beijing, China, 2010.
27. *JTG 3362—2018 Specifications for Design of Highway Reinforced Concrete and Prestressed Concrete Bridges and Culverts*; Ministry of Transport of the People's Republic of China: Beijing, China, 2018.
28. *Code for Design of Highway Reinforced Concrete and Prestressed Concrete Bridges and Culverts*; People's Communications Press: Beijing, China, 2018.
29. Wang, C.; Wei, H.; Wu, J. Experimental research and numerical simulation on bond behavior between recycled concrete and rebar. *J. Guangxi Uni. Nat. Sci. Ed.* **2013**, *4*, 218–224.
30. Liu, W.; Xu, M.; Chen, Z. Parameter calibration and verification of abaqus concrete damage plasticity model. *Ind. Build.* **2014**, *S1*, 167.
31. Zhao, W.; Xiao, J. On bond-slip Constitutive model between ribbed steel bars and concrete. *Eng. Mech.* **2011**, *28*, 164–171.
32. Xu, Y.; Shao, Z.; Shen, W. Bonding and Anchor Strength of Reinforcement and Concrete. *Architect. Sci.* **1988**, *04*, 10–16.
33. AS-3600. *Standard Concrete Structures*; Standards Australia International: Sydney, Australia, 2001.
34. ACI Committee. *ACI318-11 Building Code Requirements for Structural Concrete*; American Concrete Institute: Farmington Hills, MI, USA, 2011.
35. Yu, Q.; Xu, X.; You, G. Experimental study on bond behavior for ribbed steel bars and grout. *J. Harbin Instit. Technol.* **2017**, *49*, 91–101.
36. Shan, X. *Experimental Study on Bonding and Anchorage Properties between Deformed Steel Bar and Self-Compacting Concrete*; Harbin Institute of Technology: Harbin, China, 2008.

Article

# Future Infrastructural Replacement Through the Smart Bridge Concept

**Albert D. Reitsema [1,2], Mladena Luković [1,*], Steffen Grünewald [1,3] and Dick A. Hordijk [1,4]**

1 Engineering Structures, Delft University of Technology, Stevinweg 1, 2628 CN Delft, The Netherlands; areitsema@heijmans.nl (A.D.R.); S.Grunewald@tudelft.nl (S.G.); hordijk@adviesbureau-hageman.nl (D.G.)
2 Heijmans N.V., Graafsebaan 65, 5248 JT Rosmalen, The Netherlands
3 Department Structural Engineering and Building Materials, Ghent University, Technologiepark-Zwijnaarde 60, 9052 Ghent, Belgium
4 Adviesbureau Hageman, Polakweg 14, 2288 GG Rijswijk, The Netherlands
* Correspondence: m.lukovic@tudelft.nl

Received: 8 November 2019; Accepted: 20 December 2019; Published: 15 January 2020

**Abstract:** Most of the bridges and viaducts in the Netherlands were built in the sixties and seventies of the last century, and an increasing number of them will have to be replaced due to technical or functional reasons. The Netherlands is not an exception, many industrialized countries will face a similar replacement task in the near future. With the increased traffic intensities and the importance of mobility, the design and construction strategies for new bridges have to be different from that in the past. New methods need to ensure that traffic hindrance due to construction works and (future) maintenance activities are minimized. At the Delft University of Technology, a SMART bridge concept is being developed for fast and hindrance-free infrastructural replacement. The optimal advantage is achieved by utilizing innovative but proven technologies, and by bringing academic research into practice. A combination of recent innovations in construction technology, such as advanced cementitious materials (ACM), structural health monitoring (SHM) techniques, advanced design methods (ADM), and accelerated bridge construction (ABC) is being used. These innovations represent a step towards the next generation of infrastructure where fast construction, intelligent bridge design, sustainability, zero-energy, no/low maintenance, and aesthetics are key features.

**Keywords:** hinder-free replacement; fast construction; low maintenance; ultra high-performance concrete (UHPC); accelerated bridge construction (ABC); advanced design

## 1. Introduction

Bridges have always fascinated humanity. They symbolize links between people, communities, and nations [1]. Many bridges were built during the large infrastructural boom between the sixties and the eighties of the last century. The automotive industry has rapidly grown during this period, driving developed countries to construct many new bridges. An overview over the period from 1925 to 2004, which shows when bridges were built in the United States (US), Japan, Germany, and the Netherlands, is shown in Figure 1 [2–5].

A recent analysis [6] of data from the US Federal Highway Administration (FHWA) and the US National Bridge Inventory (NBI) indicates a strong shift in nationwide spending from building new bridges towards rehabilitation and replacement (Figure 2). Out of the 611,845 public road bridges in the US, 58,791 (9.6%) were classified as structurally deficient in 2015, and another 84,124 (13.7%) as functionally obsolete. These figures have led to claims that the US is about to experience a crisis concerning infrastructure [7]. To some extent, a similar situation is expected in many European countries. Many of the existing bridges have now reached the age of 50 years, the originally designed service life. Although this certainly does not mean that these bridges are unsafe, an increasing number will need to be replaced in the future. The replacement can be either due to deterioration or increased traffic loading, but also because of functional requirements. It is clear that, apart from rehabilitation, society will face a large infrastructural replacement task in the future.

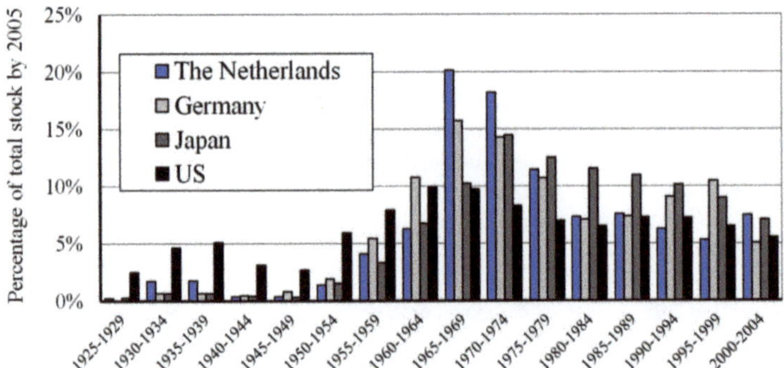


**Figure 1.** Historical overview of built bridges in Japan [2], the US [3], Germany [4], and the Netherlands [5].

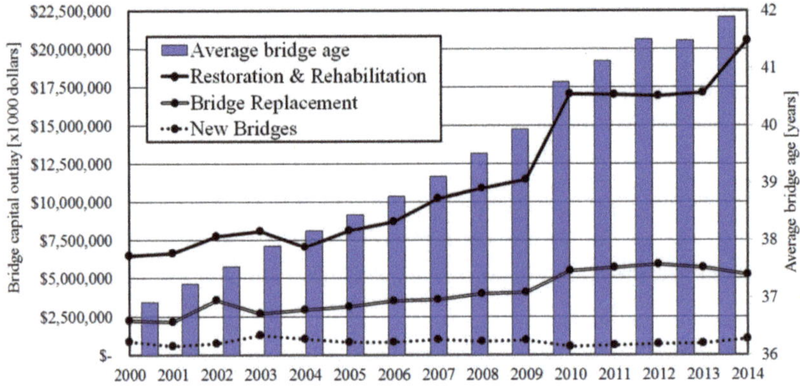


**Figure 2.** US bridges capital outlay and age (routine maintenance costs not included) [6].

A very common bridge type is the overpass (Figure 3). Traditionally, overpasses in the Netherlands were mainly built as cast in-situ reinforced concrete plate bridges. An advantage of a plate bridge with multiple intermediate supports compared to a single span bridge is the possibility to achieve higher slenderness due to a more favorable moment distribution. On the other hand, its main disadvantage is the increased on-site construction time needed for mounting the formwork, placing the reinforcement, and pouring and hardening of the concrete. In the past, traffic hindrance due to construction was not an issue, because these bridges were built in the open field, in the new highways. Nowadays, the Dutch highway network is one of the busiest in the world. With 3,075 km of highway in total and 73 km of highway per 1,000 km$^2$ [8], the Netherlands has the highest highway density in the European Union. The busiest highways accommodate over 200,000 vehicles daily. As a result, the availability of adequate infrastructure is of the utmost importance for society.

**Figure 3.** Example of a common four-span plate bridge type in the Netherlands: Bridge Zijlweg of the highway A59 in the province Brabant.

Traffic jams cause major problems in highly populated areas. Besides increased stress and time delays, an increase in traffic congestions may cause stalling of the economy. The cumulative combined costs of traffic congestion for the national economies of the US, United Kingdom (UK), France, and Germany are estimated to reach a staggering 4.4 trillion dollars by the year 2030 [9]. A report by the TNO [10] shows that Dutch companies lose around 1.1 billion euros annually due to traffic jams. These losses are equal to 18% of the Dutch annual infrastructure budget of around 5.9 billion euros. Therefore, one of the main demands of owners, contractors, and society in the modern construction industry is to minimize traffic hindrance. Therefore, for the next generation of road infrastructure, rapid and low hindrance bridge replacement will be a governing task.

## 2. Low Hindrance Bridge Replacement Method

Around 60% of the highway bridges in the Netherlands are statically undetermined three- or four-span concrete plate bridges. The bridge length of the three-span bridges is between 20 m and 40 m and of four-span bridges between 40 m and 60 m [11].

When replacing these bridges, there is a desire to have maximum freedom of space for traffic lanes. This means that spanning the total length of the existing viaduct without piers in a new bridge would be very advantageous (Figure 4, indicated with red crosses). To limit hindrance and construction time, and for economic reasons, groundwork (increasing or lowering the height of both roads) should be avoided as much as possible. However, for the new bridge, the traffic profile underneath the bridge needs to be maintained (Figure 4, green box). If the existing foundations at the abutments can be (partly) reused (Figure 4, blue box), then the requirements for an innovative, new bridge replacement concept are met.

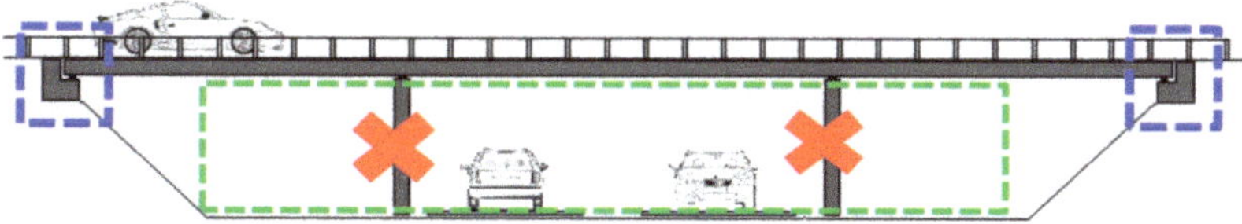

**Figure 4.** Requirements for an innovative, new bridge replacement concept for existing in-situ cast multi-span concrete plate bridges: no piers (red crosses), maintain profile (green box), no groundwork (green and blue boxes), and reuse of foundation (blue box).

With the replacement of a statically undetermined bridge (three- or four-spans) by a single span bridge, and the requirement to avoid altering the free space between the crossing road and the bridge, the height of the bridge deck cannot be increased, or it can, but only slightly. This requires a more slender bridge deck compared to the original one. It must be determined whether the conventional way of building satisfies this requirement and whether innovative materials and recently developed techniques play a role in solving this challenge. Though we realize that the above-sketched challenges could possibly be fully or partly addressed by steel or steel-concrete composite structures and/or alternative schemes, given that the Netherlands has great experience and tradition in precast concrete structures, the focus is on advanced, high-performance, prestressed concrete plates.

Looking at the stock of highway bridges in the Netherlands [11], a slenderness of $\lambda = 60$ ($\lambda$: span to bridge deck thickness) allows the replacement of 60% of the three-span plate bridges (Figure 5). Replacement of a concrete plate bridge of variable height by a new deck with a height equal to the minimal original deck height of the existing bridge plus an additional 100 mm or 200 mm, means that, with a slenderness of $\lambda = 50$, 52% or 95% of the bridges can be replaced, respectively [12]. Compared to the maximum slenderness of $\lambda \approx 30$, as applied in practice with conventional (i.e., normal strength) concrete nowadays, this represents a big step. For high slenderness, but also for reasons of transportation and crane capacities, it is essential to reduce the self-weight of precast, prestressed concrete elements. Therefore, it is essential to exploit the possibilities of newly developed advanced cementitious materials (ACMs) combined with innovative building techniques.

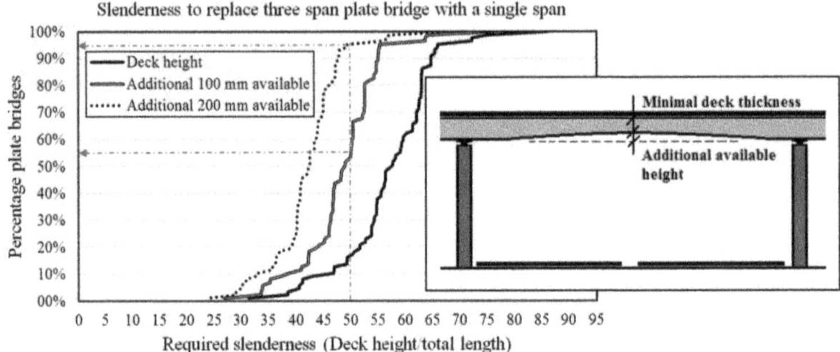


**Figure 5.** Required deck slenderness to realize a significant replacement task: processed data from Reference [11].

## 3. SMART Bridge Concept

At the Delft University of Technology, a visionary concept for new infrastructure was developed, where the main focus is on bridge engineering. The total concept is called the SMART bridge [13]. A SMART bridge utilizes existing and new technologies, as well as new design and construction methods and monitoring techniques, to develop a bridge that satisfies 21st-century demands. To exploit the potential technologies, the SMART bridge concept is developed by recognizing known shortcomings in infrastructure, such as the structural or durability deficiency of existing structures. It also addresses current requirements (i.e., sustainability, low maintenance, short construction time, and no hindrance) and future possibilities (i.e., free-form design and function integration).

Opportunities have been identified to make the concept manageable and to achieve short-term utilization. During and after the process of developing a SMART bridge, newly identified societal challenges will be integrated and research will be carried out towards developing the next, more advanced version of the bridge. This approach makes the SMART bridge concept a fast-evolving innovation platform (Figure 6).

The first SMART bridge concept combined several new technologies that incorporate new design criteria and societal demands. In the following sections, we present the experiences and future expectations of each of these technologies. We also detail steps on how to develop the first SMART bridge by application of ACMs, ABC, and the in-house developed advanced design tool (BoDeTo). Moreover, we present ways to prove its structural reliability by performing full-scale tests and by applying structural health monitoring (SHM). Attention is paid to the interactions between bridge owners, contractors, structural engineers, and building authorities (e.g., through construction tenders that incorporate societal benefits, and through considerations on how to deal with risks).

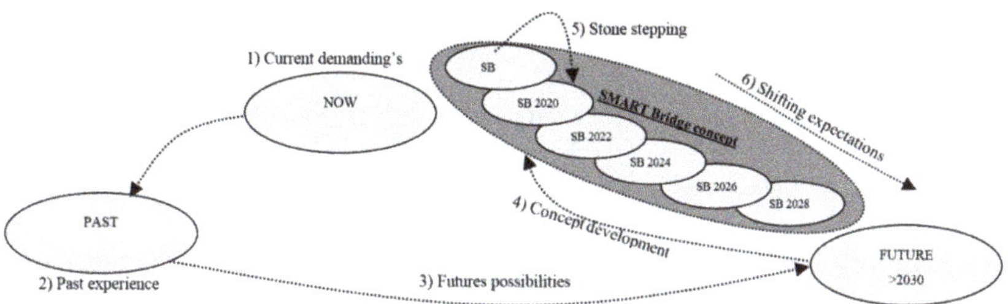


**Figure 6.** SMART bridge concept development process [13].

### 3.1. *Advanced Cementitious Materials (ACM)*

To obtain a slender bridge, a higher prestressing level has to be utilized, and therefore, ultra high-performance concrete (UHPC), a type of ACMs, must be applied. Applications of UHPC and other ACMs are still limited due to the lack of technical and economic feasibility studies, experience, design criteria, and large-scale material applications. Additionally, the costs of UHPC and its risk coverage are important factors. Since there are no design codes, the best and probably the only way to begin is to perform full-scale tests and execute pilot projects. This approach was used in Malaysia [14,15] (Figure 7a), wherein within 7 years, more than 100 road UHPC bridges were built (Figure 7b). Although in Malaysia, UHPC is mainly applied for reasons of durability and low maintenance in remote areas, the results are certainly encouraging for the application of UHPC for fast and hinder-free construction in The Netherlands.

(**a**) (**b**)

**Figure 7.** (**a**) Full-scale destructive test on a 30 m long segmental T-beam, research done in 2007–2010 [14,15]; (**b**) construction of 30-m-span Ulu Chemor bridge, stitched T-girder construction [15].

Given its low permeability and high mechanical strength, UHPC has also attained wide applications in Switzerland, in the rehabilitation of concrete bridges in zones exposed to severe environmental conditions and mechanical loading [16–18]. The superior durability of UHPC over traditional concrete is an important additional benefit for new Dutch bridges since less maintenance will lead to less out of use periods, and consequently less hinder during the exploitation period.

### 3.2. *Accelerated Bridge Construction (ABC)*

To further decrease the total project delivery time and the downtime of the roads, the ABC method can be used. For example, the lateral bridge slide-in method or self-propelled modular transporters (SPMT's) can be used to place a bridge (Figure 8a). The high slenderness and accompanying reduced self-weight of the bridge is crucial since the bridge needs to be lifted and hauled into position [19]

(a) (b)

**Figure 8.** (**a**) Illustration of placing a new bridge deck with the use of an SPMT (computer-controlled platform vehicles) with jack-ups [19]; (**b**) placing a bridge with an SPMT during the replacement project, I-84 bridges over Marion Avenue in Southington, US [20].

An example of using SPMT's for the replacement of interstate highway bridges, which typically can take months, or even a year or more using traditional construction methods, is the project I-84 bridges over Marion Avenue in Southington, Connecticut (Figure 8b). The project consisted of the replacement of two existing bridges built in 1964. The scheduled replacement of the bridges anticipated that the highway would be closed from 9 p.m. on Friday to 5 a.m. the following Monday. However, the project progressed smoothly and both bridges were opened for traffic even before the deadline [20].

### 3.3. *Advanced Design Tool (BoDeTo)*

It is expected that the application of highly prestressed UHPC concrete box girders will provide opportunities for a lightweight, slender bridge concept. Despite the relatively low height of the bridge deck, a box girder shape is regarded as beneficial due to a high torsional stiffness, vast experience in the construction industry, limited need for special edge beams and full prefabrication, which leads to higher quality and faster building. We investigated various configurations of cross-sectional areas, applied prestressing force, and compressive strength of the UHPC for various spans. For each combination of these factors, the maximum achievable slenderness needed to be identified, while conforming to the requirements related to various limit states.

A computer model named BoDeTo (Box girder Design Tool) has been developed to automate the process of hundreds of design calculations. BoDeTo is programmed with a connection between a parameter input frame (element slenderness, concrete strength, etc.), the FEM bridge design software (to determine stress distribution), and the design recommendations. As far as design codes were concerned, the maximum achievable slenderness was investigated using the Eurocode, the French recommendations for UHPC [21], and the Australian design guidelines [22]. Additionally, we investigated a configuration without shear reinforcement and with reduced concrete cover, where we did not conform to current codes in the Netherlands. Regarding dynamic response, for bridges, no additional design conditions are given in the Dutch codes. Therefore, the ratio of static deflection and first natural flexural frequency was calculated according to the New Zealand bridge design code [23].

Figure 9 shows the results of the calculations. For a chosen bridge span of 33 m, with a height of 0.9 m and prestressing strands of 15.7 mm, the optimum solution for the box girder cross-section and the prestressing configuration was determined. The obtained values show the results of all relevant design checks for the unity check. From the performed calculations, it could be seen that the moment capacity was governing, whereas the SLS criteria (e.g., stresses in service and deflection) were not critical. Note that, although generally being considered in the current approach (i.e., based on certain assumptions), for further development of the concept detailed attention has to be given to issues like dynamic response and the creep behavior of slender UHPC prestressed girders.

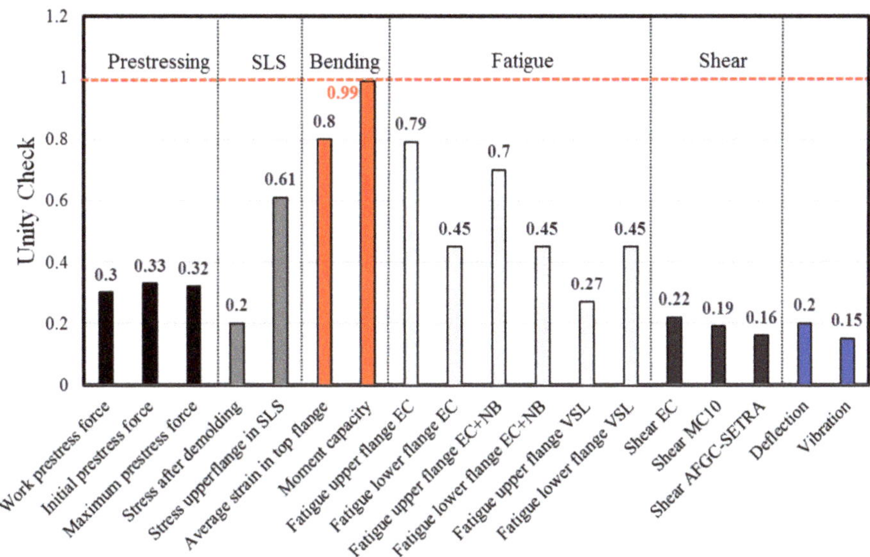


**Figure 9.** Outcome of the calculation with BoDeTo (ULS bending criteria is reached).

With the BoDeTo, the complete bridge design is performed within 90 seconds. Results are presented in Section 4.2.

### 3.4. Structural Health Monitoring (SHM) Techniques

Currently, SHM and SMART structures are popular terms in infrastructure research and asset management. At the moment, a lot of research is focused on different monitoring techniques such as radar, vibration monitoring, fiber optics, strain gauges, strain sensors, image analysis, and smart aggregates, among others. A common challenge with these techniques is detecting what is being measured and how to convert measured data to useful information for the assessment of the structural integrity of concrete bridges. At the Delft University of Technology, we believe that any monitoring that detects a change in the behavior of the bridge is very valuable. In the case of a new innovative bridge, for which limited experience is available and no design codes apply, SHM can contribute to confidence in the safe use of the bridge. It assures that we are in control and as soon as a change in behavior is detected, the cause of the change can be investigated, and action can be taken. An opportunity for monitoring and detecting degradation and changes in bridge behavior is through detecting changes in the stiffness of the structure. Laboratory and pilot projects [24] were carried out where the bending stiffness of concrete elements, measured with a limited number of displacement transducers, was used as an indicator for the structural health, with the final goal to develop a bridge monitoring model for the future [25]. Furthermore, it is believed that the new, fast, and contactless remote technique of radar interferometry can fruitfully be applied within the proposed concept. Radar-based measurements are discussed, for example, by Diaferio et al. [26] and Gentile and Bernardini [27].

### 3.5. Economical Most Advantageous Registration (EMVI)

In the past, the direct costs of bridge construction were key to bridge owners during the procurement phase. In bridge replacement, as presented in this paper, the direct costs of the bridge itself are not governing. The traditional methods of procurement and dealing with risk need to change to accommodate more innovative solutions. Aspects such as fast construction, sustainability, and low/no maintenance need to be considered in future tenders. Fortunately, developments in that direction are notable in the Netherlands. An example of such a tender is the project for the replacement of a three-span concrete plate bridge with a total span of 32 m of the Dutch highway A28 (Figure 10a).

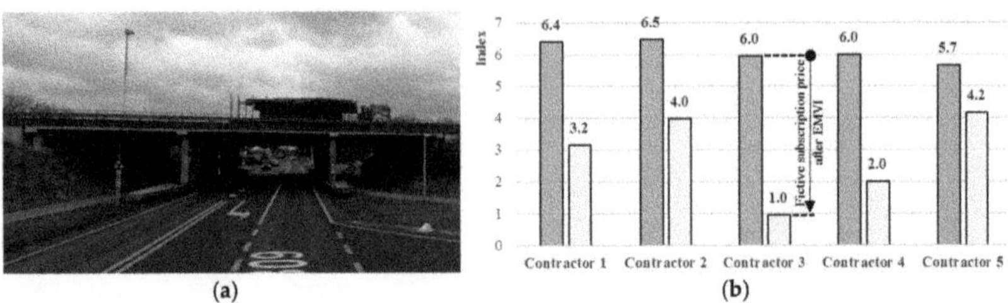


**Figure 10.** (**a**) Existing three-span plate bridge in the A28/N309; (**b**) Outcome of the tender with EMVI.

The highway crosses the underlying road N309 with three traffic lanes. The demands of the RWS (Rijkswaterstaat, the Dutch Ministry of Transportation and the Environment) were that the existing bridge should be replaced by a new bridge that will serve for the next 100 years. Additionally, the underlying road should be widened to five traffic lanes. A special feature of the project was the design and construct (D&C) contract, where a large fictive bonus, called an EMVI (Economical Most Advantageous Registration), was included for the tenderer that has the most favorable traffic model. In addition, a large bonus was available for using slender bridge girders to ensure that the underlying N309 did not need deepening to maintain the existing traffic profile under the bridge. A bonus of € 250,000 was also provided for each 50 mm in the bridge height that was reduced compared to the reference two-span bridge design (each span 24 m) consisting of 800 mm high box girders.

In the described tender procedure, the criteria for choosing the best contractor were directly related to the earned EMVI. Figure 10b shows the results of the tender for the evaluation of five contractors. The dark gray column represents the subscription price ranging from 5.7 to 6.5 (indexed). Subsequently, the fictive subscription price was calculated based on the EMVI. The white columns, ranging from 1.0 to 4.2, were the fictive subscription prices. The EMVI has a very large influence on the final result, meaning that traffic impact and infrastructure alignment were very important factors in this tender. Contractor 3 won the tender with the lowest fictive price of 1 (indexed).

Similar tenders are expected to become the standard for future replacement tasks. Furthermore, the strategy of how to reasonably quantify certain criteria has to be reconsidered. For example, sustainability is receiving increased attention in civil engineering. Concrete is generally considered a material of which the sustainability should be improved, given that per $m^3$ of concrete, 100–300 kg of $CO_2$ is emitted [28]. The $CO_2$ emission of UHPC per $m^3$ is even higher, affecting sustainability even more unless the amount of concrete used in the structure is significantly reduced. On the other hand, a report of TNO [29] shows that the $CO_2$ pollution caused by vehicles during traffic jams increases to a range of 40–70% depending on the vehicle's weight. In 2015, the total Dutch pollution linked to road transport was 29.4 billion kg of $CO_2$. Therefore, in 2015, traffic alone produced as much $CO_2$ as emitted by producing concrete for around 98,000 A28/N309 viaducts, or 25 times the total Dutch highway stock. This result highlights the urgency of reducing traffic jams in modern society.

## 4. Innovative Slender and Lightweight UHPC Bridge Concept

### 4.1. Three Types of Bridges

The basic idea of the SMART bridge concept is to replace multi-span (three or four spans) viaducts with a supported bridge structure made of UHPC, spanning the total length and with minimal changes to foundations and alignments. In this way, a reduction in construction time, traffic hindrance, and full freedom in space are achieved, while keeping the same traffic profile below the bridge (Figure 4). Despite striving for a low weight of the new single-span bridge, the forces on the foundation will increase compared to the existing bridge. Therefore, a holistic approach is adopted: from the application of innovative materials and construction techniques, through investigating existing facilities in the precast concrete factory and the use of innovative tools for the structural design, to considerations related to the reuse of the existing foundations.

One solution is to develop a new very slender deck construction to achieve the required slenderness of approximately $\lambda \approx 50$ for the bridge deck. The deck can be comprised of prestressed box girders in parallel (Figure 11, left) or it can be executed as a post-tensioned UHPC modular segmental girder bridge (Figure 11, middle). Another solution is to make use of two UHPC loadbearing girders on which the deck is hanging (Figure 11, right). Therefore, three bridge types shown in Figure 11 were examined. In the Netherlands, prestressed concrete box girders (Figure 11, left) are used widely for the construction of short- and medium-span bridges. Multiple girders are placed adjacent to each other and are connected by post-tensioning in a transverse direction, making it a fast and economical construction. As far as production of the girders is concerned, currently in the factories, the prestressing force is limited to 2250 tons (110 strands). Therefore, to achieve the required slenderness, it might be necessary to combine pretensioned prestressing with post-tensioning.

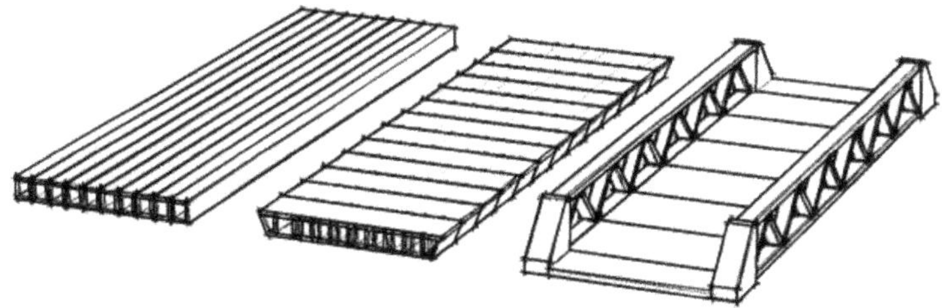

**Figure 11.** The three bridge types that are considered in the SMART bridge concept.

The advantage of building with prefabricated post-tensioned UHPC segments (Figure 11, middle) instead of girders is the ease of transport. On the other hand, the disadvantage is the need for temporary supports during construction.

An optimum has to be found between bridge stiffness, prestressing force, and self-weight, considering the maximum achievable slenderness. With high slenderness, criteria such as deflections and vibrations, similarly as in steel structures, become governing. The main benefits of the system with two UHPC girders and an in-between deck (Figure 11, right) are the increased stiffness of the bridge without affecting the free traffic profile below the bridge, as well as a shift of the main load-bearing spanning from the longitudinal to the transversal direction. As a result, the required bridge deck slenderness can easily be obtained. With the two girders beside the current road profile, the width of the bridge increases, which also provides opportunities to build additional foundations adjacent to the existing ones at the abutments. Research on the application of optimization algorithms in the design of modular girders can further reduce the bridge weight [30]. The current focus of the SMART bridge project is on bridge type 1 and UHPC precast box girders, as elaborated in the following case study.

### 4.2. Case Study Replacement Task A28/N309

The existing bridge of the A28/N309 highway in the Netherlands (Figure 10a) was used for the case study. The required length of a new single-span bridge is 48 m, with a width of 16.2 m (two traffic lanes). Different challenges related to design, construction, transport, and execution were considered. Several advanced designs and/ or calculation techniques were applied. We note that:

- Due to the limited prestressing force in the factory for the production of prefabricated girders, a combination with post-tensioning was investigated;
- The lack of design codes can be overcome by up-scaled laboratory testing;

- The increased loading on the existing foundations can be overcome by smart adjustments in the design of the supports.

The first step was to determine the maximum achievable slenderness for the current production process and geometry of the box girders. Calculations were performed using the Eurocode (EC) (Figure 12, level 1). This meant that the concrete cover was restricted to the EC's limits and stirrups were used to provide shear capacity. Besides shear, no traditional reinforcement was used. For the prestressing, Y1860 7 wire cables with a diameter of 15.7 mm were used. The maximum slenderness that could be reached with this box girder was $\lambda = 40$. Calculations showed that the application of concrete in a strength class that was higher than C130 did not result in an increased slenderness because of a limitation in the total allowable prestressing force of 2250 tons.

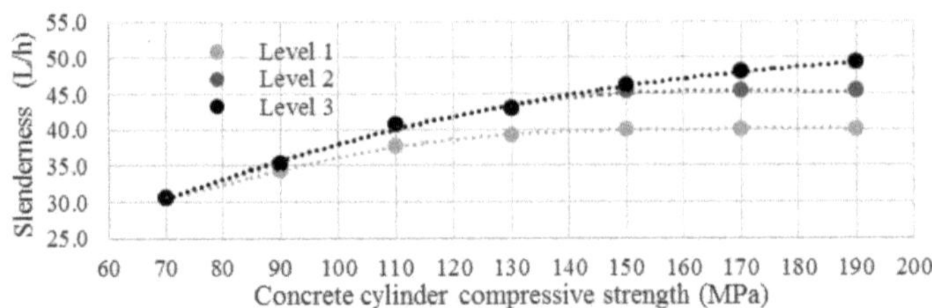


**Figure 12.** Achievable slenderness of a box girder bridge with an increasing concrete compressive strength.

In the second step, the pretensioned prestressing (110 strands) was combined with post-tensioning (Figure 12, Level 2). The maximum number of post-tensioning strands in the cross-section was determined using the design recommendation for Dywidag post-tensioning anchors [31]. The anchor edge and in-between distances that are given in this recommendation apply for concrete classes up to C45, resulting in a conservative calculation when using UHPC. The anchors were placed on the head ends of the box girder. For this situation, the maximum slenderness of $\lambda = 45$ could be achieved. A concrete strength over C150 does not increase the slenderness because of the limitation in the prestress force that can be applied.

The third step was to determine the achievable slenderness when the anchor distances given by Dywidag [31] were extrapolated towards values for UHPC, and when no shear reinforcement was used (Figure 12, Level 3). We assumed that for UHPC, smaller anchorage distances were applicable. As a result, a higher slenderness could be achieved in this case for concrete strength values above 150 MPa because the available space for the post-tensioning strands was no longer governing the design. The maximum slenderness that could be reached was $\lambda = 50$. In this case, fatigue becomes governing for design. Figure 13 shows the cross-section of the girder with the distribution of prestressing strands. Note that the box girder has a width of 1480 mm, which is in line with the production capabilities of the prefab company. In all calculations, the most unfavorable load combination was Load Model 1, in which a double-axle load (Tandem System) was applied in conjunction with a uniformly distributed load. More details of this case study and the analyses of the results can be found in Reference [12].

For the foundation, several calculations were performed with the intention of investigating whether it was possible to limit the force in the foundation piles to that similar to the existing bridge. In the FEM calculations, Menard's spring stiffness's were applied. Another option for redesigning an existing bridge foundation to increase the design capacity is by replacing a bridge girder concentrically (Figure 14b) on the abutment instead of eccentrically (Figure 14a). This removed the bending moment occurrence as a result of the abutment support reaction. This reduced the compression force in the frontal piles.

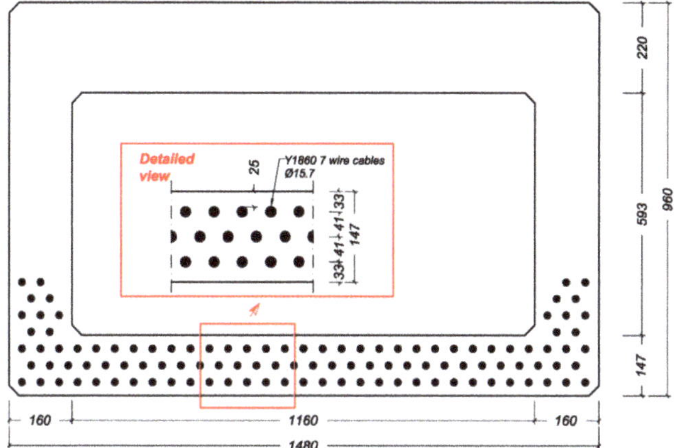


**Figure 13.** Designed cross-section of the girder with the distribution of the prestressing strands for the level 3 analysis (Figure 12).

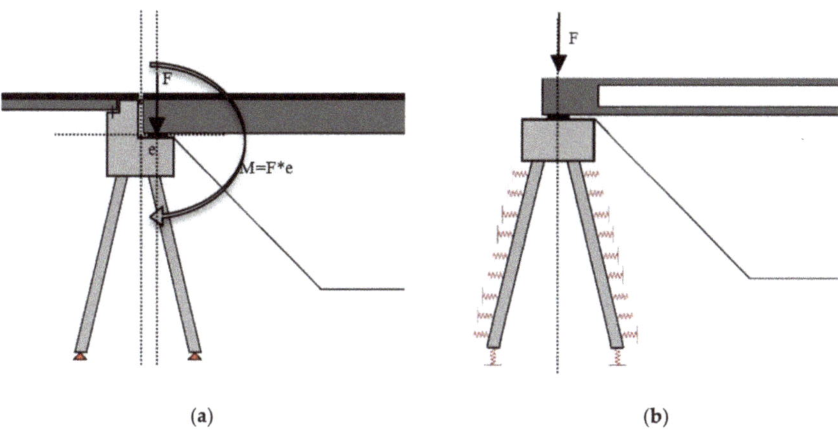


**Figure 14.** (**a**) Existing and (**b**) new bridge foundation design.

The conclusion of the case study for replacing the existing concrete plate bridge was that a slenderness of $\lambda = 50$ could be achieved by using prefabricated C190 prestressed UHPC bridge girders. This would be achieved in combination with prestressed and post-tensioned strands, while the existing foundations at the abutments could be adapted and reused.

## 5. Discussion and Potential Impact on Society

Replacement of only one plate bridge by the SMART bridge concept already has a large impact on society because of reduced hindrance. However, as shown by using the BoDeTo tool, in the Netherlands, the potential for applying this innovative technique for existing plate bridges is very high. Provided that concrete with a very high compressive strength was used, a bridge slenderness up to 60 could be achieved (Figure 15).

In Figure 16 (indicated by red color) it is shown that by designing with UHPC, according to the AFGC-SETRA recommendation [21], applying a reduced cover and taking into account fiber reinforcement and no stirrups, instead of requirements given by EC, the potential application area for bridge replacement increases from 16% to 58% when the height is kept equal to that of the existing plate bridge. The percentage increases from 52% to 96% when an additional height of 100 mm is available (indicated by green color). Since it concerns the application outside the codes, as done in Malaysia [14,15], prior full-scale testing would have to play a major role in demonstrating the structural capacity for short-term behavior. Meanwhile, monitoring, as applied in Switzerland [17], would contribute to the control of long-term structural behavior.

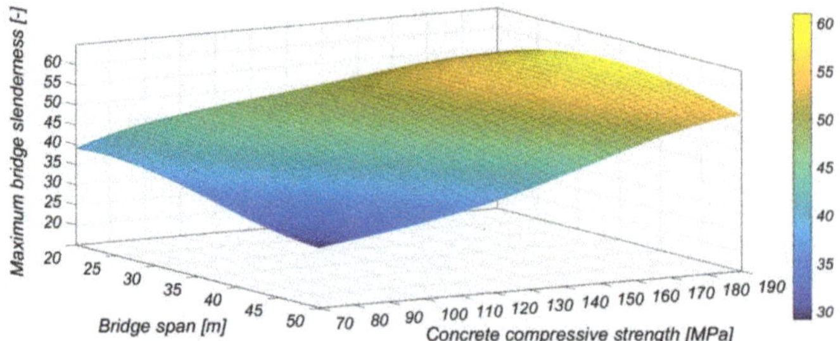


**Figure 15.** Achievable slenderness calculated with the BoDeTo.

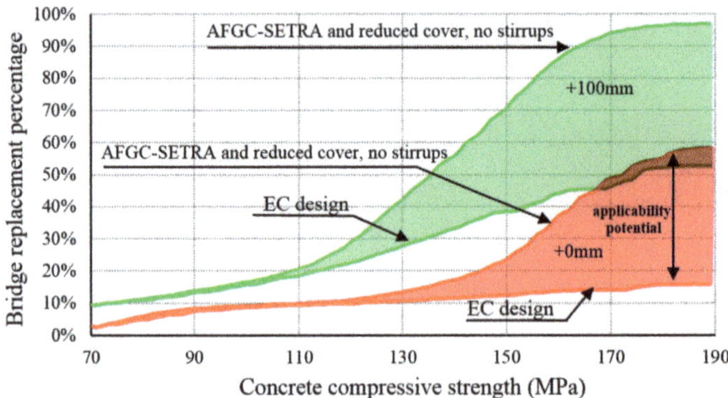


**Figure 16.** Percentage of three-span plate bridges in the Netherlands for which the SMART bridge replacement can be applied.

## 6. Conclusions

For the new generation of bridges, a paradigm shift will need to occur in bridge engineering. Especially in highly populated areas, continuous availability of infrastructure will have the highest priority and should be considered for upcoming large replacement tasks. Downtime caused by maintenance and construction works should be minimized as it has significant social and economic effects. In this paper, we presented a total concept and a holistic approach for the replacement of multi-span overpasses, of which there are many in the Netherlands and worldwide. The benefits of using the so-called SMART bridge concept for future replacement tasks are numerous, including:

- Significant reduction in traffic hindrance during construction (e.g., the 22 weeks, as is common for replacing a standard concrete plate bridge, is expected to be reduced to days).
- Full freedom in space under the bridge and no need to rebuild intermediate supports.
- No need to realign the road or reduce the available traffic profile under the bridge.
- By reusing existing foundations, an economical and fast building method is achieved.
- The enhanced durability properties of UHPC are expected to reduce maintenance needs.
- By reusing existing foundations, a sustainable solution is developed by applying durable UHPC girders and reducing traffic hindrance.

The proposed concept is innovative in many aspects. Innovation, apart from solving technical challenges, is also related to the implementation of these ideas and building awareness to sustain them. Whereas governments in the past believed that innovation has to be done by industry, given societal demands, cooperation is necessary between the government, industry, scientific institutions, and possibly other parties such as standardization organizations. Besides the costs of research efforts, acceptance criteria (no codes), and dealing with the risks related to unknown techniques, the non-financial aspects like a reduction of hindrance and improvement of sustainability play an important role. To be prepared for societal demands in the future concerning infrastructure, organizational and technical developments have to start now. For upcoming infrastructural replacement tasks, there is no time to lose.

**Author Contributions:** A.D.R., M.L., S.G., and D.A.H. performed the analysis and wrote the paper; funding acquisition, D.A.H. The presented concept won the 4th place at the World Innovation Prize in Bridge Engineering (WIBE). All authors have read and agreed to the published version of the manuscript.

**Funding:** This research received no external funding.

**Conflicts of Interest:** The authors declare no conflict of interest.

## References

1. Jensen, J. History of Bridges-A philatelic review. In Proceedings of the Fourth Symposium of Strait Crossings, Bergen, Norway, 2–5 September 2001; p. 263.
2. MLIT. *Roads in Japan*; Road Bureau, Ministry of Land, Infrastructure, Transport and Tourism: Tokyo, Japan, 2018.
3. FHWA. National Bridge Inventory in US. Available online: http://www.fhwa.dot.govbridgenbi.cfm (accessed on 21 July 2017).
4. BAST. see Bundesanstalt für Straßenwesen, Germany. Available online: http://www.bast.de/DE/Statistik/Bruecken (accessed on 21 July 2017).
5. RWS. *Inventarisatie Kunstwerken—Historical Overview Year of Construction of Dutch Bridges in Main Road Network*; RWS—Dutch Ministry of Infrastructure and the Environment: Utrecht, The Netherlands, 2007.
6. FHWA. Highway Statistics Series Publications version 2000 till 2014 in US. 2014. Available online: https://www.fhwa.dot.gov/policyinformation/statistics.cfm (accessed on 21 July 2017).
7. Kirk, R.S.; Mallett, W.J. *Highway Bridge Conditions: Issues for Congress*; Report. No. R43103; Library of Congress. Congressional Research Service: Washington, DC, USA, 2013.
8. Wegenwiki. 2017. Available online: https://www.wegenwiki.nl/Nederland (accessed on 16 September 2019).
9. Cebr. The future economic and environmental costs of gridlock in 2030, report for INRIX. 2014. Available online: https://www.ibtta.org/sites/default/files/documents/MAF/Costs-of-Congestion-INRIX-Cebr-Report%20(3).pdf (accessed on 19 September 2019).
10. Calvert, S. Schade door files groeit weer. Nieuwe economische wegwijzer TLN en EVO. *Transport en Logistiek.* **2015**, *23*, 28–31.
11. RWS. *Excel File Bridge Stock Database until Year 2014, Personal Correspondence with RWS*; Dutch Ministry of Infrastructure and the Environment: The hague, The Netherlands, 2014.
12. Reitsema, A.; Lukovic, M.; Hordijk, D. Towards slender, innovative concrete structures for replacement of existing viaducts. In Proceedings of the *fib* Symposium 2016 Performance-based Approaches for Concrete Structures, Cape Town, South Africa, 21–23 November 2016.
13. Reitsema, A.; Hordijk, D. Towards a SMART Bridge: A Bridge with Self–Monitoring, Analysing, and—Reporting Technologies. In Proceedings of the SMAR 2015: Third Conference on Smart Monitoring, Assessment & Rehabilitation of Civil Structures, Antalya, Turkey, 16–19 October 2015.
14. Voo, Y.L.; Foster, S.; Pek, L.G. Ultra-High Performance Concrete—Technology for Present and Future. In Proceedings of the High Tech Concrete: Where Technology and Engineering Meet, Maastricht, The Netherlands, 12–14 June 2017.
15. Voo, Y.L.; Foster, S.J.; Voo, C.C. Ultrahigh-performance concrete segmental bridge technology: Toward sustainable bridge construction. *J. Bridge Eng.* **2014**, *20*, B5014001. [CrossRef]

16. Brühwiler, E.; Denarié, E. Rehabilitation and strengthening of concrete structures using ultra-high performance fibre reinforced concrete. *Struct. Eng. Int.* **2013**, *23*, 450–457. [CrossRef]
17. Martín-Sanz, H.; Tatsis, K.; Dertimanis, V.K.; Avendaño-Valencia, L.D.; Brühwiler, E.; Chatzi, E. Monitoring of the UHPFRC strengthened Chillon viaduct under environmental and operational variability. *Struct. Infrastruct. Eng.* **2019**, *16*, 1–31. [CrossRef]
18. Brühwiler, E. UHPFRC technology to enhance the performance of existing concrete bridges. *Struct. Infrastruct. Eng.* **2020**, *16*, 94–105.
19. ENERPAC. Lateral Bridge Slide Jacking System—Accelerated Bridge Construction. Available online: https://www.youtube.com/watch?v=YzYCSw5mACQ (accessed on 13 January 2020).
20. Stantec. Available online: https://www.stantec.com/en/projects/united-states-projects/r/reconstruction-of-i84-bridges-over-marion-ave (accessed on 13 January 2020).
21. AFGC-SETRA. *Ultra High Performance Fibre Reinforced Concretes*; Interim Recommendations; AFGC Publication: Paris, France, 2013.
22. Gowripalan, N.; Gilbert, R. *Design Guidelines for Ductal Prestressed Concrete Beams*; The University of NSW: Sydney, Australia, 2000; p. 53.
23. NZ Transport Agency. *Bridge manual*, 3rd ed.; NZ Transport Agency: Wellington, New Zealand, 2013.
24. Mostafa, N.; Loendersloot, R.; Tinga, T.; Reitsema, A.; Hordijk, D. Monitoring dynamic stiffness that predicts concrete structure degradation. In Proceedings of the IALCCE, Delft, The Netherlands, 16–19 October 2016.
25. Westerbeek, R. Structural Health Monitoring of Concrete Bridges. Master's Thesis, Delft University of Technology, Delft, The Netherlands, 2017.
26. Diaferio, M.; Fraddosio, A.; Piccioni, M.D.; Castellano, A.; Mangialardi, L.; Soria, L. Some issues in the structural health monitoring of a railway viaduct by ground based radar interferometry. In Proceedings of the 2017 IEEE Workshop on Environmental, Energy, and Structural Monitoring Systems (EESMS), Milan, Italy, 24–25 July 2017. [CrossRef]
27. Gentile, C.; Bernardini, G. Output-only modal identification of a reinforced concrete bridge from radar-based measurements. *NDT E Int.* **2008**, *41*, 544–553. [CrossRef]
28. National Ready Mixed Concrete Association. Concrete $CO_2$ Fact Sheet. Available online: https://www.nrmca.org/sustainability/CONCRETE%20CO2%20FACT%20SHEET%20FEB%202012.pdf (accessed on 21 July 2017).
29. Ligterink, N.; Smokers, R. Uitstoot van auto's bij snelheden hoger dan 120 km/u. Available online: https://www.rijksoverheid.nl/documenten/rapporten/2016/04/11/uitstoot-van-auto-s-bij-snelheden-hoger-dan-120-km-u (accessed on 21 July 2017).
30. Shalom, I. Application of ACM's in Infrastructure. Master's Thesis, Delft University of Technology, Delft, The Netherlands, 2016.
31. Dywidag. *European Technical Approval Post-Tensioning System*; Report No. ETA-13/0815; Austrian Institute of Construction Engineering: Vienna, Austria, 2013.

*Article*

# Chemical, Physical, and Mechanical Properties of 95-Year-Old Concrete Built-In Arch Bridge

**Andrzej Ambroziak** *, **Elżbieta Haustein** and **Maciej Niedostatkiewicz**

Faculty of Civil and Environmental Engineering, Gdansk University of Technology, 11/12 Gabriela Narutowicza Street, 80-233 Gdańsk, Poland; elzbieta.haustein@pg.edu.pl (E.H.); mniedost@pg.edu.pl (M.N.)
* Correspondence: ambrozan@pg.edu.pl; Tel.: +48-(58)-347-2447

**Abstract:** This research aimed to determine the durability and strength of an old concrete built-in arch bridge based on selected mechanical, physical, and chemical properties of the concrete. The bridge was erected in 1925 and is located in Jagodnik (northern Poland). Cylindrical specimens were taken from the side ribs connected to the top plate using a concrete core borehole diamond drill machine. The properties of the old concrete were compared with the present and previous standard requirements and guidelines. The laboratory testing program consisted of the following set of tests: measurements of the depth of carbonated zone and dry density, water absorption tests, determination of concrete compressive strength and frost resistance, determination of modulus of elasticity, measurement of the pH value, determination of water-soluble chloride salt and sulfate ion content, and X-ray diffraction analyses. Large variations in the cylindrical compressive strength (14.9 to 22.0 MPa), modulus of elasticity (17,900 to 26,483 MPa), density (2064 to 2231 kg/m$^3$), and water absorption (3.88 to 6.58%) were observed. In addition to the experiments, a brief literature survey relating to old concrete properties was also conducted. This paper can provide scientists, engineers, and designers an experimental basis in the field of old concrete built-in bridge construction.

**Keywords:** structural concrete; reinforced concrete; bridge engineering; material characterization; mechanical properties; chemical properties

**Citation:** Ambroziak, A.; Haustein, E.; Niedostatkiewicz, M. Chemical, Physical, and Mechanical Properties of 95-Year-Old Concrete Built-In Arch Bridge. *Materials* **2021**, *14*, 20. https://dx.doi.org/10.3390/ma14010020

Received: 29 November 2020
Accepted: 18 December 2020
Published: 23 December 2020



## 1. Introduction

Since the second half of the 19th century when reinforced concrete was invented, there has been a rapid development of this composite material, which is made up of a combination of steel and concrete [1]. Reinforced concrete is a material used in the construction of a wide range of civil and engineering structures. Due to concrete degradation, high traffic, and the impact of high load, old concrete and reinforced concrete bridges or other types of structures require improvement, repair, and reconstruction. Before taking any action and starting the design process, it is necessary to form an expert opinion by carrying out a detailed examination and laboratory tests of the construction materials used in the old structure. In several cases, it is necessary to incorporate the scientific and engineering community to evaluate the performance of old structures. In the literature, it is possible to find many interesting descriptions related to the process of testing and repairing old concrete structures. Hellebois et al. [2,3] performed an investigation on hardened concrete samples removed from a narrow-gauge railway viaduct (Colo-Hugues viaduct) built in Belgium in 1904. Mechanical and durability performance of the 100-year-old hardened concrete samples were found to be remarkably good. Sena-Cruz et al. [4] described historical, geometrical, and damage surveys of a reinforced concrete bridge built in 1907 (the Luiz Bandeira bridge). Selected structural material properties were also determined, e.g., the strength class was found to be greater than C30/37 and the average modulus of elasticity was 30 GPa. Wolert et al. [5] investigated an 11-span flat slab reinforced concrete bridge constructed between 1914 and 1916 that goes over Barnes Slough and Jenkins Creek in the USA. The authors confirmed the overall good condition of the structure and its reserve flexural capacity. Onysyk et al. [6] described the strength of the reinforced concrete ribbed dome of Centennial Hall in Wrocław, Poland, which was built in 1911–1913. Gebauer and Harni [7] examined the composition and microstructure of the hydrated cement paste of an 84-year-old reinforced concrete bridge construction. Tests showed that the main hydration products were calcium hydroxide, fibrous calcium silicate hydrate incorporated frequently with calcium hydroxide into hexagonal plates, and calcium aluminate carbonate hydrate. Qazweeni and Daoud [8] studied the physical, mechanical, and chemical properties of a 20-year-old concrete structure in an office building. Blanco et al. [9] investigated the chemical reactions leading to the degradation of a 95-year-old concrete dam manufactured with sand–cement as a binder. Ambroziak et al. [10] determined the durability and strength of reinforced concrete continuous footing based on selected mechanical and chemical properties of a 70-year-old concrete structure in an office building. Melchers et al. [11] performed observations and analysis on a 63-year-old reinforced concrete promenade railing exposed to the harsh sea-spray environment of the North Sea in Arbroath, Scotland. Castro-Borges et al. [12] studied the physical and mechanical properties of a 60-year-old concrete pier with stainless steel reinforcement. The pier showed no visible sign of deterioration after 60 years of service. Sohail et al. [13] investigated the effects of concrete degradation in structural concrete elements in reinforced concrete structures built in the 1960s, 1970s, and 1980s in the Arabian Gulf region. The carbonation depth and chloride concentration profiles were determined from concrete core samples. Papé and Melchers [14] performed load tests on full-scale prestressed beams sampled from the 45-year-old Sorell Causeway bridge in Tasmania, Australia. The prestressing strands showed severe localized corrosion with cross-section losses between 75% and 100%. Dasar et al. [15] tested 40-year-old reinforced concrete beams exposed to real marine environments for up to 20 years at Sakata Port, Japan. The deterioration and performance reduction were investigated, and a good correlation was observed between the crack width and cross-section loss. Czaderski and Motavalli [16] performed experimental investigations on a full-scale concrete bridge girder strengthened with prestressed carbon fiber-reinforced polymer (CFRP) plates obtained from the Viadotto delle Cantine a Capolago bridge, which was constructed in 1964–1966. Usage of the gradient method for anchorage of prestressed CFRP plates on large-scale girders was confirmed. Pettigrew et al. [17] carried out experiments on 48-year-old concrete bridge girders fabricated using lightweight concrete after the decommissioning of a bridge in the USA. The designed flexural capacities were overestimated by an average of 34.0% compared to the values measured in laboratory tests. Khan et al. [18] tested reinforced concrete beams corroded by 26 years of exposure to a chloride environment. The corrosion had a significant impact on the load-carrying capacity, stiffness, and deflection of the beams. Prassianakis and Giokas [19] determined the mechanical properties of 28-year-old concrete using destructive and ultrasonic nondestructive testing methods. Chen [20] studied the dynamic mechanical properties of 10-year-old concrete exposed to high temperatures. Zhu et al. [21] studied the durability and mechanical properties of a 10-year-old crumb rubber concrete bridge deck. The investigation concluded that the deck was in good condition. Kou and Poon [22] investigated the mechanical properties of five-year-old concrete prepared with 0, 20, 50, and 100% recycled aggregates used as replacements of natural aggregates. Dasar et al. [23] studied the applicability of seawater as a mixing and curing agent in four-year-old mortar cement. The laboratory tests indicated that the effect of seawater on corrosion activity was considerably higher as a curing agent than as a mixing agent. Many engineering and scientific studies investigating the mechanical, chemical, and/or physical properties of built structures take into consideration the subject of old concrete. A proper assessment of the properties of old concrete helps determine the range of repair or reconstruction required as well as the load capacity of the investigated structure, which is needed for ensuring extended working life and the safe use of old facilities.

The present study aimed to determine selected mechanical, physical, and chemical properties of a 95-year-old concrete arch bridge. Cylindrical specimens were taken from the side ribs connected to the top plate using a concrete core borehole diamond drill machine. The drilling locations were selected based on their availability and limited interference to the bridge structure. The investigation presented in this study can be treated as part of expert opinion on the bearing capacity of the arch bridge in order to determine the possibility of the bridge carrying additional loads and help extend its working life. This paper provides scientists, engineers, and designers an experimental assessment of the mechanical, physical, and chemical properties of 95-year-old concrete.

## 2. Materials and Methods

The arch bridge investigated in this study is located above a forest canyon and the Kumiel river (watercourse) (see Figure 1). The structure of the center span bridge (about 12.95 m) is a reinforced concrete slab, monolithically connected to reinforced concrete shields supported on a plate arch. The arch bridge was built in 1925 in Jagodnik (Poland) by Karl Metzger & Co. building company (see p. 181 in [24], where a photo of the investigated arch bridge is inserted). In 1925, Jagodnik (Berendshagen) was part of the district of Elbing in Germany (present-day Elbląg in Poland). The arch bridge was erected under the guidelines of the German Committee for Structural Concrete issued in January 1916 [25] for structural use of concrete, design, and construction. This guideline was in force until September 1925, when the German standard DIN 1045 [26] was introduced.

(a) (b)

**Figure 1.** Jagodnik arch bridge before (**a**) and after (**b**) reconstruction.

Laboratory tests were carried out to determine the chemical, physical, and mechanical properties of old concrete. For this purpose, concrete specimens were taken from structural elements of the bridge, namely the side ribs connected to the top plate, using a concrete core borehole diamond drill machine. The thickness of the reinforced concrete side ribs was approximately 15 cm. Cylindrical specimens were taken from five different side ribs connected to the top plate using a concrete core borehole diamond drill machine. The spacing between the side ribs was approximately 2 m. The samples were marked as location number_specimen number (e.g., 1_2, 5_1, etc.). These denotations were used for all laboratory tests.

Two types of cylindrical samples were prepared for mechanical tests from the exploratory bore holes:

- Type 1 had a length to core diameter ratio $L/D = 1$, with diameter $D$ of approximately 100 mm ($f_{is,cycl\ 100}$); 19 samples were used for uniaxial compressive tests.
- Type 2 had a length to core diameter ratio $L/D = 1.5$, with diameter $D$ of approximately 100 mm; 7 samples were used to measure the modulus of elasticity.

The dimensions of the concrete cores for concrete compressive strength tests were determined according to the standard EN 12504-1 [27], with the preferred length/diameter ratio of 1.0. The strength results determined for the concrete cores $f_{c,cycl\ 100}$ were comparable to the cube strength $f_{c,cube}$ of 15 × 15 × 15 cm concrete specimens (i.e., $f_{c,cube} = f_{c,cycl\ 100}$).

The ASTM C469M standard [28] states that the ratio between the specimen length $L$ and the dimension $D$ should be greater than 1.50; thus, the concrete cores for the determination of the modulus of elasticity was taken as $L/D = 1.5$. In laboratory tests, the application of a greater diameter and/or greater length to core diameter ratio (e.g., $L/D = 2$ as in the ASTM C31 standard [29]) is often impossible for old concrete structures [10].

Selected chemical, physical, and mechanical properties of the 95-year-old concrete built-in arch bridge were investigated using laboratory tests. The laboratory testing program consisted of the following sets of tests.

### 2.1. Measurements of the Depth of Carbonated Zone

The depth of the carbonated zone was measured with phenolphthalein solution. Freshly fractured surfaces of old concrete were submitted under an alcoholic solution of phenolphthalein, which immediately reacted, turning to pink/purple color, indicating the presence of calcium hydroxide, except at the thin, already carbonated external layers.

### 2.2. Measurements of Dry Density

The method specified in the EN 12390-7 standard [30] was applied for determining the density of the 95-year-old concrete. The tested specimens were dried in a ventilated oven at 105 ± 5 °C until the mass changed by less than 0.2%. Before weighing, each specimen was cooled to near room temperature in a dry, airtight vessel.

### 2.3. Tests of Water Absorption

Water absorption tests were carried out following EN 13369, Annex G [31]. To measure the water uptake capacity of concrete samples, the specimens were soaked in drinking water to a constant mass and then oven-dried in a ventilated drying oven at 105 ± 5 °C to a constant mass.

### 2.4. Determination of Concrete Compressive Strength and Frost Resistance

Uniaxial compressive tests were undertaken using a computer-controlled mechanical testing machine with a constant rate of loading and a range of 0.6 MPa/s according to the EN 12390-3 standard [32]. The frost resistance of the old concrete was determined according to guidelines given by the PN-B-06250 standard [33]. A freezing chamber with a temperature- and time-controlled refrigerating and heating system was used. The freezer cycle consisted of freezing at −18 ± 2 °C for 4 h and thawing by total immersion in water at 18 ± 2 °C for 4 h.

### 2.5. Determination of Modulus of Elasticity

The ASTM C469M standard [28] guideline was used to determine the modulus of elasticity. Diamond-drilled concrete cores with a length to diameter ratio of 1.50 were used in a compressometer device to measure the static modulus of elasticity.

### 2.6. Measurement of the pH Value and Determination of Water-Soluble Chloride Salts ($Cl^-$) and Sulfate Ions ($SO_4^{2-}$)

The pH was measured according to ISO 10523 [34]. The extract with water-soluble sulfate ions and chloride ions were specified according to EN 1744-1 + A1 standard [35]. The extract with chloride ions was determined in accordance with the Volhard method. The concentration of water-soluble chloride salts and sulfate ions as well as the pH of the test samples were measured after dissolving a given amount of mass of crushed concrete in distilled water. After filtration through a mixed cellulose ester (MCE) membrane filter with a pore size of 45 μm, the obtained filtrates were tested.

### 2.7. X-ray Diffraction (XRD) Analyses

The microstructure of the cross sections of samples was studied using a JEOL JSM-7800F (Akishima, Tokyo, Japan) scanning electron microscope (SEM) equipped with an energy-dispersive X-ray spectrometer (EDAX, Octane Elite, Mahwah, NJ, USA), which allowed the element composition of the tested samples to be identified. The acceleration voltage in the X-ray tube for surface analysis of samples was 15 kV. The X-ray beam current was 5 nm. Observations were carried out to identify different phases of the microstructure.

## 3. Results and Discussion

### *3.1. Measurements of Depth of Carbonated Zone*

The chemical reaction of calcium hydroxide dissolved in pore water with atmospheric carbon dioxide (conversion into calcium carbonate, which is then mainly calcite [36]) is called carbonation of concrete. Carbon dioxide in the air penetrates into concrete and diminishes the pH value and also causes shrinkage in the concrete [37]. The depth of the carbonated zone measured with the phenolphthalein solution is illustrated in Figure 2. Freshly fractured surfaces of all specimens reacted with the alcoholic solution of phenolphthalein, immediately turning to pink/purple color, indicating the presence of calcium hydroxide, except at the thin, already carbonated external layers. Large variations in depth of the carbonated zone was observed in the investigated specimens of old concrete, ranging approximately 20 to 55 mm (see Table 1). The average depth of carbonation of the old concrete was 36 ± 2 mm. The result of the mean value is presented as the sum of mean values and standard error of the mean of the specified range.

(**a**)

(**b**)

**Figure 2.** Depth of carbonated zone: (**a**) about 5 cm; (**b**) about 2.5 cm.

**Table 1.** Measurements of depth of carbonated zone.

| **Specimen No.** | **1_1** | **1_2** | **2_1** | **2_2** | **3_1** | **3_2** | **4_1** | **4_2** | **5_1** | **5_2** | **Mean (mm)** |
|---|---|---|---|---|---|---|---|---|---|---|---|
| Cover 1 (mm) | 45 | 35 | 40 | 44 | 30 | 40 | 50 | 30 | 25 | 24 | 36 ± 2 |
| Cover 2 (mm) | 45 | 35 | 40 | 45 | 25 | 40 | 55 | 32 | 25 | 20 | |

### *3.2. Measurements of Dry Density and Water Absorption Tests*

The dry density is one of the important parameters determined for concrete. Concrete is a mixture, and its density depends on its ingredients and their proportions. The mean dry density value specified in laboratory tests was 2175 ± 7 kg/m$^3$ (see Table 2). The dry density ranged from 2000 to 2600 kg/m$^3$; thus, according to the EN 206 standard [38], the investigated old concrete could be categorized as normal concrete. The mean dry density value also fulfilled conditions for normal-weight concrete according to the ACI 318-19 standard [39] (density between 2160 and 2560 kg/m$^3$).

**Table 2.** Dry density and water absorption.

| Specimens No. | Dry Density (kg/m$^3$) | Water Absorption (%) |
|---|---|---|
| 1_1 | 2166 | 6.29 |
| 1_2 | 2183 | 6.01 |
| 1_3 | 2171 | 6.05 |
| 1_4 | 2212 | 5.68 |
| 1_5 | 2157 | 6.12 |
| 1_6 | 2186 | 6.02 |
| 1_7 | 2207 | 5.85 |
| 2_1 | 2162 | 5.94 |
| 2_2 | 2230 | 5.69 |
| 2_3 | 2132 | 6.53 |
| 2_4 | 2173 | 5.93 |
| 3_1 | 2154 | 6.47 |
| 3_2 | 2184 | 6.05 |
| 3_3 | 2155 | 5.58 |
| 3_4 | 2180 | 5.88 |
| 3_5 | 2170 | 5.84 |
| 3_6 | 2193 | 5.80 |
| 3_7 | 2179 | 5.02 |
| 4_1 | 2162 | 6.14 |
| 4_2 | 2064 | 5.98 |
| 4_3 | 2180 | 5.69 |
| 4_4 | 2231 | 5.35 |
| 5_1 | 2229 | 3.88 |
| 5_2 | 2174 | 5.27 |
| 5_3 | 2147 | 6.58 |
| 5_4 | 2191 | 6.25 |
| Mean | 2175 $\pm$ 7 | 5.84 $\pm$ 0.11 |

The laboratory tests determined water absorption of the old concrete as ranging from 3.88 to 6.58% (see Table 2). The mean value of water absorption was 5.84 $\pm$ 0.11%. According to the PN-88/B-06250 standard [33] guidelines, water absorption should not be greater than 5% for concrete exposed to atmospheric conditions and not greater than 9% for concrete protected from atmospheric conditions. PN-S-10040 [40] states that the water absorption of concrete used in bridge structures should not be greater than 5%. The mean water absorption values in our study were greater than 5%; thus, according to the International Federation for Structural Concrete (fib) report [41], the concrete quality could be categorized as poor quality.

### *3.3. Concrete Compressive Strength, Frost Resistance, and Modulus of Elasticity*

Uniaxial compressive experimental tests were carried out using the Advantest 9 C300KN mechanical testing machine. Experiments were performed on the failure of the concrete cylinder specimens (see Figure 3). The uniaxial tensile test results of compressive strength for cylindrical samples $f_{c,cycl\ 100}$ is presented in Table 3. The mean value of compressive strength of cylindrical samples $f_{c,cycl\ 100}$ was 18.8 $\pm$ 0.7 MPa. The strength results of the cylindrical samples $f_{c,cycl\ 100}$ with length/diameter ratio of 1.0 were comparable to the cube strength $f_{c,cube}$ of 15 $\times$ 15 $\times$ 15 cm concrete specimens according to the EN 12504-1 standard [27], i.e., $f_{c,cycl\ 100} = f_{c,cube}$ = 18.8 $\pm$ 0.7 MPa. The variation in compressive strength values of the 95-year-old concrete (see Table 3) can be explained by the production technology, which was probably based on portable concrete mixers with handmade proportions of concrete components. Portland cement was used as a binder in ordinary old concrete mixes. It should be noted that the new cementitious materials, e.g., geopolymer concrete (called alkali-activated materials, see e.g., [42–46]), have higher compressive strength and better durability compared to concrete mixes containing Portland cement.

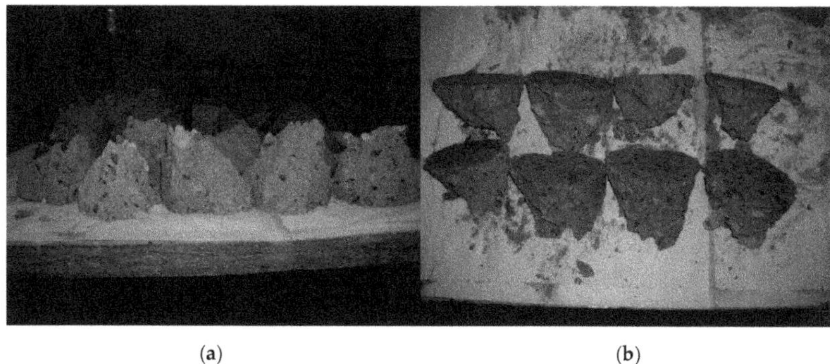

(a) (b)

**Figure 3.** Concrete specimens after uniaxial compressive tests: (**a**,**b**) views of the form of failure.

**Table 3.** Concrete compressive strength.

| Specimens No. | Compressive Strength $f_{c,cycl\ 100}$ (MPa) | Specimens No. | Compressive Strength after 50 Freezer Cycles (MPa) |
|---|---|---|---|
| 1_2 | 14.9 | 1_1 | 12.7 |
| 1_3 | 20.0 | 1_4 | 17.0 |
| 2_3 | 19.1 | 2_1 | 20.3 |
| 3_2 | 18.6 | 2_2 | 18.7 |
| 3_3 | 22.0 | 3_1 | 18.4 |
| 4_2 | 19.9 | 3_4 | 13.9 |
| 4_3 | 19.5 | 4_1 | 16.3 |
| 5_2 | 16.1 | 4_4 | 19.1 |
| 5_3 | 18.9 | 5_1 | 22.6 |
| | | 5_4 | 20.3 |
| Mean | 18.8 ± 0.7 | Mean | 17.9 ± 1.0 |

With regard to evaluation of freezing resistance, according to the PN-B-06250 standard [33], the compressive strength should not decrease by more than 20% in comparison to the base samples and the specimens should not show cracks [47]. In this study, we started the test by saturating the concrete samples with water. Then, 10 concrete samples were placed in a freezing chamber with a temperature- and time-controlled refrigerating and heating system (see Figure 4). The concrete samples were placed in the freezer compartment with a minimum 20 mm gap. A total of 50 freezer cycles were carried out consisting of freezing at $-18 \pm 2$ °C for 4 h and thawing by total immersion in water at $18 \pm 2$ °C for 4 h. After the last defrosting, a strength test was carried out. The frost resistance assessment was based on measuring the change in compressive strength. The mean value of compressive strength of the cylindrical samples after 50 freezer cycles $f_{c,cycl\ 100}^{50\ freezer\ cycles}$ was $17.9 \pm 1.0$ MPa, which was about 5% lower than the compressive strength of cylindrical samples $f_{c,cycl\ 100}$ without freezing (base samples). The difference between compressive strength $f_{c,cycl\ 100}$ and $f_{c,cycl\ 100}^{50\ freezer\ cycles}$ was small (<20% according to PN-B-06250 [33]); therefore, it could be stated that the 95-year-old concrete possessed freezing resistance.

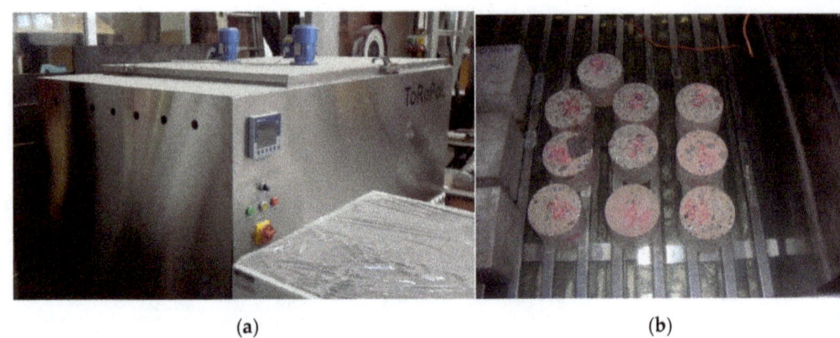

(a) (b)

**Figure 4.** Frost resistance tests: (**a**) view of freezing chamber; (**b**) view of concrete specimens inside the freezing chamber.

The two moduli of elasticity of applicable customary working stress ranged from 0 to 40% ($E_{0.0–0.4}$) and from 10 to 30% ($E_{0.1–0.3}$) were specified. Seven cylindrical specimens with a length/diameter ratio of 1.5 were stored and tested at room temperature (approximately 20 °C) in air-dry conditions. According to the laboratory tests, the modulus of elasticity ranged from 17.9 to 27 GPa for $E_{0.0–0.4}$ and from 20 to 27 GPa for $E_{0.1–0.3}$ (see Table 4). The mean values of the modulus of elasticity were 22,890 ± 1320 MPa for $E_{0.0–0.4}$ and 22,730 ± 890 MPa for $E_{0.1–0.3}$. The difference between the mean values of modulus of elasticity for $E_{0.0–0.4}$ and $E_{0.1–0.3}$ was very small (less than 1%). It should be noted that the EN 1992-1-1 standard [48] defines the modulus of elasticity as a secant value of 0 to 40% of the ultimate strength for concrete with quartzite aggregates. A limit of 10–30% should be used for limestone and sandstone aggregates. The ASTM C469M standard [28] also indicates 40% ultimate load to calculate modulus of elasticity. On the other hand, 30% of the ultimate strength is required in the ISO 1920-10 standard [49].

**Table 4.** Modulus of elasticity.

| Specimens No. | $E_{0.0–0.4}$ (MPa) | $E_{0.1–0.3}$ (MPa) |
|---|---|---|
| 1_5 | 22,567 | 21,095 |
| 1_6 | 20,613 | 21,643 |
| 1_7 | 26,483 | 24,497 |
| 2_4 | 25,369 | 22,832 |
| 3_5 | 20,264 | 22,026 |
| 3_6 | 27,027 | 27,027 |
| 3_7 | 17,900 | 20,006 |
| Mean | 22,890 ± 1320 | 22,730 ± 890 |

Hallauer [50] indicated that Hennebique seems to recommend a mixture consisting either of 1 part cement, 2 parts sand, and 4 parts gravel or 1 part cement, 3 parts sand, and 5 parts gravel (aggregate mix: sand: 0/7 mm, gravel: 7/70 mm, stone grit: 7/25 mm, stone chip: 25/70). Forecast compressive strength was about 15–18 MPa after 28 days and 18–24.5 MPa after 45 days. Note that the decrease in strength with the same cement content was due to the addition of water. The old Polish PN-B-195 standard [51] specified concrete strength equal to 0 (zero) MPa to emphasize that the amount of water should be limited and controlled in the concrete mix. The present guidelines on concrete standards state the requirements for water to cement ratio without mentioning zero-strength concrete. Wolert et al. [5] obtained compressive test results varying from 12.1 to 23.0 MPa for core samples cut out from an 11-span flat slab reinforced concrete bridge constructed between 1914 and 1916. Our laboratory tests determined the mean compressive strength of the 95-year-old concrete built-in arch bridge as $f_{c,cycl\ 100}$ = 18.8 MPa, which is similar to concrete structures build during this time period. The 95-year-old concrete also had frost resistance (after 50 freezer cycles, the compressive strength decreased only 5% to $f_{c,cycl\ 100}^{50\ freezer\ cycles}$ = 17.9 MPa). However, the compressive strength of old concrete construction sometimes varies and exhibits a wide range of compressive strengths [10]. Hellebois and Espion [3] investigated 107-year-old concrete samples taken from the Colo-Hugues viaduct, which was designed and built in 1904 in Belgium, and found different compressive strengths: 54.2 MPa for the slab, 33.3 MPa for beams, and 19.7 MPa for the column. Variation in the compressive strength of concrete is related to the type of aggregates and cement applied in concrete mixes [52] and is also affected by environmental conditions during the placement process of concrete mixes [53].

The characteristic in-situ compressive cube strength $f_{ck,is,cube}$ can be estimated according to the EN 13791 standard [54] as follows:

$$f_{ck,is,cube} = \min\left\{ \begin{array}{c} f_{m(n),is} - k_n \cdot s \\ f_{is,lowest} - M \end{array} \right\} = \min\left\{ \begin{array}{c} 18.8 - 1.96 \cdot 0.7 \\ 14.9 + 2 \end{array} \right\} = \min\left\{ \begin{array}{c} 17.43 \\ 16.9 \end{array} \right\} = 16.9 \text{ MPa} \quad (1)$$

where $f_{m(n),is}$ is the mean in-situ compressive strength of $n$ test results (in the present investigation, $n$ = 9), $f_{is,lowest}$ is the lowest in-situ compressive strength test results, $k_n$ is the factor that depends on the number of test results ($k_n$ = 1.96 for $n$ = 9 test results, see Table 6 in EN 13791 [54]), $s$ is the standard deviation of in-situ compressive strength, and $M$ is the value of margin ($M$ = 2, see Table 7 in the EN 13791 standard [54]). The characteristic in-situ cube compressive strength $f_{ck,is,cube}$ = 16.9 MPa; thus, the C12/15 compressive strength class according to the EN 206 [38] and EN 13791 [54] standards could be specified. The secant modulus of elasticity for the C12/15 concrete strength class specified in the EN 1992-2 standard [48] is 27,000 MPa. The mean modulus of elasticity $E_{0.0–0.4}$ determined by laboratory tests (see Table 4) was 22,890 MPa, which was about 15% lower than that specified in the EN 1992-2 standard [48] for design of new concrete structures. However, it should be noted that the modulus of elasticity depends not only on the strength class but also on the types and properties of the aggregates used to prepare concrete mixes. For the design of new concrete bridge structures, the EN 1992-2 standard [55] recommends the application of minimum strength classes not less than C30/37. According to the EN 206 standard [38], the C30/37 compressive strength class should have a minimum characteristic cylinder strength ($f_{ck,cyl}$) of 30 MPa (N/mm$^2$) and cube strength ($f_{ck,cube}$) of 37 MPa. The Polish standard PN-S-10042 [56] states that the C20/25 strength class may be applied for new foundations, supports, and retaining walls where the dimension of the structural elements are not less than 0.6 m thick, while the C25/30 strength class may be applied for new elements of supports and retaining walls less than 0.6 m thick and for reinforced concrete spans.

### 3.4. Chemical Properties

The pH value and the content of water-soluble chloride salts ($Cl^-$) and sulfate ions ($SO_4^{2-}$) were determined using a chemical testing program. Samples for chemical analysis were taken from the cover layer (denotated as 1s, 2s, 3s, 4s, and 5s) and center layer (denotated as 1c, 2c, 3c, 4c, and 5c) of the 95-year-old concrete samples. The cover layer samples for chemical investigations were obtained by cutting approximately 2–2.5 cm cylindrical samples from the exploratory boreholes.

The mean value of the pH was 10.6 ± 0.4 for the cover layer (range from 9.47 to 11.68) and 12.29 ± 0.04 for the center layer of samples (range from 12.14 to 12.37) (see Tables 5 and 6). The pH value of the concrete pore solution decreased as carbonation proceeded. The pH value for freshly made concrete varies from 12.5 to 13.5 [57]. The pH value is one of the most useful factors (influencing the corrosion rate [58]) for specifying the ability of concrete to protect steel rebar by the formation of a passive film protecting reinforcement from corrosion [59]. The corrosion of rebar generally occurs when the pH value is less than 9 (the passive film is not stable, see e.g., [60]). Despite the large depth of the carbonated zone (see Table 1), the pH of the old concrete was still in the safety range.

**Table 5.** pH values of concrete specimens and content of chloride ions and sulfate ions in concrete as a percentage of dry weight for the cover layer.

| Samples | pH | $Cl^-$ (%) | $SO_4^{2-}$ (%) |
|---|---|---|---|
| 1s | 9.47 | 0.036 | 0.670 |
| 2s | 9.98 | 0.020 | 0.298 |
| 3s | 10.54 | 0.014 | 0.658 |
| 4s | 11.37 | 0.014 | 0.127 |
| 5s | 11.68 | 0.012 | 0.203 |
| Mean | 10.6 ± 0.4 | 0.019 ± 0.004 | 0.39 ± 0.11 |

**Table 6.** pH values of concrete specimens and content of chloride ions and sulfate ions in concrete as a percentage of dry weight for the center layer.

| Samples | pH | $Cl^-$ (%) | $SO_4^{2-}$ (%) |
|---|---|---|---|
| 1c | 12.32 | 0.017 | 0.089 |
| 2c | 12.37 | 0.018 | 0.019 |
| 3c | 12.14 | 0.014 | 0.022 |
| 4c | 12.28 | 0.013 | 0.011 |
| 5c | 12.34 | 0.010 | 0.016 |
| Mean | 12.29 ± 0.04 | 0.014 ± 0.001 | 0.031 ± 0.014 |

The content of water-soluble chloride salts and sulfate ions was determined as a percentage of dry weight (see Tables 5 and 6). The water-soluble chloride salt values ranged from 0.012 to 0.036% dry weight for the cover layer and from 0.010 to 0.018% dry weight for the center layer of samples. The mean values of water-soluble chloride salt were 0.019 ± 0.004% and 0.014 ± 0.001% for the cover layer and the center layer, respectively. The sulfate ion values ranged from 0.127 to 0.670% dry weight for the cover layer and from 0.011 to 0.089% dry weight for the center layer of samples. The mean values of sulfate ions were 0.39 ± 0.11% and 0.031 ± 0.014% for the cover layer and the center layer, respectively.

In 1925, the cement content was usually set at 300 kg/m$^3$; it could be 270 kg/m$^3$ for buildings without the influence of moisture (see e.g., [50]). Assuming usage of 300 kg/m$^3$ cement and specified dry density of 2175 kg/m$^3$ (see Table 2), it was possible to convert the percentage content of chloride ions and sulfate ions by mass of cement (see Table 7). The chloride content of the concrete samples expressed as a percentage of chloride ions by mass of cement did not exceed 0.2 and 0.15%, which are limits for reinforced concrete stated by the EN 206 standard [38] and the ACI 318 code [61], respectively. The 95-year-old concrete built-in arch bridge was not exposed to chloride attack. A chloride content in concrete of over 0.2–0.3% of cement weight can be treated as being exposed to chloride attack.

**Table 7.** Mean content of chloride ions and sulfate ions in concrete as a percentage of mass of cement.

| Layer | $Cl^-$ (%) | $SO_4^{2-}$ (%) |
|---|---|---|
| Cover | 0.14 | 2.83 |
| Center | 0.10 | 0.22 |

The mean percentage content of sulfate ions $SO_4^{2-}$ by mass of cement was 2.83 for the cover layer and 0.22 for the center layer of concrete samples (see Table 7). The concrete samples did not exceed 4%, which is the limit by mass of cement based on the total acid-soluble sulfate given by the BS 8110-1:1985 standard [62] (this guideline was excluded in the new edition of the BS 8110-1 standard [63]). The low concentration of sulfate ions in the concrete samples indicated that the low contamination was due to external sources.

### 3.5. X-ray Diffraction Analyses

Based on the specified chemical properties, samples from 1s and 1c concrete specimens were chosen for XRD analyses to assess the presence of crystalline compounds and identify them. For the tests, specimens measuring 10 × 10 × 10 mm were taken from the cover layer and the center layer (about 7.5 cm below the surface of the concrete) of the tested concrete element. Microstructure tests were carried out for four samples (two samples from the cover and two samples from the center layer of old concrete specimens). Element composition tests (EDS) were carried out for four samples (three samples from the cover and three samples from a depth of about 7.5 cm below the surface). The results of the microstructure (SEM images) and element composition of the tested samples are shown in Figures 5 and 6.

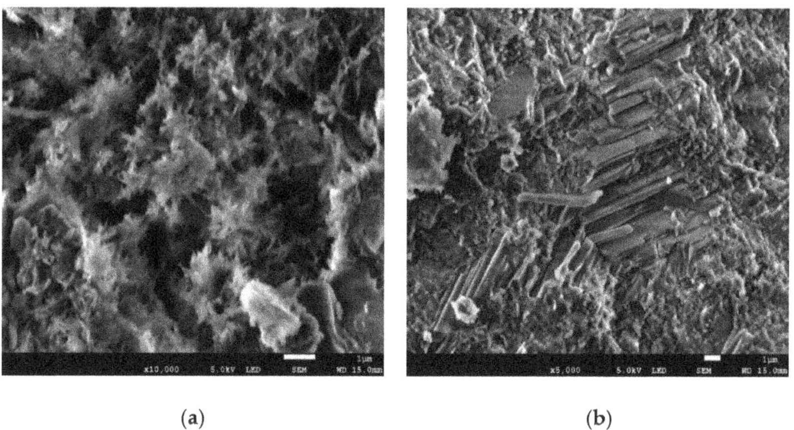

(a) (b)

**Figure 5.** Scanning electron microscope (SEM) images of the samples taken from the cover layer of the concrete structure: (**a**) ×10,000; (**b**) ×5000.

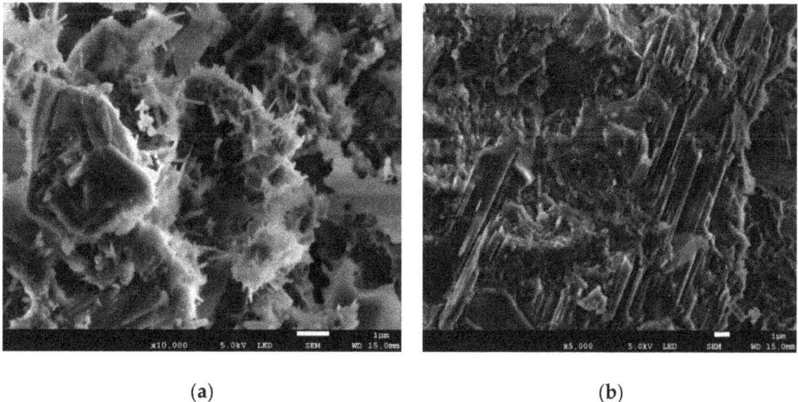

(a) (b)

**Figure 6.** SEM images of the samples taken from center layer of the concrete structure: (**a**) ×10,000; (**b**) ×5000.

The results of element composition tests are presented in Tables 8 and 9. They indicated the dominance of two main elements in the microstructure of the tested samples: Si (silica) and Ca (calcium) (see Figure 7). The content of these elements for samples taken from the cover layer was 29.4 to 34.7% for silica and 54.1 to 57.5% for calcium. For the specimen samples taken from center layer, the content was 25.6 to 36.8% for silica and 51.6 to 60% for calcium. The content of other elements, such as Al, Fe, Mg, K, and Na, was definitely lower compared to the Si and Ca content in all samples.

**Table 8.** Element compositions of samples taken from the cover layer of the concrete structure as determined by EDS.

| Type of Element | Sample 1 | Sample 2 | Sample 3 | Mean (%) |
|---|---|---|---|---|
| Mg | 1.2 | 1.0 | 1.4 | 1.2 |
| Al | 4.8 | 5.0 | 6.2 | 5.3 |
| Si | 36.8 | 34.7 | 25.6 | 32.4 |
| S | 1.3 | 1.3 | 1.8 | 1.5 |
| K | 1.0 | 1.6 | 1.0 | 1.2 |
| Na | - | 0.6 | - | 0.6 |
| Ca | 51.7 | 51.6 | 60.0 | 54.4 |
| Fe | 3.3 | 3.3 | 3.9 | 3.5 |
| F | - | 1.0 | - | 1.0 |

**Table 9.** Element compositions of samples taken from the center layer of the concrete structure as determined by EDS.

| Type of Element | Sample 1 | Sample 2 | Sample 3 | Mean (%) |
|---|---|---|---|---|
| Mg | 1.1 | 1.3 | 1.2 | 1.2 |
| Al | 4.3 | 5.3 | 5.7 | 5.1 |
| Si | 34.7 | 28.9 | 29.4 | 31.0 |
| S | 1.1 | 0.8 | 1.0 | 0.9 |
| K | 1.7 | 1.7 | 1.7 | 1.7 |
| Ca | 54.1 | 58.0 | 57.5 | 56.6 |
| Fe | 3.2 | 3.4 | 3.5 | 3.4 |

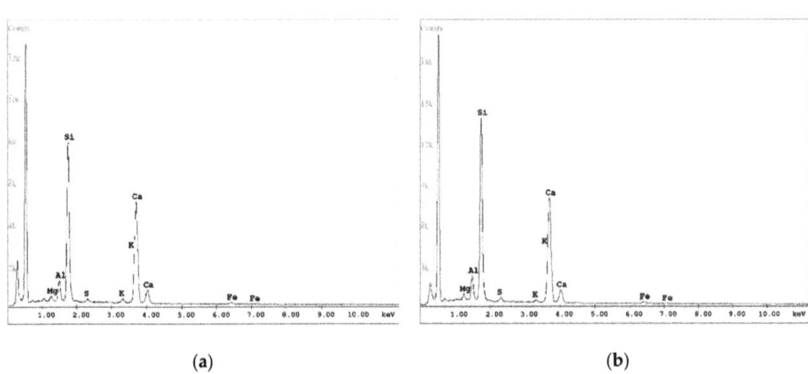


(a) (b)

**Figure 7.** Energy-dispersion X-ray spectrometer (EDS) spectra of samples taken from the (**a**) cover layer; (**b**) center layer.

## 4. Conclusions

The main objective of the present research was to determine the durability and strength of a 95 year-old concrete built-in arch bridge in Jagodnik (northern Poland) based on selected mechanical, physical, and chemical properties. Based on the experimental investigation, the following conclusions can be drawn:

- The old concrete could be categorized as normal concrete according to the EN 206 standard [38], with a mean dry density value of 2175 ± 7 kg/m$^3$ (see Table 2).
- The average depth of carbonation of the old concrete was 36 ± 2 mm (see Table 1). The old concrete had large variations in depth of the carbonated zone ranging from 20 to 55 mm. Despite the large depth of the carbonated zone, the pH of the old concrete was still in the safety range.
- The mean values of water absorption was 5.84 ± 0.11% (see Table 2), so the quality of the old concrete could be categorized as poor quality.
- The mean value of the compressive strength of cylindrical samples $f_{c,cycl\ 100}$ was 18.8 ± 0.7 MPa. Estimated characteristic in-situ compressive cube strength $f_{ck,is,cube}$ was 16.9 MPa (according to the EN 13791 standard [54]). The old concrete was categorized as C12/15 compressive strength class according to EN 206 [38]. However, the 95-year-old concrete possessed good freezing resistance.
- The mean value of the modulus of elasticity $E_{0.0\text{–}0.4}$ (see Table 4) was 22,890 MPa, which was about 15% lower than that specified in the EN 1992-2 standard [48] for C12/15 concrete strength class.
- The mean value of the pH was 10.6 ± 0.4 and 12.29 ± 0.04 for the cover layer and the center layer of samples, respectively (see Tables 5 and 6). The pH values for the old concrete indicated that there was no corrosion of the steel rebars. Generally, when the pH value decreases below 9–9.5, corrosion of the reinforcing steel may be indicated.
- The chloride content of the old concrete did not exceed 0.2% by mass of cement; thus, the old concrete arch bridge was not exposed to chloride attack. The low concentration of sulfate ions in the old concrete samples indicated that the low contamination was due to external sources.
- The element composition tests indicated the dominance of two main elements (Si and Ca) in the microstructure of the old concrete (see Tables 8 and 9). The significant content of the two elements suggested complete reaction of the clinker phases of the cement over the time the structure of concrete was exposed to the environment.
- Portlandite crystals were found in the tested samples, shaped most often as hexagonal platelets. Both isolated and clustered occurrence was noted. The visible portlandite occurred in the form of large pseudohexagonal crystals, forming columnar conglomerates arranged in parallel. In both cases, the resulting products (phases CSH and CH) were due to cement hydration reactions with water. The microstructure of the concrete has a significant influence on its properties, such as its strength and durability.

The design team of the Jagodnik arch bridge made a decision on the possibility of reconstructing the 95-year-old arch bridge and designing it for car traffic to a limited tonnage. On top of the existing old structure, a new deck slab with pavement covers was made. The entire surface of the old concrete was protected with repair mortars.

Deterioration of old concrete due to corrosion is a serious problem. Proper assessment of the properties of old concrete is critical to make decisions regarding the reconstruction and repair of old concrete structures. Inappropriate or incorrect estimation of the properties of old concrete can have catastrophic consequences. Reconstruction and renovation of old civil and building structures often require incorporation of the scientific and engineering community in order to evaluate the performance of old structures and give them "new life" and extended service. The authors are hopeful that this investigation sparks interest among a wide group of engineers and scientists to take into consideration the subject of old concrete structures.

**Author Contributions:** Conceptualization, A.A.; methodology, A.A.; validation, A.A. and E.H.; formal analysis, A.A.; investigation, A.A. and E.H.; resources, A.A.; data curation, A.A. and E.H.; writing—original draft preparation, A.A., E.H., and M.N.; writing—review and editing, A.A.; visualization, A.A. and E.H.; supervision, A.A.; project administration, A.A. All authors have read and agreed to the published version of the manuscript.

**Funding:** This research received no external funding.

**Institutional Review Board Statement:** Not applicable.

**Informed Consent Statement:** Not applicable.

**Data Availability Statement:** All laboratory test results (data) are included in Tables 1–9 in the present paper. No new data were created or analyzed in this study. Data sharing is not applicable to this article.

**Conflicts of Interest:** The authors declare no conflict of interest.

## References

1. Moussard, M.; Garibaldi, P.; Curbach, M. The invention of Reinforced concrete (1848–1906). In *High Tech Concrete: Where Technology and Engineering Meet*; Springer: Cham, Switzerland, 2017; pp. 2785–2794.
2. Hellebois, A.; Launoy, A.; Pierre, C.; De Lanève, M.; Espion, B. 100-year-old Hennebique concrete, from composition to performance. *Constr. Build. Mater.* **2013**, *44*, 149–160. [CrossRef]
3. Hellebois, A.; Espion, B. Concrete properties of a 1904 Hennebique reinforced concrete viaduct. *WIT Trans. Built Environ.* **2011**, *118*, 589–600. [CrossRef]
4. Sena-Cruz, J.; Ferreira, R.M.; Ramos, L.F.; Fernandes, F.; Miranda, T.; Castro, F. Luiz bandeira bridge: Assessment of a historical reinforced concrete (RC) bridge. *Int. J. Archit. Herit.* **2013**, *7*, 628–652. [CrossRef]
5. Wolert, P.J.; Kolodziejczyk, M.K.; Stallings, J.M.; Nowak, A.S. Non-destructive Testing of a 100-Year-Old Reinforced Concrete Flat Slab Bridge. *Front. Built Environ.* **2020**, *6*, 31. [CrossRef]
6. Onysyk, J.; Biliszczuk, J.; Prabucki, P.; Sadowski, K.; Toczkiewicz, R. Strengthening the 100-year-old reinforced concrete dome of the Centennial Hall in Wrocław. *Struct. Concr.* **2014**, *15*, 30–37. [CrossRef]
7. Gebauer, J.; Harnik, A.B. Microstructure and composition of the hydrated cement paste of an 84 year old concrete bridge construction. *Cem. Concr. Res.* **1975**, *5*, 163–169. [CrossRef]
8. Qazweeni, J.; Daoud, O. Concrete deterioration in a 20-years-old structure in Kuwait. *Cem. Concr. Res.* **1991**, *21*, 1155–1164. [CrossRef]
9. Blanco, A.; Segura, I.; Cavalaro, S.H.P.; Chinchón-Payá, S.; Aguado, A. Sand-Cement Concrete in the Century-Old Camarasa Dam. *J. Perform. Constr. Facil.* **2016**, *30*, 04015083. [CrossRef]
10. Ambroziak, A.; Haustein, E.; Kondrat, J. Chemical and Mechanical Properties of 70-Year-Old Concrete. *J. Mater. Civ. Eng.* **2019**, *31*. [CrossRef]
11. Melchers, R.E.; Li, C.Q.; Davison, M.A. Observations and analysis of a 63-year-old reinforced concrete promenade railing exposed to the North Sea. *Mag. Concr. Res.* **2009**, *61*, 233–243. [CrossRef]
12. Castro-Borges, P.; De Rincón, O.T.; Moreno, E.I.; Torres-Acosta, A.A.; Martínez-Madrid, M.; Knudsen, A. Performance of a 60-year-old concrete pier with stainless steel reinforcement. *Mater. Perform.* **2002**, *41*, 50–55.
13. Sohail, M.G.; Kahraman, R.; Ozerkan, N.G.; Alnuaimi, N.A.; Gencturk, B.; Dawood, M.; Belarbi, A. Reinforced Concrete Degradation in the Harsh Climates of the Arabian Gulf: Field Study on 30-to-50-Year-Old Structures. *J. Perform. Constr. Facil.* **2018**, *32*, 04018059. [CrossRef]
14. Papé, T.M.; Melchers, R.E. The effects of corrosion on 45-year-old pre-stressed concrete bridge beams. *Struct. Infrastruct. Eng.* **2011**, *7*, 101–108. [CrossRef]
15. Dasar, A.; Hamada, H.; Sagawa, Y.; Yamamoto, D. Deterioration progress and performance reduction of 40-year-old reinforced concrete beams in natural corrosion environments. *Constr. Build. Mater.* **2017**, *149*, 690–704. [CrossRef]
16. Czaderski, C.; Motavalli, M. 40-Year-old full-scale concrete bridge girder strengthened with prestressed CFRP plates anchored using gradient method. *Compos. Part B Eng.* **2007**, *38*, 878–886. [CrossRef]
17. Pettigrew, C.S.; Barr, P.J.; Maguire, M.; Halling, M.W. Behavior of 48-Year-Old Double-Tee Bridge Girders Made with Lightweight Concrete. *J. Bridg. Eng.* **2016**, *21*, 04016054. [CrossRef]
18. Khan, I.; François, R.; Castel, A. Structural performance of a 26-year-old corroded reinforced concrete beam. *Eur. J. Environ. Civ. Eng.* **2012**, *16*, 440–449. [CrossRef]
19. Prassianakis, I.N.; Giokas, P. Mechanical properties of old concrete using destructive and ultrasonic non-destructive testing methods. *Mag. Concr. Res.* **2003**, *55*, 171–176. [CrossRef]
20. Chen, C.; Chen, X.; Li, X. Dynamic Compressive Behavior of 10-Year-Old Concrete Cores after Exposure to High Temperatures. *J. Mater. Civ. Eng.* **2020**, *32*, 04020076. [CrossRef]
21. Zhu, H.; Duan, F.; Shao, J.; Shi, W.; Lin, Z. Material and durability study of a 10-year old crumb rubber concrete bridge deck in Tianjin China. *Mag. Concr. Res.* **2019**, 1–39. [CrossRef]
22. Kou, S.-C.; Poon, C.-S. Mechanical properties of 5-year-old concrete prepared with recycled aggregates obtained from three different sources. *Mag. Concr. Res.* **2008**, *60*, 57–64. [CrossRef]
23. Dasar, A.; Patah, D.; Hamada, H.; Sagawa, Y.; Yamamoto, D. Applicability of seawater as a mixing and curing agent in 4-year-old concrete. *Constr. Build. Mater.* **2020**, *259*, 119692. [CrossRef]
24. Lockemann, T. *Elbing*; Deutscher Architectur und Industrie Verlag: Berlin, Germany, 1926.
25. *Bestimmungen für die Ausführung von Bauwerken aus Eisenbeton*; DAfStB (Deutscher Ausschuss für Stahlbeton): Berlin, Germany, 1916.

26. Deutsche Institut für Normung. *DIN 1045:1925-09 Bestimmungen für Ausführung von Bauwerken aus Eisenbeton*; Deutsche Institut für Normung: Berlin, Germany, 1925.
27. *EN 12504-1 Testing Concrete in Structures—Part 1: Cored Specimens—Taking, Examining and Testing in Compression*; CEN (European Committee for Standardization): Brussels, Belgium, 2009.
28. *ASTM C469M-14 Standard Test Method for Static Modulus of Elasticity and Poisson's Ratio of Concrete in Compression*; ASTM International (American Society for Testing and Materials): West Conshohocken, PA, USA, 2014.
29. *ASTM C31/C31M—18b Standard Practice for Making and Curing Concrete Test Specimens in the Field*; ASTM International (American Society for Testing and Materials): West Conshohocken, PA, USA, 2018.
30. *EN 12390-7 Testing Hardened Concrete—Part 7: Density of Hardened Concrete*; CEN (European Committee for Standardization): Brussels, Belgium, 2019.
31. *EN 13369 Common Rules for Precast Concrete Products*; CEN (European Committee for Standardization): Brussels, Belgium, 2001; ISBN 0000105058.
32. *EN 12390-3 Testing Hardened Concrete. Compressive Strength of Test Specimens*; CEN (European Committee for Standardization): Brussels, Belgium, 2019.
33. *PN-88/B-06250 Normal Concrete*; PKN (Polish Committee for Standardization): Warsaw, Poland, 1988.
34. *ISO 10523 Water Quality—Determination of pH*; ISO (International Organization for Standardization): Geneva, Switzerland, 2008.
35. *EN 1744-1:2009+A1 Tests for Chemical Properties of Aggregates—Part 1: Chemical Analysis*; CEN (European Committee for Standardization): Brussels, Belgium, 2009.
36. Huang, N.M.; Chang, J.J.; Liang, M.T. Effect of plastering on the carbonation of a 35-year-old reinforced concrete building. *Constr. Build. Mater.* **2012**, *29*, 206–214. [CrossRef]
37. Dhir, R.K.; de Brito, J.; Silva, R.V.; Lye, C.Q. Deformation of Concrete Containing Recycled Concrete Aggregate. In *Sustainable Construction Materials*; Elsevier: Amsterdam, The Netherlands, 2019; pp. 283–363.
38. *EN 206:2013+A1:2016 Concrete—Specification, Performance, Production and Conformity*; CEN (European Committee for Standardization): Brussels, Belgium, 2016.
39. *ACI 318-19 Building Code Requirements for Structural Concrete*; ACI (American Concrete Institute): Farmington Hills, MI, USA, 2019; ISBN 9781641950565.
40. *PN-S-10040 Bridges—Concrete, Reinforced Concrete and Prestressed Concrete Structures—Requirements and Testings*; PKN (Polish Committee for Standardization): Warsaw, Poland, 1999.
41. *Report No. 192. Diagnosis and Assessment of Concrete Structures—State-of-Art*; CEB-FIP (Euro-International Committee for Concrete-International Federation for Pre-stressing): Lausanne, Switzerland, 1989.
42. Steinerova, M.; Matulova, L.; Vermach, P.; Kotas, J. The Brittleness and Chemical Stability of Optimized Geopolymer Composites. *Materials* **2017**, *10*, 396. [CrossRef]
43. Rovnaník, P.; Šafránková, K. Thermal Behaviour of Metakaolin/Fly Ash Geopolymers with Chamotte Aggregate. *Materials* **2016**, *9*, 535. [CrossRef] [PubMed]
44. Provis, J.L. Alkali-activated materials. *Cem. Concr. Res.* **2018**, *114*, 40–48. [CrossRef]
45. Xiao, R.; Jiang, X.; Zhang, M.; Polaczyk, P.; Huang, B. Analytical investigation of phase assemblages of alkali-activated materials in $CaO-SiO_2-Al_2O_3$ systems: The management of reaction products and designing of precursors. *Mater. Des.* **2020**, *194*, 108975. [CrossRef]
46. Xiao, R.; Ma, Y.; Jiang, X.; Zhang, M.; Zhang, Y.; Wang, Y.; Huang, B.; He, Q. Strength, microstructure, efflorescence behavior and environmental impacts of waste glass geopolymers cured at ambient temperature. *J. Clean. Prod.* **2020**, *252*, 119610. [CrossRef]
47. Glinicki, M.A.; Jaskulski, R.; Dąbrowski, M. Design principles and testing of internal frost resistance of concrete for road structures—Critical review. *Roads Bridg. Drog. Most.* **2016**, *15*, 21–43. [CrossRef]
48. *EN 1992-1-1 Eurocode 2: Design of Concrete Structures—Part 1-1: General Rules and Rules for Buildings*; CEN (European Committee for Standardization): Brussels, Belgium, 2004.
49. ISO (International Organization for Standardization). *ISO 1920-10 Testing of Concrete—Part 10: Determination of Static Modulus of Elasticity in Compression*; ISO (International Organization for Standardization): Geneva, Switzerland, 2010.
50. Hallauer, O. Die Entwicklung der Zusammensetzung von Beton für Wasserbauten. *Mtteilungsblaetter Bundesanst. Wasserbaut.* **1989**, *65*, 39–56.
51. *PN-B-195 Concrete and Reinforced Concrete Structures. Structural Analysis and Design*; PKN (Polish Committee for Standardization): Warsaw, Poland, 1945.
52. Petrounias, P.; Giannakopoulou, P.; Rogkala, A.; Stamatis, P.; Lampropoulou, P.; Tsikouras, B.; Hatzipanagiotou, K. The Effect of Petrographic Characteristics and Physico-Mechanical Properties of Aggregates on the Quality of Concrete. *Minerals* **2018**, *8*, 577. [CrossRef]
53. Ambroziak, A.; Ziolkowski, P. Concrete compressive strength under changing environmental conditions during placement processes. *Materials* **2020**, *13*, 4577. [CrossRef] [PubMed]
54. *EN 13791 Assessment of In-Situ Compressive Strength in Tructures and Precast Concrete Components*; CEN (European Committee for Standardization): Brussels, Belgium, 2019.
55. *EN 1992-2 Eurocode 2: Design of Concrete Structures—Concrete Bridges—Design and Detailing Rules*; European Committee for Standardization: Brussels, Belgium, 2005.

56. *PN-S-10042 Bridges—Concrete, Reinforced Concrete and Prestressed Concrete Structures—Design*; PKN (Polish Committee for Standardization): Warsaw, Poland, 1991.
57. Duffó, G.S.; Farina, S.B.; Giordano, C.M. Characterization of solid embeddable reference electrodes for corrosion monitoring in reinforced concrete structures. *Electrochim. Acta* **2009**, *54*, 1010–1020. [CrossRef]
58. Stojanović, G.; Radovanović, M.; Krstić, D.; Ignjatović, I.; Dragaš, J.; Carević, V. Determination of pH in powdered concrete samples or in suspension. *Appl. Sci.* **2019**, *9*, 3257. [CrossRef]
59. Abdulrahman, A.S.; Mohammad, I.; Mohammad, S.H. Corrosion inhibitors for steel reinforcement in concrete: A review. *Sci. Res. Essays* **2011**, *6*, 4152–4162. [CrossRef]
60. Hansson, C.M.; Poursaee, A.; Jaffer, S.J. Corrosion of Reinforcing Steel Bars in Concrete. *Masterbuilder* **2012**, *12*, 106–124.
61. *ACI 318-14 Building Code Requirements for Structural Concrete*; ACI (American Concrete Institute): Farmington Hills, MI, USA, 2014.
62. *BS 8110-1:1985 Structural Use of Concrete. Code of Practice for Design and Construction*; BSI (British Standards Institution): London, UK, 1985.
63. *BS 8110-1:1997 Structural Use of Concrete—Part 1: Code of Practice for Design and Construction*; BSI (British Standards Institution): London, UK, 1997.

*Case Report*

# A 95-Year-Old Concrete Arch Bridge: From Materials Characterization to Structural Analysis

**Andrzej Ambroziak** * and **Maciej Malinowski**

Faculty of Civil and Environmental Engineering, Gdansk University of Technology, 11/12 Gabriela Narutowicza Street, 80-233 Gdańsk, Poland; maciej.malinowski@pg.edu.pl
* Correspondence: ambrozan@pg.edu.pl; Tel.: +48-58-347-2447

**Abstract:** The structural analysis of a 95-year-old concrete arch bridge located in Jagodnik (Poland) is performed in this paper, in order to check its behavior under today's traffic loads. The mechanical properties of both the concrete and the reinforcement are investigated by testing cores and bar stubs extracted from the bridge. Structural analysis confirms that the bridge meets today's load requirements in terms of bearing capacity, serviceability state, and that the adopted structural improvements (a new deck slab on top of the existing structure and a layer of mortar to protect the surface of the old concrete) are effective. In this way, the 95-year-old arch bridge was given a new life. The structural improvements show how combining numerical modelling and laboratory tests can contribute to the preservation of an old—though fairly simple—and valuable structure, otherwise destined to demolition, with both environmental and economic benefits.

**Keywords:** structural analysis; bridge engineering; reinforced concrete; mechanical properties

**Citation:** Ambroziak, A.; Malinowski, M. A 95-Year-Old Concrete Arch Bridge: From Materials Characterization to Structural Analysis. *Materials* **2021**, *14*, 1744. https://doi.org/10.3390/ma14071744

Academic Editors: Eva O.L. Lantsoght and F. Pacheco Torgal

Received: 6 March 2021
Accepted: 30 March 2021
Published: 1 April 2021

**Publisher's Note:** MDPI stays neutral with regard to jurisdictional claims in published maps and institutional affiliations.



## 1. Introduction

Arch bridges are one of the most popular types of bridges. At present, there are over 40 concrete arch bridges in the world with a span of greater than 200 m [1]. Concrete application in the development of arch bridges has a long and interesting history [2]. In most countries, there are old bridges that require maintenance, renovation, or reconstruction [3]. In the literature, it is possible to find many interesting investigations related to the process of testing and repairing old concrete bridges [4–15] or to the structural analysis of old bridges [16–22]. Reconstruction and renovation of old bridge structures and adaptation to new traffic loads are complex issues often requiring not only the experience of civil engineers, but also that of the scientific community. Before dealing with the technical conditions and bearing capacity of any given bridge structure, a detailed inspection is a must. The code provisions concerning the original materials like mechanical, chemical, and also physical properties are required for proper assessment of conditions of old structures to reflect their real technical conditions. The scope of material tests should be adapted to the specificity of the construction and location of the bridge structure. The design team often faces the problem of limited or even lacking original documentation. It is then necessary to make a detailed inspection which is instrumental in defining the scope of the reconstruction or repair process, as well as in formulating a numerical model. New technics, like laser scanning, photogrammetry techniques and ground penetrating radar [23–29], have been increasingly used in terms of monitoring, inventory control and structural inspection. New tools and techniques are very helpful in the technical and theoretical assessment of old bridge structures.

The present study is aimed at the structural analysis of the 95-year-old concrete arch bridge based on mechanical properties measured by means of laboratory tests. Structural analysis was a part of an expert opinion, required to check whether the old arch bridge has an adequate bearing capacity face to today's traffic loads, in order to extend its service life. The present paper supplements and extends the investigations performed by Ambroziak et al. [30] on the design stage. In the present paper, new concrete samples taken from the old concrete arch bridge during its reconstruction were tested in a laboratory and the results of the tests on the original steel reinforcement are presented. The results about the internal forces under design loads are determined and the maximum stresses in both the concrete and the reinforcement are evaluated and compared with the design stresses, and the displacements are checked with reference to those specified for the service limit state. The paper provides scientists, engineers, and designers the example of structural analysis results and experimental assessment of the 95-year-old concrete arch bridge.

## 2. Materials and Methods

The old bridge investigated (see Figure 1) is an arch bridge built in 1925 close to the city of Elbląg (village Jagodnik in Poland) above the Kumiel river. Karl Metzger & Co. building company [31] was responsible for the construction of the bridge, which consists of a reinforced-concrete slab monolithically connected to a reinforced-concrete arch with a span equal to 12.95 m.

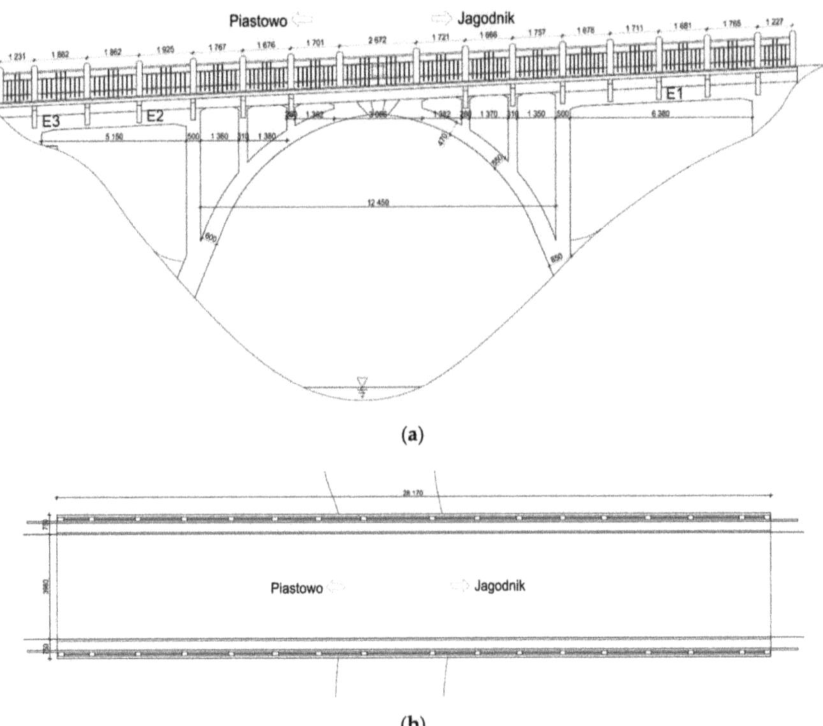


**Figure 1.** Jagodnik arch bridge before reconstruction (dimension units—mm): (**a**) Side view; (**b**) view from above.

Before the reconstruction process of the 95-year-old concrete arch bridge (see Figure 2) a few fragments of the old concrete were delivered to a laboratory (see Figure 3a) and seven concrete cores were extracted by means of a borehole diamond drill machine (see Figure 3b). After finishing, the length-to-diameter ratio L/D was 1 (L = D = 100 mm; see EN 12504-1 [32] standard), as the thickness of the old structural concrete members was close to 15–16 cm. The cores were marked with the indication of their location and specimen number (location number_specimens number, e.g., 1_2, 2_1). Stubs of steel bars were tested as well. The concrete fragments E1, E2, and E3 (Figure 3a) were taken from locations indicated in Figure 1a.

(a)

(b)

(c)

**Figure 2.** Jagodnik arch bridge: (**a**) Before; (**b**) during; and (**c**) after reconstruction.

The concrete cylinders were tested in compression to the failure according to EN 12390-3 standard [33]. Concrete dry density was determined according to method guidelines in EN 12390-7 standard [34], after drying the specimens in a ventilated oven (T = 105 ± 5 °C) until mass stabilization (not more than 0.2% mass variation with respect to the original mass). The density was derived after cooling down to room temperature in dry conditions. The tests in uniaxial tension of steel reinforcements were performed in accordance with

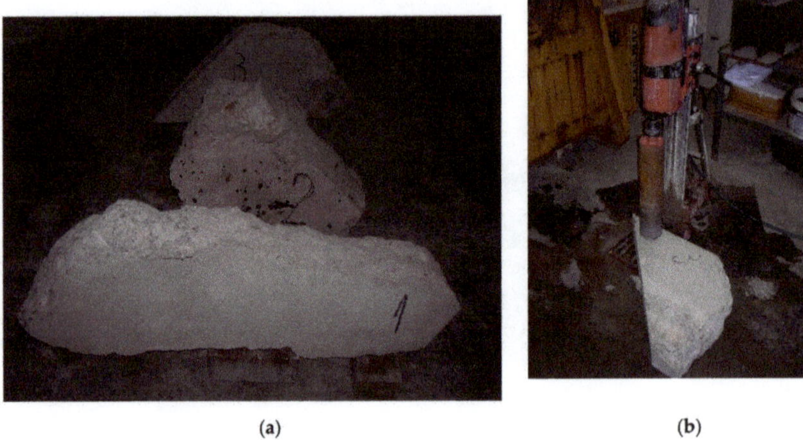

(a) (b)

**Figure 3.** Preparation of the concrete cores: (**a**) Fragments of old concrete; (**b**) drilling process.

## 3. Laboratory Test Results and Discussion

### *3.1. Reinforced Steel Tensile Tests*

The reinforcement smooth steel bars of a 6 ± 0.1 mm diameter were taken from the old arch bridge and subjected to uniaxial tension by means of the computer-controlled Zwick Z400 testing machine (ZwickRoell GmbH & Co. KG, Ulm, Germany), see Figure 4. The length of the sample between the grips is 100 mm and the displacement rate is 5 mm/min. All tests were performed at room temperature (about 20 °C) and were carried out up to specimen failure. Three specimens were chosen for uniaxial tensile tests. During tensile tests, the results were recorded sampling every 10 μm (traverse displacement interval), 20 ms (time interval) and 1 N (force interval). The tests in tension on the bar stubs were carried out in accordance with ISO 6892-1 standard [35]. The engineering strain at rupture show range from 15% to near 25%, while the ultimate tensile strength covers the 374–380 MPa interval. The investigated steel rebars are characterized by clear yield strength, strain hardening, and necking range at the stress–strain curves, see Figure 5a. The yield strength was determined in laboratory tests equals 291 ± 7 MPa. The yield strength is defined as the lowest value of stress during plastic yielding, ignoring any initial transient effects. The stress is obtained by dividing the force by the original cross-sectional area of the steel bars.

The Regulations on the Construction and Maintenance of Road Bridges [36] approved by the Polish Minister of Public Works (the ordinance of 9.XI.1925 no. XIII-1386) state the yield strength of steel rebars not to be less than 294 MPa (3000 kg/cm$^2$). The yield strength (291 ± 7 MPa) specified in laboratory tests corresponds to the guidelines issued in the arch bridge construction time. The Post-Second World War standard PN-B-195 [37] made it possible to apply three types of steel bars of variable yield strengths equals: 196 MPa (2000 kg/cm$^2$), 235 MPa (2400 kg/cm$^2$), and 353 MPa (3600 kg/cm$^2$). The S500 steel grade of 500 MPa characteristic yield strength is an abundant concrete reinforcement on bridge and building sites today.

Additionally, two bar stubs were tested at a displacement rate equal to 10 mm/min (Figure 5b). The elastic-viscous behavior of the material is evident (the higher the strain rate, the higher the strength at yielding [38]). In Figure 5b, the initial S-shaped loading branch was probably due to grip sliding, while the subsequent mostly-linear branch (stress comprised between 150 and 350 MPa) is related to concrete linear-elastic behavior. Finally, nonlinearity starts at 350 MPa.

(**a**)

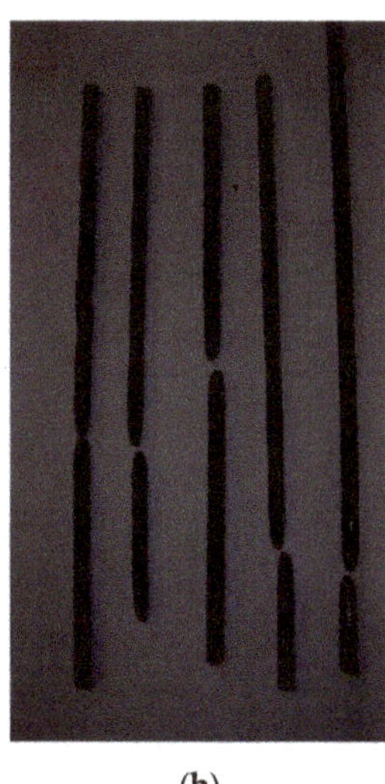

(**b**)

**Figure 4.** Uniaxial tensile tests: (**a**) Test setup; (**b**) steel bars after failure.

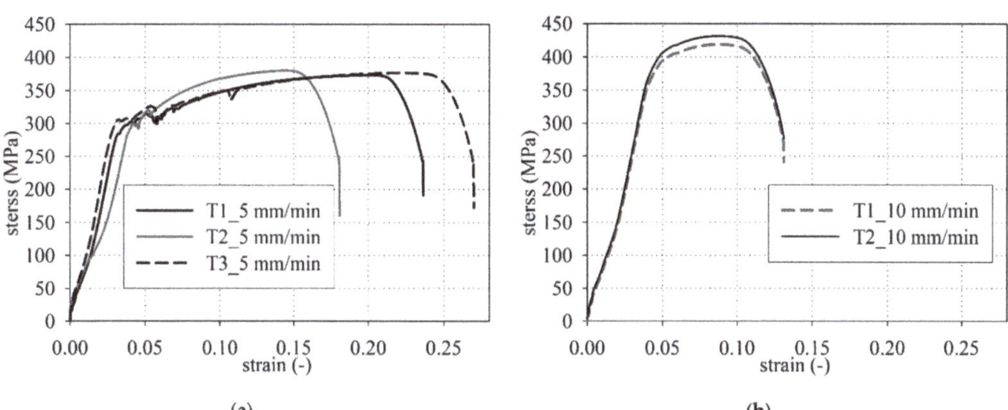


**Figure 5.** Stress–strain curves in tension: Test repeatability for two different displacement rates, 5 mm/s (**a**) and 10 mm/s (**b**).

### *3.2. Uniaxial Compressive Tests and Dry Density*

The uniaxial compressive experimental tests were conducted on the Advantest 9 C300 KN mechanical testing machine. The experiments were performed to the failure of the concrete cylinder specimens and were used at a constant rate of loading with a range of 0.6 MPa/s according to EN 12390-3 [33]. The uniaxial compression test results of compressive strength for cylindrical samples are presented in Table 1. These results are accompanied by the results of the previous investigation. The mean compressive strength of cylindrical samples is equal to 20.5 $\pm$ 1 MPa while the median is equal to 19.7 MPa. The strengths of normal-weight concrete determined on cored specimens with a diameter of 100 mm have no different from those for standardized cube specimens with a 150 mm side length [39] as opposed to lightweight aggregate concrete [40]. It is worth noting that concrete tends to behave as a homogeneous material as long as the sample size is a multiple of the maximum aggregate size, which implies for the diameter of concrete cores to be at least three times larger than the maximum aggregate size. The strength results determined for the concrete cores $f_{ck,is,cycl\ 100}$ according to standard EN 12504-1 [32] are identical to the cube strength of 15 × 15 × 15 cm concrete specimens, thus $f_{ck,is,cube} = f_{ck,is,\ cycl\ 100} = 20.5 \pm 1$ MPa. The mean compressive strength of old concrete cylindrical samples slightly exceeds the value presented by Ambroziak et al. [30] (18.8 ± 0.7 MPa) in early investigations.

**Table 1.** Concrete compressive strength and dry density.

| Specimens No. | Compressive Strength MPa | Dry Density kg/m$^3$ |
|---|---|---|
| Results Obtained by Ambroziak et al. [30] | 14.9 | 2166 |
| | 20.0 | 2171 |
| | 19.1 | 2132 |
| | 18.6 | 2184 |
| | 22.0 | 2155 |
| | 19.9 | 2064 |
| | 19.5 | 2180 |
| | 16.1 | 2174 |
| | 18.9 | 2147 |
| 1_1 | 24.3 | 2157 |
| 1_2 | 21.6 | 2186 |
| 1_3 | 24.5 | 2207 |
| 2_1 | 17.1 | 2173 |
| 2_2 | 25.3 | 2170 |
| 3_1 | 29.7 | 2193 |
| 3_2 | 17.0 | 2179 |
| mean | 20.5 ± 1 | 2165 ± 8 |
| median | 19.7 | 2172.1 |

The concrete had large variations in compressive strength ranging from 14.9 MPa to 29.7 MPa, see Table 1. To properly perform the structural analysis of old concrete structures, it is necessary to evaluate the old concrete compressive strength. The characteristic in-situ compressive cube strength $f_{ck,is,cube}$ according to EN 13791 [41] standard can be determined as:

$$f_{ck,is,cube} = \min\left\{ \begin{array}{c} f_{m(n),is} - k_n \cdot s \\ f_{is,lowest} + M \end{array} \right\} = \min\left\{ \begin{array}{c} 20.5 - 1.81 \cdot 1 \\ 14.9 + 2 \end{array} \right\} = 16.9 \text{ MPa} \quad (1)$$

where $f_{m(n),is}$ is the mean in-situ compressive strength of $n$ = 16 test results, $f_{is,lowest}$ is the lowest in-situ compressive strength test results, $k_n$ is the factor depends on the number of tests results ($k_n$ = 1.81 for tests results equal to 16, see EN 13791 [41], $s$ = 1 MPa is the standard deviation of in situ compressive strength, $M$ = 2 MPa is the value of margin depend on value of $f_{is,lowest}$ (see EN 13791 [41], 12 MPa $\geq f_{is,lowest} <$ 16 MPa). Ambroziak et al. [30] in their earlier research for old concrete set the same value of the characteristic in-situ compressive cube strength equal to 16.9 MPa. The decisive condition for determining the characteristic in-situ compressive cube strength is governed by the value of the lowest in-situ compressive strength. The differences in single cylindrical compressive strength exhibit a non-homogenous distribution of concrete strength in the old concrete arch bridge.

High compressive strength variation among individual concrete specimens is produced by impurities that were identified after uniaxial compressive tests, see Figure 6a–f. Parts of timber, piece of clay, coarse aggregates (large stones) with cavities and pores were detected in some concrete cores. The maximum aggregate size used in the old concrete mix is up to 20 mm. In a single individual case, the maximum aggregate size was up to 50 mm (see Figure 6e). Additionally, a wide scatter in compressive strength may be affected by the proportion of cement and aggregate (sand to gravel volumetric ratios) for the old concrete mix preparation. Concrete strength may also be affected by different climatic conditions in the course of placement [42]. The specimens were crushed or got separated along a slanted surface and columnar vertical cracking through both ends with no well-formed cones was observed. Generally, the failure mode of core specimens (see Figure 6a,b) was typical and fulfills requirements guidelines in EN 12390-3 [33] standard.

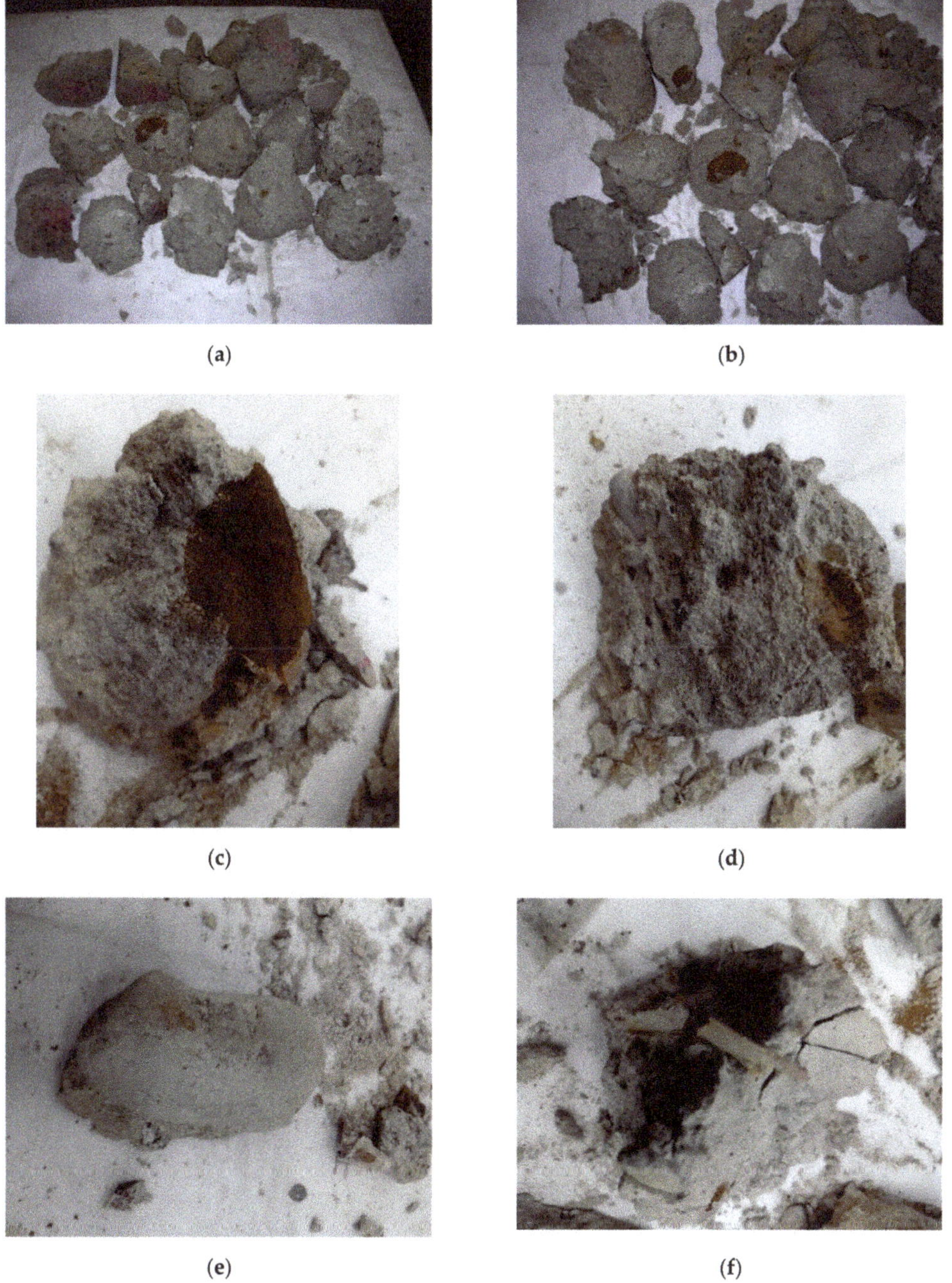

**Figure 6.** Concrete specimens after uniaxial compressive tests: (**a**,**b**) Form of specimens failure; (**c**) dirty fine aggregate; (**d**) clay inclusion; (**e**) large stone; (**f**) timber inclusions.

The reinforced concrete structural guideline [43] issued in January 1916 by the German Committee for Structural Concrete specified two main concrete strength classes 14.7 MPa (150 kg/cm$^2$) and 17.7 MPa (180 kg/cm$^2$) for the erection of concrete structures. This guideline has been applicable till September 1925, i.e., while German standard DIN 1045 [44] was introduced. On the other hand, while the 15–18 MPa compressive strength is required the Hennebique recommends a mixture consisting either of a single part cement, two parts sand and four parts gravel or of one part cement, three parts sand and five parts of gravel [45]. The Regulations on the Construction and Maintenance of Road Bridges [36] made it possible to forecast the cube compressive strength of concrete related to the amount of cement to 1 m$^3$ aggregate in concrete mixes, see Table 2. According to these regulations the amount of water in the concrete mix should be appropriate to locate the mixed concrete in the formwork, or to hand-knead the compacted concrete ball. In the hand-mixing case the amount of cement should be increased by 5%, while consistency of a liquid mix is regarded, the 10% increment is anticipated.

**Table 2.** Concrete strength depending on the amount of cement in 1 m$^3$ aggregate (according to guidelines given in [36]).

| The Amount of Cement (kg) in 1 m$^3$ of Aggregate | Forecast Compressive Strength MPa (kg/cm$^2$) |
|---|---|
| 500 | 19.6 (200) |
| 400 | 16.7 (170) |
| 300 | 13.7 (140) |
| 200 | 9.8 (100) |
| 100 | 5.9 (60) |

The Polish standard PN-B-195 [37] specified the forecast compressive strength of concrete with regard to the amount of cement in a 1 m$^3$ concrete mix, the volume ratio of aggregates and the consistency of the ready concrete mix, see Table 3.

**Table 3.** Concrete strength in MPa (kg/cm$^2$) depending on the amount of cement in 1 m$^3$ of ready concrete on the degree of liquidity and the volume ratio of aggregate (according to guidelines given in [37]).

| Volume Ratios | The Amount of Cement (kg) in 1 m$^3$ of Concrete Mix | Liquid Consistency | Plastic Consistency | Rammed Consistency |
|---|---|---|---|---|
| Sand to gravel 1:1 or sand to stone gravel 1:0.8 | 200 | 0 (0) | 2.9 (30) | 5.9 (60) |
| | 300 | 4.9 (50) | 8.8 (90) | 11.8 (120) |
| | 400 | 9.8 (100) | 13.7 (140) | 15.7 (160) |
| Sand to gravel 1:2 or sand to stone gravel 1:1.6 | 200 | 3.9 (40) | 8.8 (90) | 11.8 (120) |
| | 300 | 9.8 (100) | 13.7 (140) | 15.7 (160) |
| | 400 | 13.7 (140) | 17.7 (180) | 19.6 (200) |

According to the PN-B-195 standard [37] the characteristic strength of concrete, equal 19.6 MPa (200 kg/cm$^2$) may be achieved by use of 400 kg of cement with a 1:2 ratio of sand to gravel parts in 1 m$^3$ of finished concrete of a rammed concrete consistency. The standard PN-B-195 [46] in its early version of 1934 specified the strength equal to 16.7 MPa (170 kg/cm$^2$) with the same amount of cement. The lack of clear and detailed water dosage guidelines produced variable compressive strengths of old concrete structures. The standard PN-B-195 [37] emphasized that the amount of water should be limited in ready concrete mixes of 0 MPa (0 kg/cm$^2$, see Table 3) concrete strength class when the liquid concrete mix consistency is assumed. The water-to-cement ratio is defined in the present standards and guidelines regarding concrete mixes, specifying the proper amount of water in concrete mix for a prescribed concrete strength class.

The method specified in EN 12390-7 [34] standard is applied for determining the dry density of 95-year-old concrete. The tested specimens were dried in a ventilated oven at

105 ± 5 °C until the mass relative decrement reaches 0.2%. Before weighing each specimen was cooled to near room temperature in a dry airtight vessel. The mean dry density value is equal to 2164 ± 9 $kg/m^3$ while the median is equal to 2173.5 $kg/m^3$, see Table 1. According to the EN 206 standard [47] and the ACI 318-19 code [48], the investigated old concrete satisfies the conditions for the normal-weight concrete category.

### 3.3. Modulus of Elasticity and Durability

The secant modulus of elasticity in the range 0 to 40% of the ultimate strength (according to EN 1992-1-1 [49] standard) is assumed 22,890 MPa (mean value of the modulus of elasticity, see Ambroziak et al. [30]) with regard to structural analysis of old arch concrete bridge. The ASTM C469 M standard [50] guideline was used to determine the modulus of elasticity. Diamond-drilled concrete cores with a length to diameter ratio of 1.50 were used in a compressometer device to measure the static modulus of elasticity. The regulations [36], applicable in time of the bridge erection indicated that the concrete compressive strength greater or equal to 13.73 MPa corresponds to the modulus of elasticity 14,715 MPa (150,000 $kg/cm^2$). The modulus of elasticity assumed for investigated old concrete is 1.55 times greater than the value specified in regulations [36], possibly due to various aggregates types in concrete mixes.

Based on the early experimental investigation performed by Ambroziak et al. [30] it can be concluded that the 95-year-old concrete has good freezing resistance. The chloride content of the old concrete did not exceed 0.2% by mass of cement; thus, the old concrete arch bridge was not exposed to chloride attack. The pH values for the old concrete indicated that there is no corrosion of the steel rebar's. Nevertheless, the old concrete has large variations in depth of the carbonated zone ranging from 20 to 55 mm. Despite the large depth of the carbonated zone, the pH of the old concrete is still in the safety range.

## 4. Structural Analysis

### 4.1. Description of FEM Model

The three-dimensional finite element model of the 95-year-old concrete arch bridge is built (see Figure 7), followed by numerical calculations and structural analysis. The SOFiSTiK structural engineering system is applied in numerical calculations. This software is frequently used in the design and analysis of bridge structures [51–55]. In the present structural analysis, the traffic loads class C are assumed, according to standard PN-S-10030 [56]. This standard was applicable in Poland during the reconstruction design process of the old bridge. The traffic load is composed of q and K types of loads, see Figure 8. The load values of q and K type are equal to 2 kPa and 400 kN (8.P = 8.50 kN = 400 kN), respectively. The distribution of q load (arbitrary distribution) and location of K load have to produce the largest responses in analyzed structural elements. Besides this traffic load, permanent loads are also imposed (dead-weight and bridge equipment's loads). The shell (SH3 D) and 3 D beam (B3 D) finite elements (FE) are applied in the FEM model of the arch bridge. The finite elements adopted in this study are 4-node isoparametric shell finite elements of DKMQ type and 2-node 3 D beam elements of the Timoszenko type, C0 class with linear shape functions. The mesh independence study of the bridge finite element model is carried out to ensure that the results of an analysis are not affected by changing the size of the mesh. The FEM model of the old concrete arch bridge is meshed by 45,075 shell elements (modeled slab and concrete arch) and 776 3 D beam elements (girders), the model includes 42,223 nodes with and 940 support constraints (Figure 7). It should be noted, that the authors' experiences in the design and monitoring of bridge structures are also exploited during the construction of the FEM model [57,58]. It can be pointed, that more detailed numerical models are created to investigate the individual structural elements of bridges and comparisons them with laboratory tests [59,60].

(a) (b)

**Figure 7.** FEM model: (**a**,**b**) 3 D views.

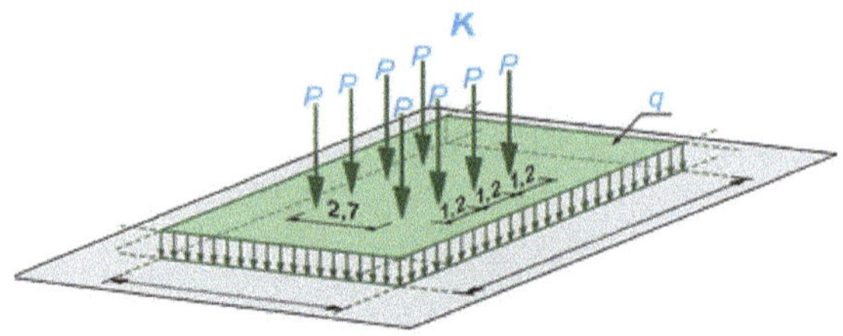


**Figure 8.** Traffic loads class C (q and K types) according to standard PN-S-10030 [56].

### 4.2. Results of Structural Analysis

The numerical analysis allows to evaluate the internal forces acting on the sections and specifically their largest values, which in turns make it possible to define the geometry of the members and later to proceed with the verifications at the Serviceability Limit State (SLS) under the service loads, and at the Ultimate Limit State (ULS) under the ultimate loads. The maps of bending moments in the deck slab and the concrete arch under the dead weight of the structure and loads of the bridge equipment are determined in Figures A1 and A2 respectively. On the other hand, for the moving loads q and K of class C (see Figure 8), the envelope maps of characteristic values of bending moments in the deck slab (see Figure A3) and the envelope maps of characteristic values of forces in the concrete arch (see Figure A4) are presented. In Figures A5 and A6 the characteristic values of bending moments for inner and outer girders are shown. The largest bending moments and axial forces at the chosen, critical cross-sections of a superstructure correspond to the most unfavorable combinations of permanent and live loads according to PN-S-10030 [56], these values are presented in Table 4. The lack of symmetry in the results of internal forces between the right and left span girders results from different spans (see Figure 1). The results for the outer girder and the inner girder are similar in the longer span (right span, see Table 4) where they have a similar height, while in the shorter span (left span) the results differ significantly between the outer girder and inner girder because the girders have different constructional heights.

The site inspection and inventory of old arch bridge make it possible to assess the amount of reinforcement in structural elements, see Table 5. Concrete detector for rebar localization, depth measurement, and size estimation are necessary for the proper estimation of steel reinforcements in the existing old concrete bridge. New technologies, methods, and systems are still developed for the detection of steel rebars in concrete structures [61–64].

**Table 4.** Characteristic and design value of internal forces under designed loads.

| Loads | Right Span | | Left Span | | Slab and Arch | | |
|---|---|---|---|---|---|---|---|
| | Outer Girder | Inner Girder | Outer Girder | Inner Girder | Center Deck Plate | Arch | |
| | $M$ kNm | $M$ kNm | $M$ kNm | $M$ kNm | $M$ kNm | $M$ kNm | $N$ kN |
| L1—characteristic load by dead-weight | 9.8 | 10.0 | 8.4 | 3.7 | 1.4 | 56.3 | 560.4 |
| L2—characteristic load by bridge equipment's | 5.1 | 6.1 | 4.5 | 2.5 | 0.9 | 16.2 | 115.0 |
| L3—characteristic live load, standard load q type | 1.5 | 1.9 | 1.4 | 0.8 | 0.5 | 6.8 | 37.4 |
| L4—characteristic live load, standard load K typewithout the dynamic factor ϕ | 17.0 | 15.3 | 16.0 | 6.5 | 9.2 | 38.7 | 196.8 |
| Sum: L1 ÷ L4 | 33.4 | 33.3 | 30.3 | 13.5 | 12 | 118 | 909.6 |
| dynamic factor ϕ | 1.318 | | 1.324 | | 1.325 | 1.284 | |
| design value with dynamic factor ϕ | 55.1 | 53.8 | 50.4 | 22.3 | 21.7 | 172.6 | 1280.3 |

**Table 5.** Materials characteristic and calculated extreme stresses in concrete and steel rebar's.

| Properties | Left Span | | Right Span | | Slab and Arch | |
|---|---|---|---|---|---|---|
| | Outer Girder | Inner Girder | Outer Girder | Inner Girder | Deck Plate | Arch |
| Steel rebar's (bottom) | 6#20 | | 6#20 | | 14#12/m | 5#12/m |
| Design yield strength of reinforcement | $f_{yk}$= 291 MPa, thus $f_{yd} = \frac{f_{yk}}{\gamma_s} = \frac{291}{1.15} = 257$ MPa | | | | | |
| Design compressive strength of concrete | $f_{ck,is,cube}$= 16.9 MPa, thus $f_{cd} = \alpha_{cc}\frac{f_{ck}}{\gamma_c} = 1.0\frac{0.8\cdot 16.9}{1.4} = 9.7$ MPa | | | | | |
| Determined largest stress in steel rebar's, MPa | 32.7 | 31.9 | 38.1 | 16.8 | 78.6 | 145.6 |
| Determined highest stress in concrete, MPa | 2.2 | 2.2 | 3.0 | 1.3 | 3.3 | 9.3 |

The design yield strength of reinforcement ($f_{yd}$) and design compressive strength of old concrete ($f_{cd}$) are determined according to specified characteristic yield strength of reinforcement ($f_{yk}$), characteristic compressive strength of concrete ($f_{ck}$), partial safety factors for reinforcement ($\gamma_s$) and concrete ($\gamma_c$) material properties. The partial safety factors $\gamma_s$ and $\gamma_c$ are assumed according to Table NA.2 guidelines in National Annex to PN-EN 1992-1-1 [65]. The characteristic yield strength of steel rebars and the characteristic compressive strength of concrete are specified in laboratory tests.

The largest stresses in reinforced steel and old concrete in representative structural elements of the arch bridge are determined and collected in Table 5. The highest stress in steel rebar's equals 145.6 MPa, approximately equal to 57% of the design yield strength of reinforcement, $f_{yd}$ = 257 MPa, see Table 5. The specified maximal compressive stress in old concrete arch bridge is 9.3 MPa, approximately equal to 96% of the design compressive strength of concrete $f_{cd}$ = 9.7 MPa.

The largest displacement under the SLS combination of the traffic loadings (see Figure A7) is equal to about 1.3 mm. According to the assumptions of PN-S-10042 [66] standard, the permissible deflection of reinforced concrete beam elements in continuous systems (in this case the largest spans of the analysed bridge, see Figures 1a and A7) is $f_{perm} = L/1000 = (6380 + 2\cdot 250)/1000 = 6.88$ mm. Determined largest deflections fulfils the SLS condition $f_{max} = 1.32$ mm $< f_{perm} = 6.88$ mm.

Structural analysis indicated that the 95-year old concrete arch bridge structure satisfies SLS and ULS requirements under the traffic loads class C according to PN-S-10030 [56] standard. Sometimes the car traffic on old bridges is not permitted and after reconstructions are able to meet the requirements for footbridges only [67]. The structural analysis confirmed the validity of the adopted design solutions. On top of the existing old deck structure, a new deck slab with pavement covers is designed and the entire surface of the old concrete is protected with repair mortars. The task of the new deck slab is not only to strengthen the structure but also to increase the resistance to present environmental influences and also the shear capacity of the deck slab. The higher strength class of concrete is designed and used to fulfill the EN 206 [47] standard requirement of the present exposure classes.

All newly erected bridges and old rebuilt bridges need a detailed assessment and dynamic analysis. In this paper, this analysis is not included but the dynamic analysis is performed to properly assess the properties of the 95-year-old arch concrete bridge. It should be noted, that improper design of bridge structure cause the risk of excessive structural vibrations throughout the operation [68–73].

## 5. Conclusions

A 95 year-old concrete arch bridge in Jagodnik (northern Poland) is examined in this paper, from the properties of the materials—by testing samples in a lab—to the structural behavior—by finite-element modeling, with the following results:

- The investigated core samples of the 95-year-old concrete exhibit a high compressive strength scatter from 14.9 MPa to 29.7 MPa, see Table 1. These results serve as evidence of the very non-homogenous distribution of concrete strength in the old concrete arch bridge.
- The concrete specimens tested in compression show that there is a large number of impurities in the old concrete (i.e., pieces of timber and clay, coarse aggregates, large stones) with cavities and soft pockets. The scatter in the mechanical properties is, therefore, due to either very low quality-control in the building phase or very poor concrete technology based on portable concrete mixers and hand-made proportioning of concrete constituents.
- The mean dry density of the original concrete equals 2165 $\pm$ 8 kg/m$^3$, see Table 1. The old concrete satisfies the conditions for normal-weight concrete category according to the EN 206 standard [47] and the ACI 318-19 code [48].
- The in-situ compressive cube strength $f_{ck,is,cube}$ equals 16.9 MPa, according to EN 13791 [41] standard. Calculated and applied in structural analysis, the design compressive strength of 95-year-old concrete $f_{cd}$ equals 9.7 MPa.
- The reinforced steel tensile tests confirm that the characteristic yield strength specified in laboratory tests ($f_{yk}$ = 291 $\pm$ 7 MPa) corresponds to the guidelines [36] issued in the time of the 95-year-old arch bridge construction.
- The highest stress in steel rebars of structural cross-section is approximately equal to 57% of the design yield strength of reinforcement, the maximum compressive stress in the original parts of the arch bridge is approximately equal to 96% of the design compressive strength of concrete.
- The largest displacement under the SLS combination of the traffic loadings satisfies the SLS condition according to o PN-S-10042 [66] standard. The determined largest deflections are five times less than the permissible deflection.
- The structural analysis confirmed the validity of the adopted design solutions. On top of the existing old deck structure, a new deck slab with pavement covers is designed and the entire surface of the old concrete is protected with repair mortars. The task of the new deck slab is not only to strengthen the structure but also to increase the resistance to present environmental influences and also the shear capacity of the deck slab. The higher strength class of concrete is designed and used to fulfill the EN 206 [47] standard requirement of the present exposure classes.
- The 95-year-old arch bridge satisfies bearing capacity (ULS) and SLS conditions, capable of carrying designed traffic loads class C according to PN-S-10030 [56] in extended working life.

This paper provides scientists, engineers, and designers with experimental and structural assessments of the 95-year-old concrete. The study confirms the old arch bridge ability to carry newly designed traffic loads and helped give him a new structural life and extend his working life. Lots of existed old and historic bridge structures require restoration and reconstruction, which require not only a sound financial plan and the expertise of civil engineers, but—more often than not—the expertise of the scientific community, for an appropriate assessment of structural safety. Old bridge structures should return like the Phoenix from their ashes and recovery their past appearance. The protection of old bridges coincides with the preservation of their cultural heritage.

**Author Contributions:** Conceptualization, A.A.; methodology, A.A. and M.M.; validation, A.A. and M.M.; formal analysis, A.A. and M.M.; investigation, A.A. and M.M.; resources, A.A. and M.M.; data curation, A.A. and M.M.; writing—original draft preparation, A.A. and M.M.; writing—review and editing, A.A.; visualization, A.A.; supervision, A.A.; project administration, A.A. All authors have read and agreed to the published version of the manuscript.

**Funding:** This research received no external funding.

**Institutional Review Board Statement:** Not applicable.

**Informed Consent Statement:** Not applicable.

**Data Availability Statement:** All laboratory test results (data) are included in Table 1 in the present paper.

**Conflicts of Interest:** The authors declare no conflict of interest.

## Appendix A FEM Calculation Results

The chosen results of internal forces used in the verification and design process (see Table 4) are presented in Figures A1–A6. In Figure A7 extreme displacement under the SLS combination of traffic loadings is given.

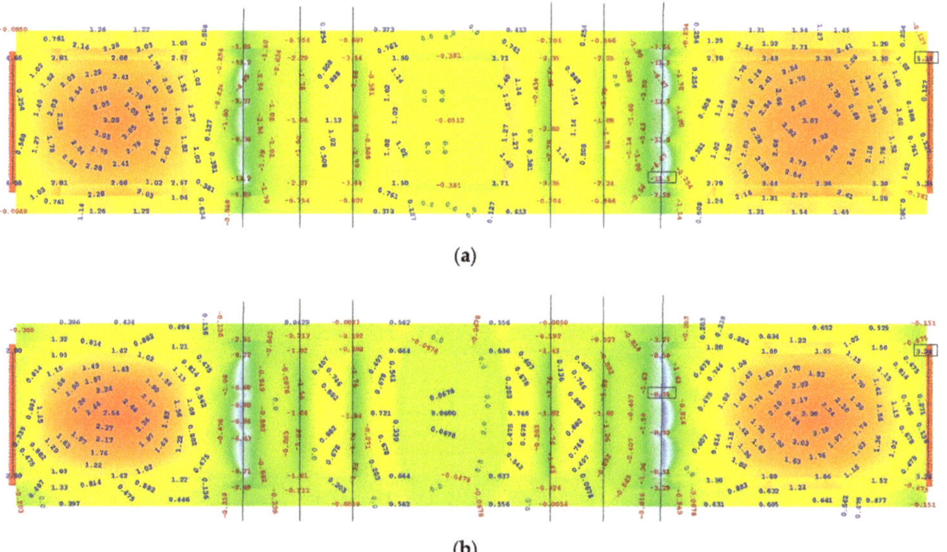

(a)

(b)

**Figure A1.** Maps of characteristic values of bending moments *M*xx (kNm) in the deck slab: (**a**) Dead weight; (**b**) bridge equipment loads.

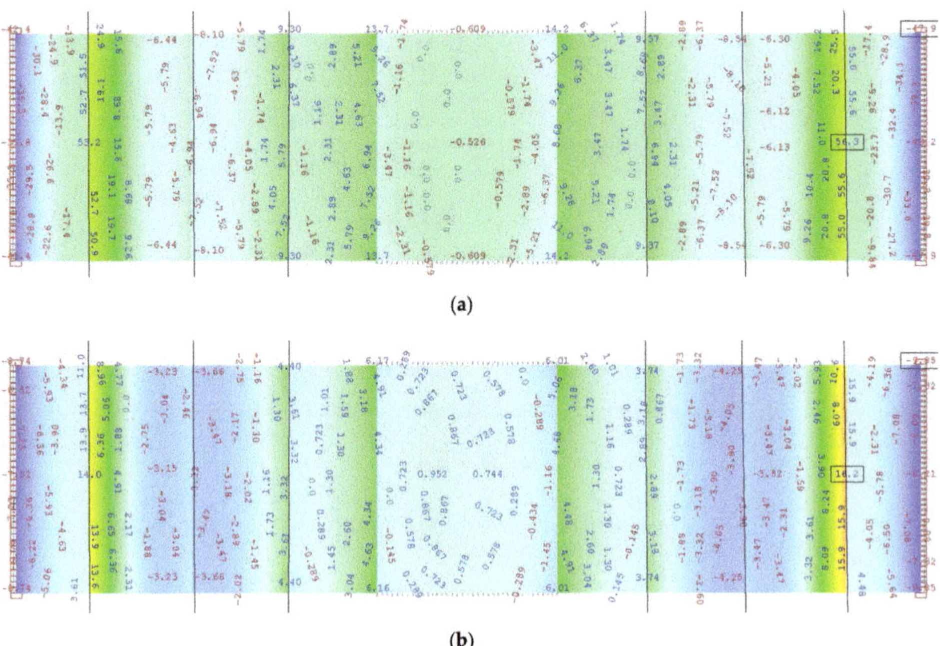

Figure A2. Maps of characteristic values of bending moments $Mxx$ (kNm) in the concrete arch: (a) Dead weight; (b) bridge equipment loads.

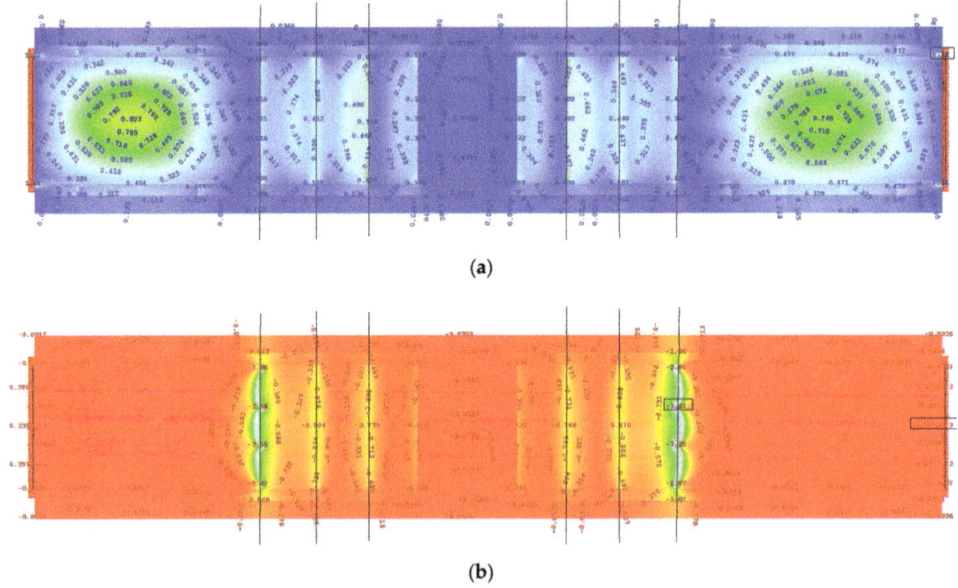

Figure A3. Cont.

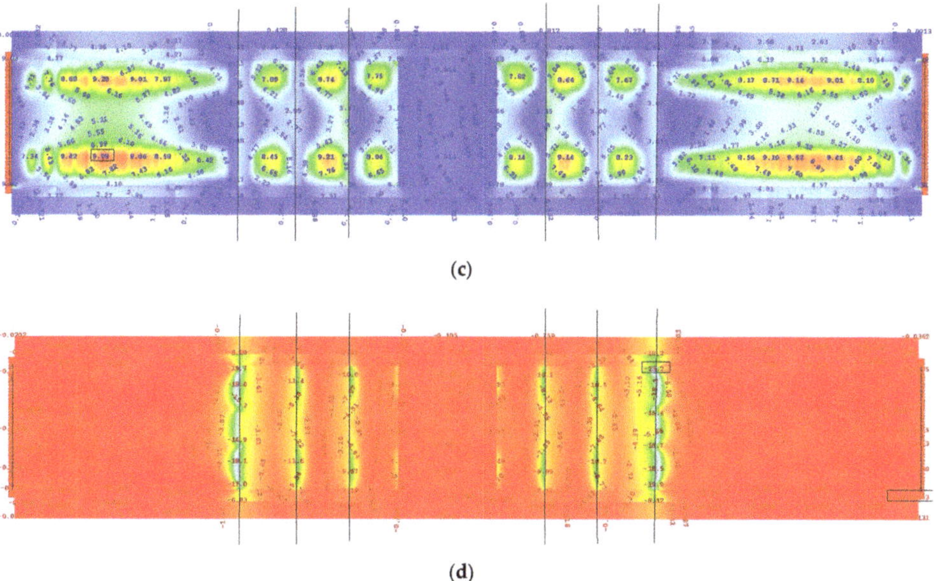

(c)

(d)

**Figure A3.** The envelope maps of characteristic values of bending moments *M*xx (kNm) in the deck plate: (**a**) Max Mxx (kNm) under the moving load q of class C; (**b**) min Mxx (kNm) under the moving load q of class C; (**c**) max Mxx (kNm) under the moving load K of class C; (**d**) min Mxx (kNm) under the moving load K of class C.

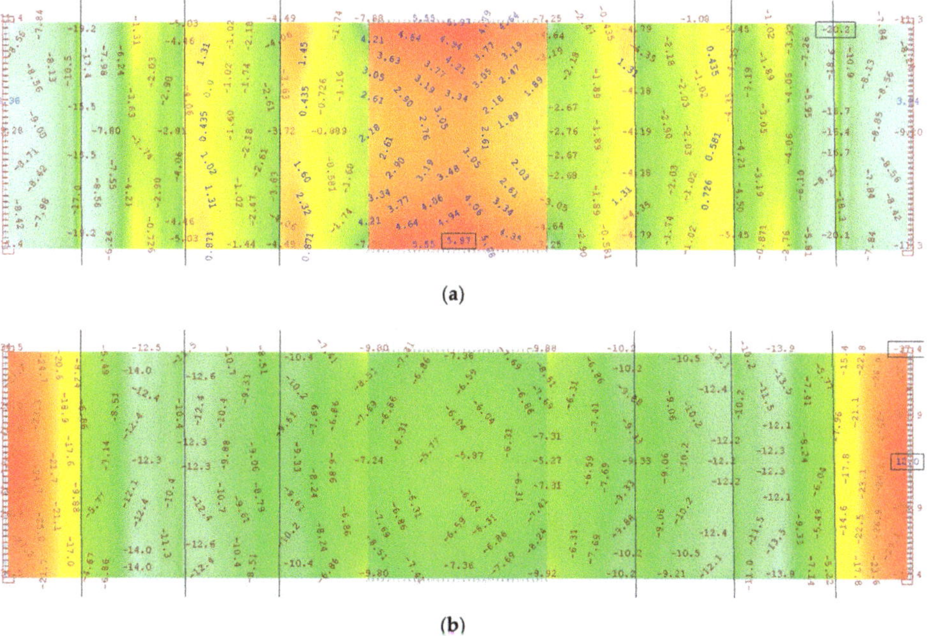

(a)

(b)

**Figure A4.** *Cont.*

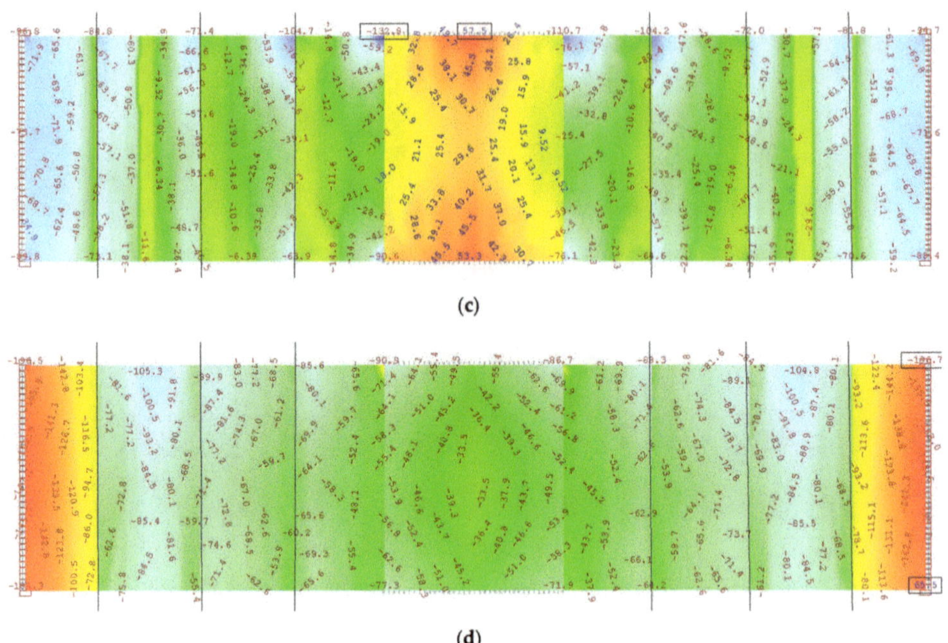

**(c)**

**(d)**

**Figure A4.** Envelope maps of characteristic values of forces *N*xx (kN) in the concrete arch: (**a**) Max Nxx (kN) under the moving load q of class C; (**b**) min Nxx (kN) under the moving load q of class C; (**c**) max Nxx (kN) under the moving load K of class C; (**d**) min Nxx (kN) under the moving load K of class C.

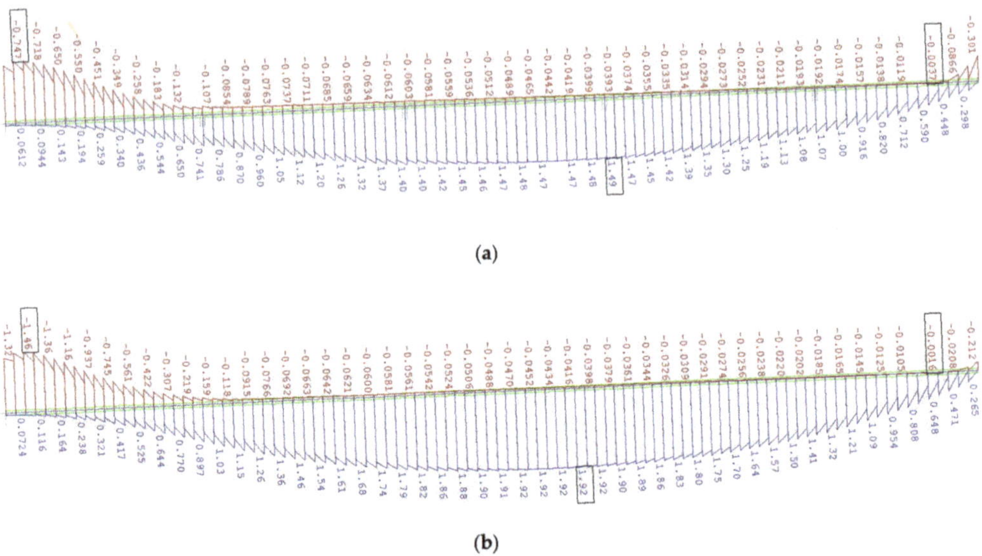

**(a)**

**(b)**

**Figure A5.** Characteristic values of bending moments *M* (kNm) under the moving load q of class C in the right span: (**a**) Outer girder; (**b**) inner girder.

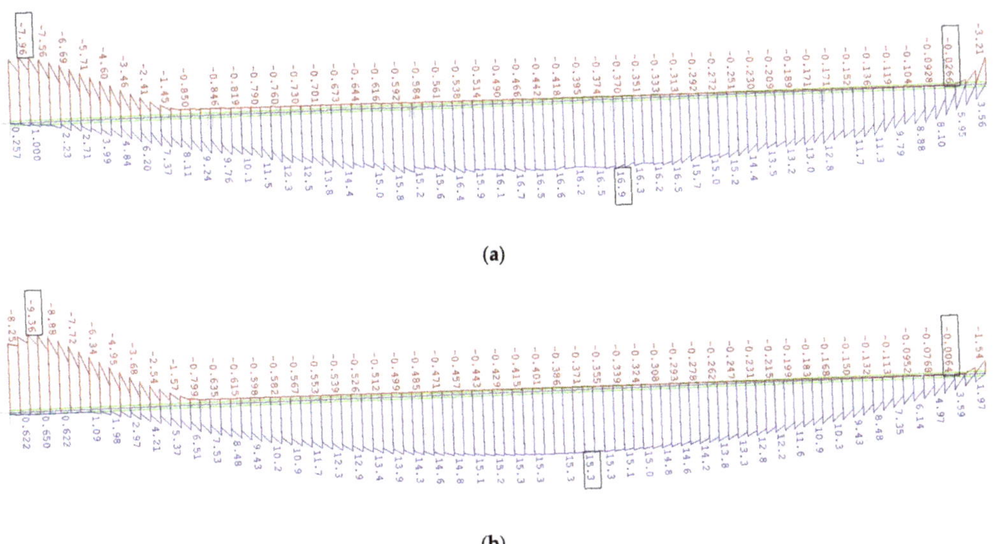

(a)

(b)

**Figure A6.** Characteristic values of bending moments $M$ (kNm) under the moving load K of class C in the right span: (**a**) Outer girder; (**b**) inner girder.

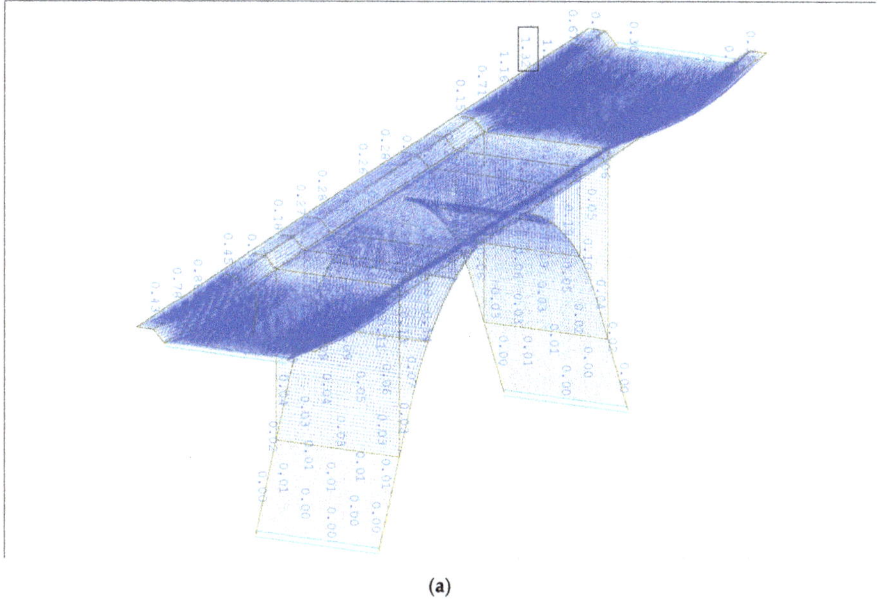

(a)

**Figure A7.** *Cont.*

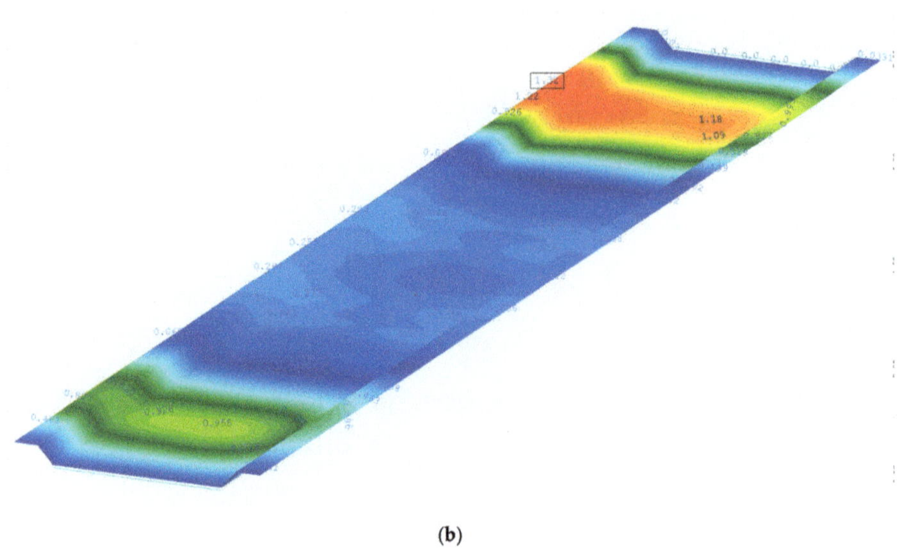

(**b**)

**Figure A7.** Extreme displacement *f* (mm) under the SLS combination of traffic loadings class C according to PN-S-10030 [56] standard: (**a**) 3 D structure view; (**b**) displacement map on the deck slab.

## References


1. Chen, B.; Su, J.; Lin, S.; Chen, G.; Zhuang, Y.; Tabatabai, H. Development and application of concrete arch bridges in China. *J. Asian Concr. Fed.* **2017**, *3*, 12–19. [CrossRef]
2. Radić, J.; Kindij, A.; Mandić, A. History of Concrete Application in Development of Concrete and Hybriad Arch Bridges. In *Proceedings of the Chinese-Croatian Joint Colloquium on Long Arch Bridges, Brijuni Islands, Croatia, 10–14 July 2008*; Radić, J., Chen, B., Eds.; SECON HDGK: Zagreb, Croatia, 2008; Volume LONG ARCH, pp. 9–118.
3. Omar, T.; Nehdi, M. Condition Assessment of Reinforced Concrete Bridges: Current Practice and Research Challenges. *Infrastructures* **2018**, *3*, 36. [CrossRef]
4. Paulík, P.; Bačuvčík, M.; Brodňan, M.; Koteš, P.; Vičan, J. Reconstruction of the Oldest Reinforced Concrete Bridge in Slovakia in Krásno nad Kysucou. *Procedia Eng.* **2016**, *156*, 334–339. [CrossRef]
5. Witzany, J.; Zigler, R. Rehabilitation Design of a Historic Concrete Arch Bridge in Prague from the Early 20th Century. *J. Perform. Constr. Facil.* **2020**, *34*, 04020056. [CrossRef]
6. Hellebois, A.; Espion, B. Concrete properties of a 1904 Hennebique reinforced concrete viaduct. *WIT Trans. Built Environ.* **2011**, *118*, 589–600. [CrossRef]
7. Sena-Cruz, J.; Ferreira, R.M.; Ramos, L.F.; Fernandes, F.; Miranda, T.; Castro, F. Luiz bandeira bridge: Assessment of a historical reinforced concrete (RC) bridge. *Int. J. Archit. Herit.* **2013**, *7*, 628–652. [CrossRef]
8. Wolert, P.J.; Kolodziejczyk, M.K.; Stallings, J.M.; Nowak, A.S. Non-destructive Testing of a 100-Year-Old Reinforced Concrete Flat Slab Bridge. *Front. Built Environ.* **2020**, *6*, 1–12. [CrossRef]
9. Czaderski, C.; Motavalli, M. 40-Year-old full-scale concrete bridge girder strengthened with prestressed CFRP plates anchored using gradient method. *Compos. Part B Eng.* **2007**, *38*, 878–886. [CrossRef]
10. Pettigrew, C.S.; Barr, P.J.; Maguire, M.; Halling, M.W. Behavior of 48-Year-Old Double-Tee Bridge Girders Made with Lightweight Concrete. *J. Bridg. Eng.* **2016**, *21*, 04016054. [CrossRef]
11. Gebauer, J.; Harnik, A.B. Microstructure and composition of the hydrated cement paste of an 84 year old concrete bridge construction. *Cem. Concr. Res.* **1975**, *5*, 163–169. [CrossRef]
12. Papé, T.M.; Melchers, R.E. The effects of corrosion on 45-year-old pre-stressed concrete bridge beams. *Struct. Infrastruct. Eng.* **2011**, *7*, 101–108. [CrossRef]
13. Dasar, A.; Hamada, H.; Sagawa, Y.; Yamamoto, D. Deterioration progress and performance reduction of 40-year-old reinforced concrete beams in natural corrosion environments. *Constr. Build. Mater.* **2017**, *149*, 690–704. [CrossRef]
14. Khan, I.; François, R.; Castel, A. Structural performance of a 26-year-old corroded reinforced concrete beam. *Eur. J. Environ. Civ. Eng.* **2012**, *16*, 440–449. [CrossRef]
15. Shu, J.; Bagge, N.; Nilimaa, J. Field Destructive Testing of a Reinforced Concrete Bridge Deck Slab. *J. Bridg. Eng.* **2020**, *25*, 04020067. [CrossRef]
16. Turker, T. Structural evaluation of Aspendos (Belkis) Masonry Bridge. *Struct. Eng. Mech.* **2014**, *50*, 419–439. [CrossRef]

17. Orlando, M.; Spinelli, P.; Vignoli, A. Structural analysis for the reconstruction design of the old bridge of Mostar. *Adv. Archit.* **2003**, *15*, 617–626.
18. Domede, N.; Sellier, A.; Stablon, T. Structural analysis of a multi-span railway masonry bridge combining in situ observations, laboratory tests and damage modelling. *Eng. Struct.* **2013**, *56*, 837–849. [CrossRef]
19. Pipinato, A.; Modena, C. Structural Analysis and Fatigue Reliability Assessment of the Paderno Bridge. *Pract. Period. Struct. Des. Constr.* **2010**, *15*, 109–124. [CrossRef]
20. Cardinale, G.; Orlando, M. Structural Evaluation and Strengthening of a Reinforced Concrete Bridge. *J. Bridg. Eng.* **2004**, *9*, 35–42. [CrossRef]
21. Richard, B.; Epaillard, S.; Cremona, C.; Elfgren, L.; Adelaide, L. Nonlinear finite element analysis of a 50 years old reinforced concrete trough bridge. *Eng. Struct.* **2010**, *32*, 3899–3910. [CrossRef]
22. Zhang, J.; Li, C.; Xu, F.; Yu, X. Test and Analysis for Ultimate Load-Carrying Capacity of Existing Reinforced Concrete Arch Ribs. *J. Bridg. Eng.* **2007**, *12*, 4–12. [CrossRef]
23. Lubowiecka, I.; Armesto, J.; Arias, P.; Lorenzo, H. Historic bridge modelling using laser scanning, ground penetrating radar and finite element methods in the context of structural dynamics. *Eng. Struct.* **2009**, *31*, 2667–2676. [CrossRef]
24. Lõhmus, H.; Ellmann, A.; Märdla, S.; Idnurm, S. Terrestrial laser scanning for the monitoring of bridge load tests—Two case studies. *Surv. Rev.* **2018**, *50*, 270–284. [CrossRef]
25. Valença, J.; Puente, I.; Júlio, E.; González-Jorge, H.; Arias-Sánchez, P. Assessment of cracks on concrete bridges using image processing supported by laser scanning survey. *Constr. Build. Mater.* **2017**, *146*, 668–678. [CrossRef]
26. Riveiro, B.; González-Jorge, H.; Varela, M.; Jauregui, D.V. Validation of terrestrial laser scanning and photogrammetry techniques for the measurement of vertical underclearance and beam geometry in structural inspection of bridges. *Measurement* **2013**, *46*, 784–794. [CrossRef]
27. Binczyk, M.; Kalitowski, P.; Szulwic, J.; Tysiac, P. Nondestructive Testing of the Miter Gates Using Various Measurement Methods. *Sensors* **2020**, *20*, 1749. [CrossRef] [PubMed]
28. Ziolkowski, P.; Szulwic, J.; Miskiewicz, M. Deformation Analysis of a Composite Bridge during Proof Loading Using Point Cloud Processing. *Sensors* **2018**, *18*, 4332. [CrossRef] [PubMed]
29. Rashidi, M.; Mohammadi, M.; Sadeghlou Kivi, S.; Abdolvand, M.M.; Truong-Hong, L.; Samali, B. A Decade of Modern Bridge Monitoring Using Terrestrial Laser Scanning: Review and Future Directions. *Remote Sens.* **2020**, *12*, 3796. [CrossRef]
30. Ambroziak, A.; Haustein, E.; Niedostatkiewicz, M. Chemical, Physical, and Mechanical Properties of 95-Year-Old Concrete Built-In Arch Bridge. *Materials* **2020**, *14*, 20. [CrossRef]
31. Lockemann, T. *Elbing*; Deutscher Architectur und Industrie—Verlag: Berlin-Halensee, Germany, 1926.
32. *EN 12504-1 Testing Concrete in Structures—Part 1: Cored Specimens—Taking, Examining and Testing in Compression*; CEN (European Committee for Standardization): Brussels, Belgium, 2009.
33. EN 12390-3 Testing Hardened Concrete. *Compressive Strength of Test Specimens*; CEN (European Committee for Standardization): Brussels, Belgium, 2019.
34. *EN 12390-7 Testing Hardened Concrete—Part 7: Density of Hardened Concrete*; CEN (European Committee for Standardization): Brussels, Belgium, 2019.
35. *ISO (International Organization for Standardization) ISO 6892-1 Metallic Materials—Tensile Testing—Part 1: Method of Test at Room Temperature*; ISO: Geneva, Switzerland, 2019.
36. *Rules for the Construction and Maintenance of Road Bridges (Przepisy o Budowie i Utrzymaniu Mostów Drogowych)*; Pomorska Drukarnia Rolnicza, S.A.: Warszawa, Poland, 1926.
37. PN-B-195 Concrete and Reinforced Concrete Structures. *Structural Analysis and Design*; PKN (Polish Committee for Standardization): Warsaw, Poland, 1945.
38. Ambroziak, A. Application of elasto-viscoplastic Bodner-Partom constitutive equations in finite element analysis. *Comput. Assist. Mech. Eng. Sci.* **2007**, *14*, 405–429.
39. Brunarski, L.; Dohojda, M. An approach to in-situ compressive strength of concrete. *Bull. Polish Acad. Sci. Tech. Sci.* **2016**, *64*, 687–695. [CrossRef]
40. Domagała, L. Size Effect in Compressive Strength Tests of Cored Specimens of Lightweight Aggregate Concrete. *Materials* **2020**, *13*, 1187. [CrossRef]
41. *EN 13791 Assessment of In-Situ Compressive Strength in Tructures and Precast Concrete Components*; CEN (European Committee for Standardization): Brussels, Belgium, 2019.
42. Ambroziak, A.; Ziolkowski, P. Concrete compressive strength under changing environmental conditions during placement processes. *Materials* **2020**, *13*, 4577. [CrossRef]
43. *Bestimmungen für die Ausführung von Bauwerken aus Eisenbeton*; DAfStB (Deutscher Ausschuss für Stahlbeton): Berlin, Germany, 1916.
44. Deutsche Institut für Normung. *DIN 1045:1925-09 Bestimmungen für Ausführung von Bauwerken aus Eisenbeton*; Deutsche Institut für Normung: Berlin, Germany, 1925.
45. Hallauer, O. Die Entwicklung der Zusammensetzung von Beton für Wasserbauten. *Mtteilungsblaetter Bundesanstalt für Wasserbaut.* **1989**, *1988*, 39–56.

46. PN-B-195 Concrete and Reinforced Concrete Structures. *Structural Analysis and Design*; PKN (Polish Committee for Standardization): Warsaw, Poland, 1934.
47. *EN 206:2013 + A1:2016 Concrete—Specification, Performance, Production and Conformity*; CEN (European Committee for Standardization): Brussels, Belgium, 2016.
48. *ACI 318-19 Building Code Requirements for Structural Concrete*; ACI (American Concrete Institute): Farmington Hills, MI, USA, 2019; ISBN 9781641950565.
49. *EN 1992-1-1 Eurocode 2: Design of Concrete Structures—Part 1-1: General Rules and Rules for Buildings*; CEN (European Committee for Standardization): Brussels, Belgium, 2004.
50. *ASTM C469M-14 Standard Test Method for Static Modulus of Elasticity and Poisson's Ratio of Concrete in Compression*; ASTM International (American Society for Testing and Materials): West Conshohocken, PA, USA, 2014.
51. Żółtowski, K.J. Bridge over Vistula river in Kiezmark. Practical application of nonlinear shell FEM system. In *Shell Structures Theory and Applications*; Pietraszkiewicz, W., Szymczak, C., Eds.; Taylor & Francis Group: Abingdon, UK, 2005; pp. 521–525.
52. Siwowski, T.; Rajchel, M.; Kulpa, M. Design and field evaluation of a hybrid FRP composite—Lightweight concrete road bridge. *Compos. Struct.* **2019**, *230*, 111504. [CrossRef]
53. Siwowski, T.; Wysocki, A. Horizontal Rotation via Floatation as an Accelerated Bridge Construction for Long-Span Footbridge Erection: Case Study. *J. Bridg. Eng.* **2015**, *20*, 05014014. [CrossRef]
54. Erdenebat, D.; Waldmann, D. Application of the DAD method for damage localisation on an existing bridge structure using close-range UAV photogrammetry. *Eng. Struct.* **2020**, *218*, 110727. [CrossRef]
55. Skokandić, D.; Mandić Ivanković, A. Value of additional traffic data in the context of bridge service-life management. *Struct. Infrastruct. Eng.* **2020**, 1–20. [CrossRef]
56. PN-S-10030 Bridges. *Loads*; PKN (Polish Committee for Standardization): Warsaw, Poland, 1985.
57. Malinowski, M.; Banas, A.; Jeszka, M.; Sitarski, A. Imaginative footbridge in Mikolajki, Poland. *Stahlbau* **2018**, *87*, 248–255. [CrossRef]
58. Malinowski, M.; Banas, A.; Cywiński, Z.; Jeszka, M.; Sitarski, A. Zur Wiedergeburt einer historischen Gitterbrücke. *Stahlbau* **2017**, *86*, 789–796. [CrossRef]
59. Imperatore, S.; Lavorato, D.; Nuti, C.; Santini, S.; Sguerri, L. Numerical Modeling of Existing RC Beams Strengthened in Shear with FRP U-sheets. In Proceedings of the 6th International Conference on FRP, Composites in Civil Engineering, CICE 2012, Rome, Italy, 13–15 June 2012; pp. 1–8.
60. El-Gamal, S.E. Finite Element Analysis of Concrete Bridge Slabs Reinforced with Fiber Reinforced Polymer Bars. *J. Eng. Res.* **2014**, *11*, 50. [CrossRef]
61. Ahmed, H.; La, H.M.; Tran, K. Rebar detection and localization for bridge deck inspection and evaluation using deep residual networks. *Autom. Constr.* **2020**, *120*, 103393. [CrossRef]
62. Erdogmus, E.; Garcia, E.; Amiri, A.S.; Schuller, M. A Novel Structural Health Monitoring Method for Reinforced Concrete Bridge Decks Using Ultrasonic Guided Waves. *Infrastructures* **2020**, *5*, 49. [CrossRef]
63. Asadi, P.; Gindy, M.; Alvarez, M. A Machine Learning Based Approach for Automatic Rebar Detection and Quantification of Deterioration in Concrete Bridge Deck Ground Penetrating Radar B-scan Images. *KSCE J. Civ. Eng.* **2019**, *23*, 2618–2627. [CrossRef]
64. Kobaka, J.; Katzer, J.; Ponikiewski, T. A Combined Electromagnetic Induction and Radar-Based Test for Quality Control of Steel Fibre Reinforced Concrete. *Materials* **2019**, *12*, 3507. [CrossRef] [PubMed]
65. *PN-EN 1992-1-1 Eurocode 2: Design of Concrete Structures—Part 1-1: General Rules and Rules for Buildings*; PKN (Polish Committee for Standardization): Warsaw, Poland, 2008.
66. *PN-S-10042 Bridges -Concrete, Reinforced Concrete and Prestressed Concrete Structures—Design*; PKN (Polish Committee for Standardization): Warsaw, Poland, 1991.
67. Karas, S. Unique Hennebique Bridges in Lublin, Poland. *Am. J. Civ. Eng. Archit.* **2013**, *1*, 47–51. [CrossRef]
68. Banas, A.; Jankowski, R. Experimental and Numerical Study on Dynamics of Two Footbridges with Different Shapes of Girders. *Appl. Sci.* **2020**, *10*, 4505. [CrossRef]
69. Nowak, A.S.; Eom, J.; Sanli, A. Control of Live Load on Bridges. *Transp. Res. Rec. J. Transp. Res. Board* **2000**, *1696*, 136–143. [CrossRef]
70. Kilikevičius, A.; Bačinskas, D.; Jurevičius, M.; Kilikevičienė, K.; Fursenko, A.; Jakaitis, J.; Toločka, E. Field testing and dynamic analysis of old continuous truss steel bridge. *Balt. J. Road Bridg. Eng.* **2018**, *13*, 54–66. [CrossRef]
71. Podworna, M.; Klasztorny, M. Vertical vibrations of composite bridge/track structure/high-speed train systems. Part 3: Deterministic and random vibrations of exemplary system. *Bull. Polish Acad. Sci. Tech. Sci.* **2014**, *62*, 305–320. [CrossRef]
72. Law, S.S.; Zhu, X.Q. Dynamic behavior of damaged concrete bridge structures under moving vehicular loads. *Eng. Struct.* **2004**, *26*, 1279–1293. [CrossRef]
73. Biliszczuk, J.; Barcik, W.; Onysyk, J.; Toczkiewicz, R.; Tukendorf, A. The two largest Polish concrete bridges—Design and construction. *Proc. Inst. Civ. Eng. Bridg. Eng.* **2016**, *169*, 298–308. [CrossRef]

 materials

MDPI

Article

# Activated Ductile CFRP NSMR Strengthening

**Jacob Wittrup Schmidt [1,*], Christian Overgaard Christensen [2], Per Goltermann [2] and José Sena-Cruz [3]**

[1] Department of Built Environment, University of Aalborg, 9220 Aalborg, Denmark
[2] Department of Civil Engineering, Technical University of Denmark, 2800 Kongens Lyngby, Denmark; coch@byg.dtu.dk (C.O.C.); pg@byg.dtu.dk (P.G.)
[3] Department of Civil Engineering, ISISE/IB-S, University of Minho, 4800-058 Guimarães, Portugal; jsena@civil.uminho.pt
* Correspondence: jws@build.aau.dk; Tel.: +45-994-08484

**Abstract:** Significant strengthening of concrete structures can be obtained when using adhesively-bonded carbon fiber-reinforced polymer (CFRP) systems. Challenges related to such strengthening methods are; however, the brittle concrete delamination failure, reduced warning, and the consequent inefficient use of the CFRP. A novel ductile near-surface mounted reinforcement (NSMR) CFRP strengthening system with a high CFRP utilization is introduced in this paper. It is hypothesized that the tailored ductile enclosure wedge (EW) end anchors, in combination with low E-modulus and high elongation adhesive, can provide significant strengthening and ductility control. Five concrete T-beams were strengthened using the novel system with a CFRP rod activation stress of approximately 980 MPa. The beam responses were compared to identical epoxy-bonded NSMR strengthened and un-strengthened beams. The linear elastic response was identical to the epoxy-bonded NSMR strengthened beam. In addition, the average deflection and yielding regimes were improved by 220% and 300% (average values), respectively, with an ultimate capacity comparable to the epoxy-bonded NSMR strengthened beam. Reproducible and predictable strengthening effect seems obtainable, where a good correlation between the results and applied theory was reached. The brittle failure modes were prevented, where concrete compression failure and frontal overload anchor failure were experienced when failure was initiated.

**Keywords:** CFRP; strengthening; ductility; post-tensioning; concrete structures

**Citation:** Schmidt, J.W.; Christensen, C.O.; Goltermann, P.; Sena-Cruz, J. Activated Ductile CFRP NSMR Strengthening. *Materials* **2021**, *14*, 2821. https://doi.org/10.3390/ma14112821

Academic Editor: Eva O. L. Lantsoght

Received: 22 April 2021
Accepted: 19 May 2021
Published: 25 May 2021



## 1. Introduction

Pultruded carbon fiber-reinforced polymers (CFRPs) for near-surface mounted reinforcement (NSMR), as a means to strengthen concrete structures, has been used for several decades (e.g., [1–10]). The strengthening often results in a significant amount of benefits, namely lower additional permanent loads (CFRP is ~80% lighter than steel), possibility of mobilizing higher tensile strength (CFRP systems are approximately five times stronger than steel), excellent durability properties, installation easiness, among others.

However, several additional failure modes, from those already known from un-strengthened concrete structures, may arise when using CFRP strengthening, such as IC (intermediate crack) de-bonding [11–16] and concrete cover separation [17,18]. Consequently, CFRP strengthening is often at the expense of a significantly reduced ductility compared to the un-strengthened concrete structure. These applications are; thus, often challenged by a more general discussion related to lack of failure warning and expected safety margin.

A way to further utilize the CFRP as well as increase the ductility of the strengthened elements may be based on the use of more ductile anchorage zones along the CFRP laminate [19–22]. However, these systems are more related to CFRP plate and sheet systems. Similar for these systems is; however, still the low magnitude of ductility provided by the systems compared to the un-strengthened reference beam. In addition, limited solutions are commercially available in the field of CFRP plate and sheet strengthening and seems even less available for CFRP rod strengthening systems [23,24].

CFRP rods can experience different types of failure modes when mechanically anchored for NSMR CFRP strengthening systems [25–27]. These failures are due to the anisotropic nature of the CFRP material, which provides weaker properties in the transverse direction of the fibers as well as brittleness. Premature mechanical anchorage failure modes, due to crushing of the CFRP, soft slip, power slip, cutting of the fibers, bending of fibers, frontal overload, and fiber failure [27], are thus often experienced. Fiber failure and power slip are typically reached of higher load magnitudes. Control of the stress development along the anchorage length seems to be the key aspect to increase utilization of the CFRP [21].

Mechanical anchoring and activation of a CFRP NSMR system in a bonded configuration have shown to provide high material utilization (e.g., [28,29]). Activation of 50% (approximately 980 MPa) and 70% (approximately 1415 MPa) of the tensile capacity of the CFRP rod was applied before adhesively bonding the system to the concrete structure. Consistent results as well as high levels of strengthening efficiency were obtained, where CFRP rupture was experienced at approximately 3300 MPa (for the 70% activated configuration).

This stress magnitude corresponds to an approximate ultimate strain magnitude of 0.02 mm/mm. Consequently, the method provided better utilization of the CFRP compared to un-activated freely-bonded CFRP strengthening systems and seemed beneficial. It is; however, still an open question whether this CFRP strain magnitude combined with brittle failure modes still provides a sufficient safety margin and failure warning. Increased ductility seems possible if introducing a flexible system which allows for increased elongation in the bonded zone of the CFRP rod and at the anchorage zones. In an unbonded configuration, the CFRP rod is free to elongate along its entire length, thus resulting in increased ductility [30].

However, in such a system the CFRP material is exposed, losing some of the advantages of the NSMR; moreover, the anchorage zones are subjected to the full load regime during the structural service life.

This paper presents a strengthening system where a novel ductile anchorage [30] is used to anchor 8 mm CFRP rods in an NSMR strengthening setup. Flexible adhesive is used, replacing typical stiff epoxy adhesives, thus facilitating a larger elongation provided by the system. It is hypothesized that the system provides a combination of a high strengthening magnitude and significantly increased ductility to the concrete element (in this case T-section reinforced concrete (RC) beams). The system is additionally deemed to enable a wider stress distribution in the adhesively-bonded interface, thus reducing or even preventing a brittle behavior. Finally, a tailorable response control provided by the system components is assumed to enable a high control of stresses and thus a good result consistency.

Consequently, it is hypothesized that the ductile system response provides significantly increased ductility compared to conventional epoxy-bonded NSMR strengthening systems. An additional scope is to provide a strengthening system that provides high CFRP utilization but with a more controlled response and non-brittle failure mode compared to other CFRP strengthening systems. According to the authors' best knowledge, the researched system has not been proposed before and constitutes; therefore, a further major step towards providing ductility and non-brittle failure modes to RC elements strengthened with CFRP materials.

An experimental program which includes five concrete T-beams strengthened with the novel system is; thus, presented in the paper, where a CFRP rod activation stress of approximately 980 MPa is applied. The unique ductile CFRP system behavior and mounting procedure, as well as the ductile anchor behavior during beam loading, is evaluated experimentally and compared to analytical theoretical predictions.

## 2. Experimental Program

### 2.1. RC T-Section Beams and Test Setup

Figures 1 and 2 shows details about the T-section RC beams used for investigating the activated ductile CFRP NSMR strengthening. The beam length was 6.4 m with a support distance of 5 m and a distance of 1.4 m between the two loading points in the four-point-bending configuration. Six beams were quasi-statically loaded until failure with a deformation-controlled load rate of 2 mm/min. Concrete strain gauges, Strain gauge 1 (SG1) and Strain gauge 2 (SG2), (type: 1-LY41-50/120) were placed on the top flange surface at the transverse center line. Two steel strain gauges, Strain gauge 3 (SG3) and Strain gauge 4 (SG4), (type: 1-LY416/120) were placed on the exposed bottom reinforcement steel in the groove. Seven strain gauges (Strain gauge 5 (SG5) to Strain gauge 11 (SG11)) were placed on the CFRP rod using three different types (Types: 1-LD20-6/120, 1-LD20-6/350, and 1-LD2010/350). Linear Variable differential transformers (LVDTs) (Novotechnik position transducer; repeatability ±0.002 mm, Novotechnik, Southborough, MA, USA) were used to measure deformations at the quarter points, at the supports, and at the center location. A data acquisition frequency of 1 Hz was used. Figure 1 depicts the cross section geometry, the reinforcement location, and where the anchor blocks with the ductile anchorage were installed, approximately 1800 mm from the center line.

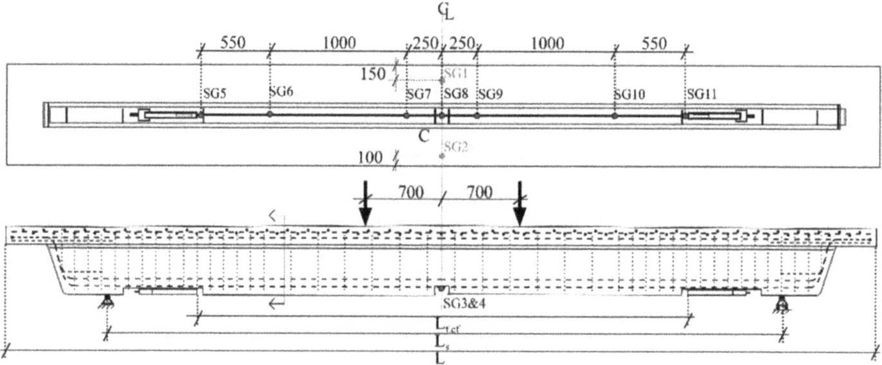


**Figure 1.** Concrete T-beam dimensions, load setup, and applied monitoring. Note: Units in mm.

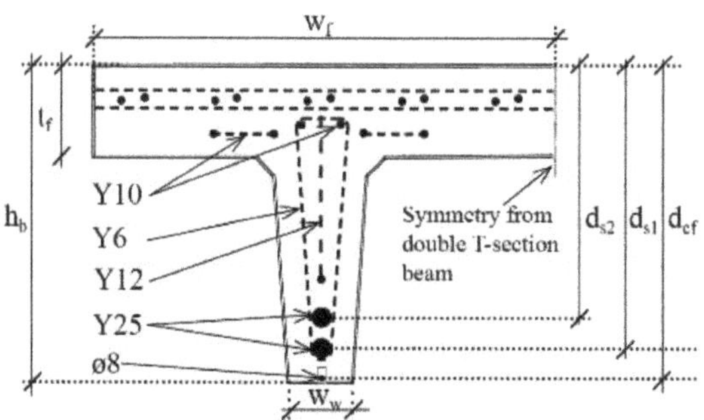


**Figure 2.** Cross section measures and reinforcement location.

The longitudinal tensile reinforcement consisted of 2Ø25 deformed steel reinforcement bars, and the top reinforcement consisted of longitudinal Ø10 bars as well as two reinforcement grids. The Ø6 stirrups were spaced by 100 mm and with a double spacing after 1.4 m from the extremities until the beam middle (see Figure 1). More detailed information concerning the beam can be found in [29]. However, relevant geometrical parameters of the T-beams are presented in Table 1.

**Table 1.** Geometrical parameters of the T-beams, See Figure 2.

| Parameter | Value (m) |
|---|---|
| Length of the beam, L | 6.4 |
| Distance between supports, $L_s$ | 5.0 |
| Free length of the carbon fiber reinforced polymer (CFRP) rod, $L_{f.cf}$ | 3.9 |
| Height of the beam, $h_b$ | 0.47 |
| Height of beam flange, $t_f$ | 0.12 |
| Width of beam web, $w_w$ | 0.10 |
| Width of beam flange, $w_f$ | 0.74 |
| Distance from top surface to outer reinforcement, $d_{s1}$ | 0.42 |
| Distance from top surface to inner reinforcement, $d_{s2}$ | 0.37 |
| Distance from top surface to CFRP rod, $d_{cf}$ | 0.46 |

Three different grooves were cut in the concrete beam before applying the ductile CFRP NSMR strengthening system: (i) Anchor block groove (Figure 3a), (ii) groove in which the NSMR rod was mounted (Figure 3b), and (iii) notch located at the mid-span (Figure 3c).

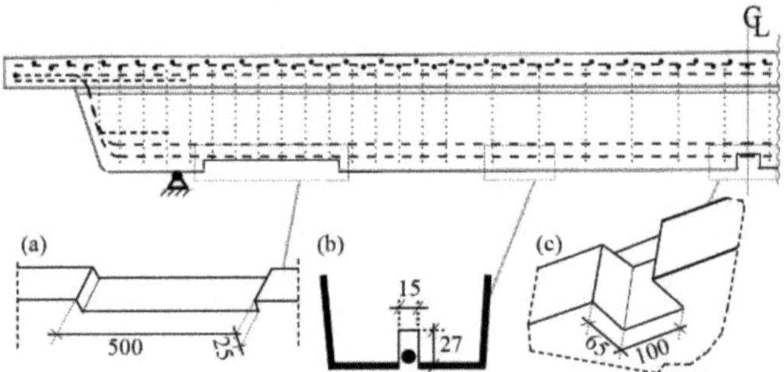


**Figure 3.** Groove dimensions used for (**a**) anchorage zone, (**b**) Near surface mounted reinforcement (NSMR), and (**c**) mid-span notch. Note: units in mm.

The mid-span notch had a width of 100 mm and a depth of 65 mm, which ensured exposure of the internal Y25 steel reinforcement and thus enabled mounting of strain gauges. In addition, a groove width of 15 m with a depth of 27 mm was cut into the web bottom concrete cover, along the beam length, in order to allow NSMR mounting. The depth was ensured relatively large to provide a good support for the CFRP rod mounted into the more flexible low E-modulus adhesive, and allow for a more desirable distribution of stresses. Finally, the anchor block groove ensured a 4 mm cover layer on the CFRP rod, using a width and depth of 500 and 25 mm, respectively.

### *2.2. Activation of the System and Mounting Procedure*

Figure 4 depicts the installed anchor block at the ends of a CFRP NSMR strengthening system. The assembled anchor system consists of the following components: (a) Enclosure wedge (EW) anchor, (b) response control pin (RCP), (c) ductile mechanism, (d) anchor block, (e) high strength M16 threaded activation bar, and (f) tensioning tray [29]. The clearance hole in the anchor block allows the movement of the tensioning tray thus enabling activation of the anchored NSMR system. The existing clearance in the tensioning tray provides the space for the response control pin to deform with the ductile mechanism and thus control the deformation.

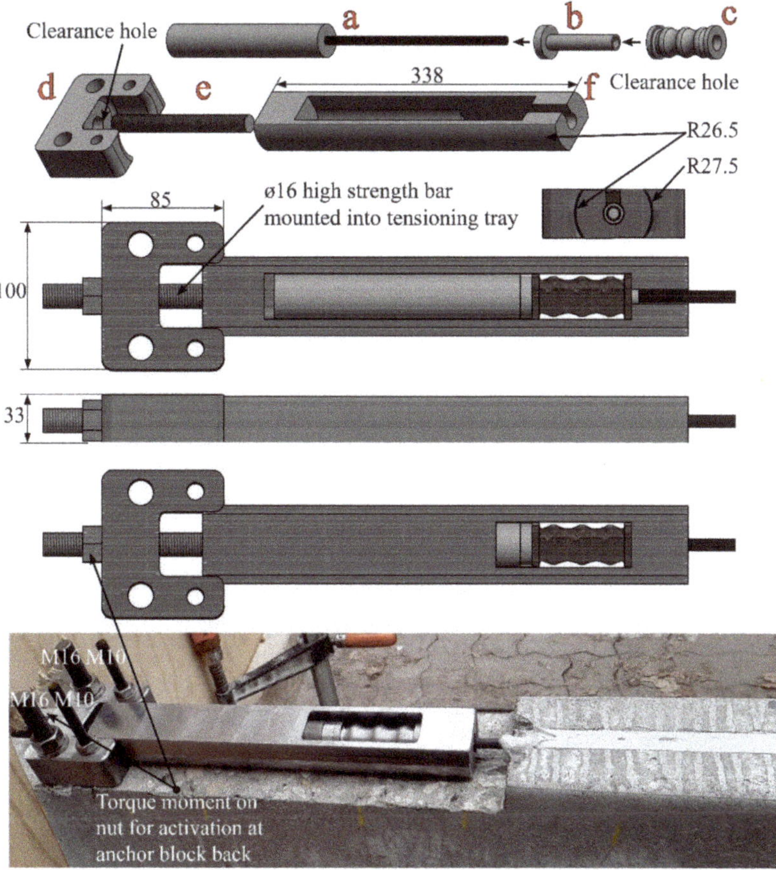


**Figure 4.** Anchor block used for the ductile Enclosure Wedge (EW) anchor. Note: Units in mm.

In the activated ductile system, the 100 × 85 × 33 mm$^3$ anchor block was mounted on the concrete beam using two M16 and M10 adhesively-anchored M8.8 threaded bars. The anchor block provides the base for the high strength M16 activation bar mounted into the 338 × 53 mm$^2$ tensioning tray. The threaded bar, on which the activation nut was mounted, penetrates through an Ø17 clearance hole in the anchor block and additionally stabilizes the tensioning procedure. In addition, this stabilization is improved by using circular side faces of the tensioning tray, which fits into the indent in the anchor block. By tightening the activation nut, the activation of the ductile-anchored NSMR CFRP strengthening system is enabled.

The strengthening system mounting procedure involves: (i) Cutting of grooves for anchor block and CFRP NSMR slit into the concrete beam, (ii) installation of the anchor block on the RC beam, (iii) mounting the ductile EW anchor on the CFRP rod, (iv) placing of the mounted anchor in the tensioning tray, (v) injection of adhesive in the slit bottom, (vi) inserting the tension tray into the anchor block, (vii) activation of the CFRP rod through a torque moment on the nut applied to the M16 threaded high strength bar, and (viii) final injection of the adhesive to fully cover the CFRP rod in the NSMR system.

## 2.3. Ductile Anchor System

Figure 5 shows the response of the fully assembled system (i.e., the combined tailored ductile mechanism, EW anchor, and CFRP rod) along with the individual responses of the ductile mechanism and CFRP rod. The combined response generates an initial linear elastic response and develops into a yielding regime with an approximate threshold at approximately 115 kN. This threshold was chosen to ensure that significant yielding occurred before anchor failure. When constructing the ductile mechanism, this threshold can; thus, be chosen to a desired value based on the anchorage response.

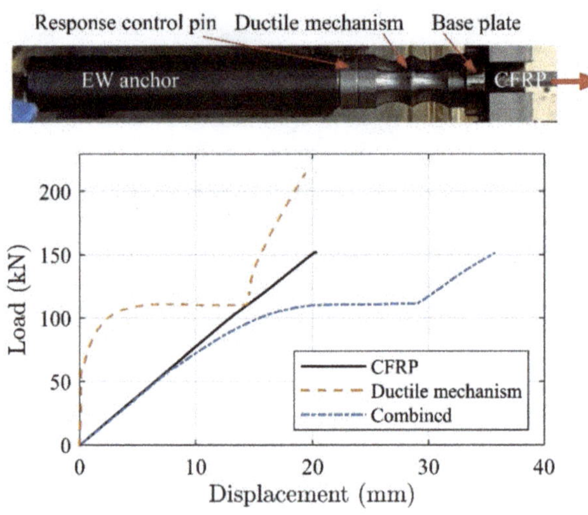


**Figure 5.** Ductile EW anchorage and related response [30].

The high strength, linear elastic, and brittle behavior of the CFRP rod was converted into a similar high tensile strength response; however, with significant ductility provided by the ductile anchor [30]. The anchor response; thus, seems to provide a good basis for stress control when combined with an NSMR strengthening system. It is; however, a prerequisite that the applied ductile anchor system is activated, during the structural loading procedure, in order to ensure the desired strengthening effect. This activation does; however, not seem possible when an epoxy adhesive is used for the NSMR, since anchorage is provided by the adhesive bond line. The use of a flexible adhesive is; thus, hypothesized to provide a more even distribution of the stresses along the bond line and thus reduce the possibility of IC debonding. In addition, it is deemed to enable activation of the ductile anchor system. Consequently, two flexible adhesives were used to address the anchorage activation.

## 2.4. Material Properties

Material properties related to the characteristics of the beams and end-anchor parts are presented in Table 2. Further details about the material properties can be found in [29,30].

Table 2. Material properties related to the beam and end-anchor parts.

| Parameter | Value (MPa) |
|---|---|
| Mean E-modulus of concrete **, Ec | 39,000 |
| Mean concrete cylinder compressive strength **, fcm | 62 |
| Mean yield strength of steel reinforcement **, fym | 565 |
| Mean E-modulus steel reinforcement **, Es | 200,000 |
| CFRP: Recommended ult. stress */Ult. mean stress **, fcfm | 2200/3300 |
| E-modulus CFRP tendon *, Ecf | 165,000 |
| Steel yield/tension strength of the anchor block parts *, fy/fu | 235/340 |
| Steel yield/tension strength of the threaded bar used for activation *, fy,bar/fu,bar | 900/1000 |

Notes: * Data provided by manufacturer; ** data obtained from testing at the lab.

Flexible adhesives (PU1 and PU2) were explored in this work since they are deemed to have a significant effect on the shear-slip behavior in the bond between the CFRP and concrete adherents. Material parameters of the adhesives, PU1 and PU2, are shown in Table 3. The adhesives were chosen due to the low E-moduli and large extension at failure.

Table 3. Material properties of the flexible adhesive.

| Adhesive | Parameter | Values |
|---|---|---|
| PU1 * | Shrinkage | 3–4% [31] |
| | Hardness | 55–60, Shore A |
| | Extension at failure | 400% [32] |
| | E-modulus | 1.1 MPa [32] |
| | Elasticity | ±20% |
| | Temperature resistance | −40 to +90 °C |
| | Curing | 2–3 mm/day (20 °C, AH50%) |
| PU2 * | Hardness | 60, Shore A [33] |
| | Failure stress | 3.0 N/mm$^2$ [34] |
| | E-modulus | 3.0 MPa |
| | Extension at failure | 200% [34] |
| | Elasticity | ±20% |
| | Curing | 2 mm/day (Environment dependent) |

Note: * Approximate PU material data, provided by manufacturer.

The curing days for the PU beams, when tested, ranged from 7–11 days. This was due to project time limitations and laboratory availability. Nevertheless, the manufacturer curing time data indicate that the adhesive is cured around the CFRP rod. However, with the relatively large NSMR notch in mind, this could still result in adhesives with reduced stiffness and strength in some locations (compared to the fully cured state).

PU adhesive dog bone specimens were; thus, cast in order to provide an understanding concerning the elongation of these materials, as well as the time-dependent curing influence on the material tensile behavior. This investigation was additionally done to get an indication of different behaviors in the used PU adhesives at a certain curing time. A behavior difference was important to potentially defer between these and was considered (due to the time limitations) acceptable to evaluate the ductile anchored NSMR system bonded with flexible adhesives.

The slit dimension in the concrete beams was used as a reference for the dog bone specimens, where the tested minimum cross section area was approximately 10 × 15 mm$^2$. Molds made of plywood were used to cast the specimens and are depicted on Figure 6.

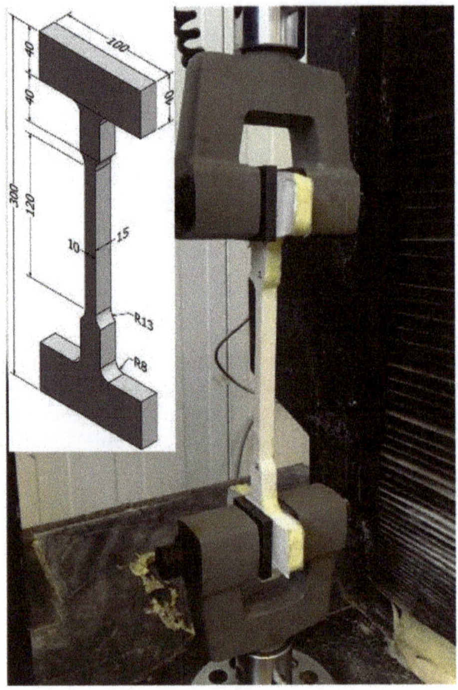


**Figure 6.** Injection of PU adhesive and related dog bone test specimen. Note: Units in mm.

At least three specimens per adhesive were tested in tension at a load rate of 10 mm/min The load–deformation curves related to each specimen are depicted in Figure 7 together with the related curing time. Note that, in test PU2-3, two curves were combined due to slippage and reloading, which do not seem to have an effect on the response.

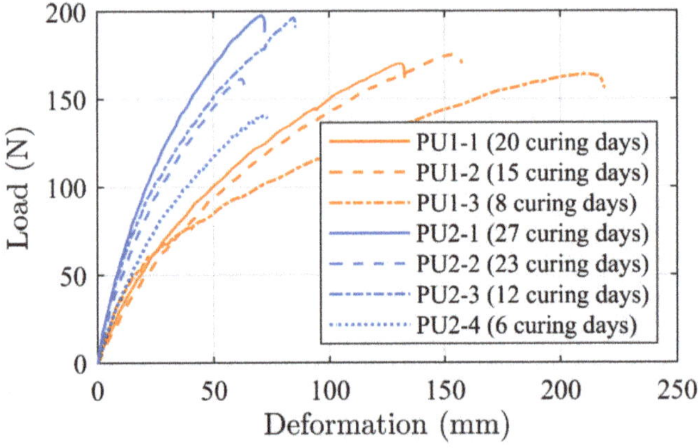


**Figure 7.** PU dog bone specimen, time related, load–deformation curves.

It is seen that the curve developments provide grouped formations. This indicates that the testing method seems to provide consistency in relation to the material properties as well as the preparation method. The PU1 tests provide a maximum capacity in the interval from 140–200 N. In addition, a significant stiffness increase is seen when comparing the specimen which was cured for 20 days compared to 8 days of curing. PU1-2, which cured for 15 days, had a response curve which was located in between, but closer to the PU1-1 specimen, as expected.

The stiffness increase seems; however, to be at the expense of a reduced elongation where PU1-3 elongates more than 200 mm, while PU1-1 and PU1-2 elongate 130 and 160 mm, respectively. PU2 specimens shows more deviations in the results where two specimens provide a higher tensile value than PU1 but also an identical and lower value. However, the stiffness increases as the curing days increases. The adhesive seems to enable large elongations which, consequently, may have the desirable influence on the shear-slip bond behavior, thus enabling large deformations. It should; however, be noted that the curves provide indicational information only, since more tests at each curing day are needed to provide a verified evaluation background.

### 2.5. Test Beam Configuration and Activation

The experimental program consisted of six beams with three different strengthening configurations. In addition, an un-strengthened beam (REF) response from another experimental program [29] was included to show the actual strengthening effect. The test program and curing days related to each beam are presented in Table 4. One reference beam (PCF) was strengthened with conventional activated CFRP NSMR using epoxy resin identical to the beams tested and presented in [29]. PU1 was used as the NSMR adhesive for two beams, whereas PU2 was used for three beams. The CFRP rod in all the beams had an activation level of approximately 50% (980 MPa) of the guaranteed manufacturer CFRP strength. It shall be noted that beam PU1-1 had an activation to 44%, which was due to initial technical issues.

**Table 4.** Test program and activation magnitudes.

| Beam | Activation (%) | Curing Days |
|---|---|---|
| REF (1) | - | - |
| PCF | 50 | 6 |
| PU1-1 | 44 | 10 |
| PU1-2 | 50 | 7 |
| PU2-1 | 50 | 11 |
| PU2-2 | 50 | 7 |
| PU2-3 | 50 | 7 |

(1) Beam tested in the scope of the work [29].

Figure 8 shows the stress development until testing. It should be advised that the activation took place prior to the curing of the adhesive in order to avoid initial stresses in the bonded connection.

The activation procedure was performed with the beams in an upside-down position, on the T-flange top surface. The change from activation stress did not reduce significantly during curing, which indicates a stable activation system. Thus, an additional average stress increase in the CFRP of about 21 MPa due to the beam dead load occurred in five beams when placed in its final position for testing. However, a reduction of stress occurred in the CFRP of 10 and 14 MPa PCF and PU-3, respectively. Generally, the obtained stress discrepancies from activation to before testing are small, thus not evaluated further. It should; however, be noted that future research will include further investigations of optimal adhesives used in conjunction with the ductile strengthening system.

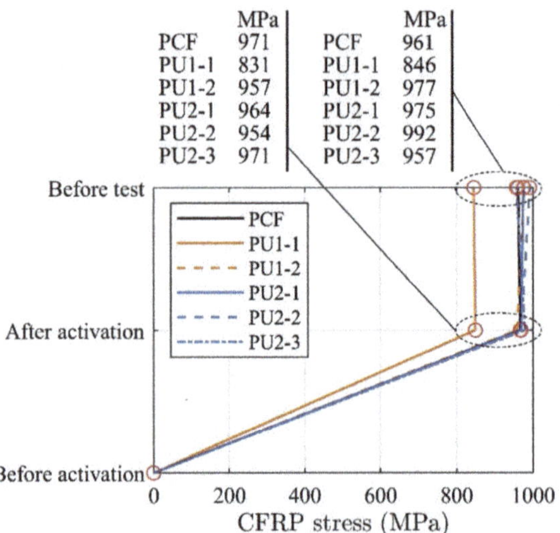


**Figure 8.** Evolution of the activation values until testing.

## 3. Test Results

Figure 9 shows the moment–deflection curves related to the tested configurations. The results are compared to a representative un-strengthened beam (REF) [29].

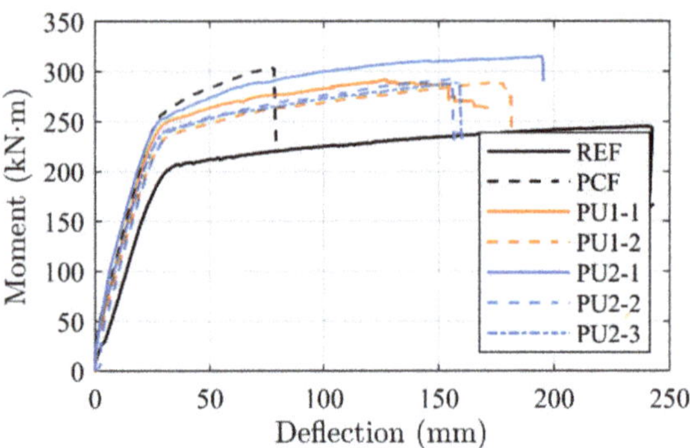


**Figure 9.** Moment–deformation test results.

Crack initiation seems to occur at approximately 35 kNm for all strengthened configurations. The linear elastic cracked stage of the PU configurations continue until magnitudes of 235 to 250 kNm, and 250 kNm for PCF, where the internal reinforcement yielding point is reached. Stiffness in, what seems to be, the linear elastic cracked regime of the PU1 and PU2 configurations seems unchanged, when comparing to the conventionally epoxy-bonded NSMR strengthened reference beam (PCF).

An inclined yielding regime is achieved at this branch, which terminates at 280 to 320 kNm, where the ultimate capacity is reached. It is seen that, for all configurations, significant strengthening is gained both in the serviceability and ultimate limit state along with an increased beam stiffness. However, while for the conventional CFRP NSMR strengthened reference beam (PCF) the failure occurred at approximately 300 kNm and a deflection of 76.2 mm, in the case of PU configurations, failure took place at deflections spanning from 155 to 195 mm, corresponding to an increase up to 256% of the PCF. In addition, when isolating the yielding regime alone, an increase up to 357% was detected for the beams strengthened with the PU1 and PU2 adhesives. The brittle failure modes, which are usually seen for CFRP strengthening configurations (end or intermediate debonding), were mitigated and the governing failure mode for the PU strengthened beams was a mix of concrete crushing (CCF) in the compressive zone of the top flange of the beam and frontal overload (FO) anchorage failure (MCFO failure).

Figure 10 shows some of the obtained failure modes. Large deflections were obtained when failure occurred. It seemed that the final stage, just before failure, reached a state which provided an equilibrium where there was an identical possibility to initiate concrete compression failure and frontal overload in the anchor. This seem to be supported by the concrete strain gauges placed at the beams center location, where all measured values at ultimate reached approximately $2.3 \times 10^{-3}$. Consequently, the failure mode is described as MCFO (CCF or FO), to indicate this state together with the observed failure mode after testing. It should be noted that the configuration PU2-1 seems to provide a more desirable strengthening magnitude. However, one of the rolling supports was identified to have some friction during testing, which seems to result in a contributing arch effect. This effect was prevented in the other test configurations.

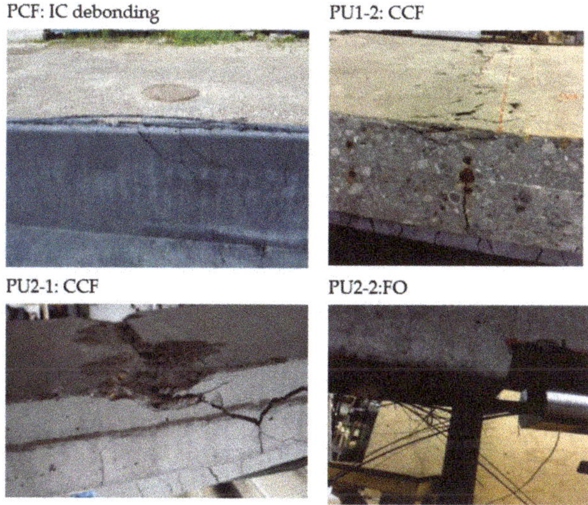


**Figure 10.** Examples of experienced failure modes.

Figure 11 shows the visual deflection of a representative overturned PU strengthened beam after testing. The permanent deformation with a large permanent deflection indicates that the yielding regime of the steel reinforcement was reached, thus providing significant warning before failure. A good correlation between the beam responses is seen. This indicates that a tailored and thus reproducible strengthening effect may be provided when applying the ductile strengthening system.

**Figure 11.** Example of deflected beam after testing.

Table 5 shows the increase in deformation and yielding regime extension magnitude, when comparing the PU configurations to the PCF configuration. It is seen that a significant deformation and yielding increase are obtained when applying the ductile strengthening system.

**Table 5.** Deformation and yielding regime extension compared to PCF configuration.

| Test Configuration | Deformation (%) | Yield Branch (%) |
|---|---|---|
| PU1-1 | 203 | 271 |
| PU1-2 | 234 | 320 |
| PU3-1 | 256 | 357 |
| PU3-2 | 205 | 273 |
| PU3-3 | 209 | 279 |

It could be argued that the values should be compared with the un-strengthened beam. However, since a mixed concrete crushing and anchorage frontal overload failure is obtained, the concrete failure at the higher load magnitudes provide a limitation to such an evaluation. Consequently, a comparison between the well-known epoxy-bonded NSMR strengthening method used in the REF beam configurations and the PU strengthened configuration seems more suitable.

### 3.1. *Activation of the Ductile Anchorage*

It is hypothesized that the ductile mechanism provides significant control of the strengthening effect. A prerequisite for the desired response seems to be the interaction between the PU adhesives and the ductile EW anchors. A tailorable response is; thus, deemed to be provided by the ductile anchorages if the flexible NSMR adhesive ensures sufficient elongation and stiffness.

Figure 12 depicts a typical ductile mechanism deformation history. Four deformation magnitudes are chosen, correlating somewhat with (i) test initiation, (ii) approximate yielding point of the concrete beam, (iii) deformation at a point in the yielding regime, and (iv) ultimate capacity. It is seen that the anchorage is significantly activated, with a final deformation of 15.8 mm, when beam failure occurs. Figure 13 depicts these magnitude points and related load levels on the ductile mechanism response curve. It is seen that the ductile mechanism behaved as desired (i.e., yielding without exhausting its plastic capacity).

Table 5 shows all ductile mechanism deformation magnitudes at the left and right beam anchorage locations, as well as the summarized deformation of both ductile mechanisms from each beam configuration. It is seen from Figure 12 and Table 6 that the ductile mechanism can provide even more deformation to the strengthening system, which is not needed in the present case where MCFO failure is achieved.

**Table 6.** Ductile mechanism deformation compared to ultimate moment and deflection.

| Configuration | $D_{left}$ (mm) | $D_{right}$ (mm) | $D_{total}$ (mm) | Utilization (%) | $D_{beam}$ (mm) | $M_{ult}$ (kN·m) |
|---|---|---|---|---|---|---|
| PCF | - | - | - | - | 76.2 | 303.3 |
| PU1-1 | 13.0 | 15.7 | 28.7 | 47.8 | 154.8 | 291.1 |
| PU1-2 | 17.1 | 15.8 | 32.9 | 54.8 | 178.4 | 288.5 |
| PU2-1 | 16.7 | 23.7 | 40.4 | 67.3 | 194.8 | 315.1 |
| PU2-2 | 12.1 | 12.7 | 24.8 | 41.3 | 156.1 | 292.2 |
| PU2-3 | 12.3 | 14.0 | 26.3 | 43.8 | 159.2 | 288.7 |

$D_{left}$: Ductile mechanism deformation, at left hand side; $D_{right}$: Ductile mechanism deformation, at right hand side; $D_{total}$: Total deformation of both ductile mechanisms; $D_{beam}$: Maximum beam deflection, identified in test; $M_{ult}$: Ultimate capacity moment, identified in test.

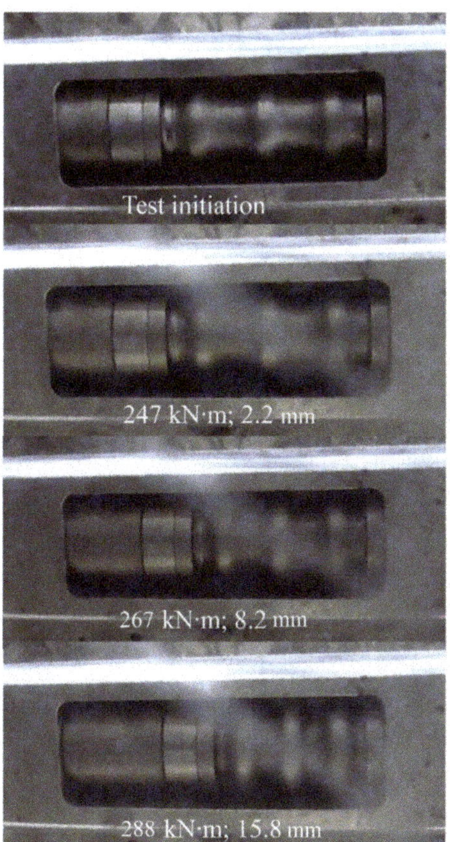


**Figure 12.** Deformation magnitudes during test of beam PU1-2.

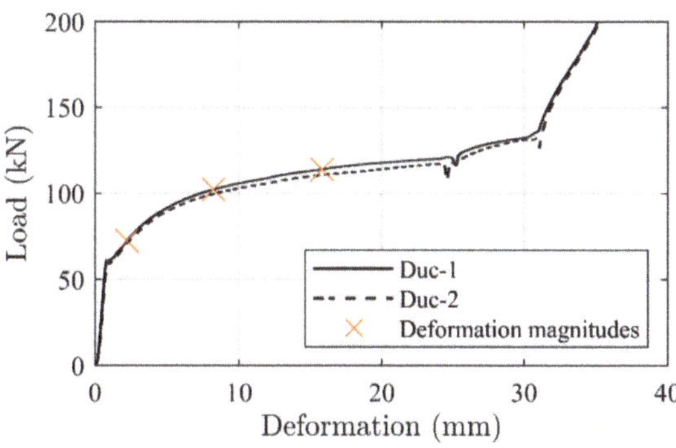


**Figure 13.** Typical complete ductile mechanism response versus ductile mechanism response in the beam PU1-2 for different stages (crosses in orange).

*3.2. Stress Development*

Typical stress variations (without considering the activation component) at different load levels along the CFRP rod, measured by the applied strain gauges, are shown in Figure 14. A comparison to beam PCF* in [29] is included. In this case the values of the stress variations at the anchorage locations and at mid span of an activated epoxy adhesively-bonded NSMR strengthened beam are provided. Stresses at moment levels of 240 and 260 kNm and at the approximate ultimate capacity of 280 kNm are chosen to demonstrate changes in the stress development. For the case of PU beams, it is seen that the stress along the CFRP rod is almost constant until 240 kNm, but changes into a more inclined stress development at 260 kNm, increasing towards 280 kNm.

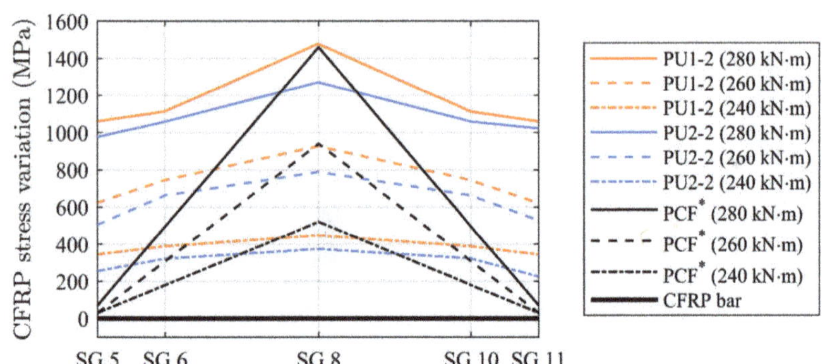


**Figure 14.** Measured stress transfer at different moment magnitudes.

At ultimate, both flexible adhesive configurations seem to enable stress transfer in the adhesive interface between the CFRP rod and concrete. The stress transfer to the concrete beam adherent; thus, ensures an increased stress at mid span, of approximately 380 and 220 MPa for the PU1 and PU2, respectively (difference between midspan CFRP stress and stress at the anchor location).

In the case of the beam PCF*, as expected, the stress variations are marginal in the CFRP rod at the anchorage zone, since stiff epoxy does not have the same ability of stress transfer along the bond line.

Figure 15 shows the stress development in the steel reinforcement. The ductile strengthening system seems to support the internal reinforcement well. When comparing to the un-strengthened beam REF*, the yielding thresholds are postponed and the tailored response seem to affect stresses in a controlled way until failure initiation. In addition, the strengthening support can be identified in all regimes of the beam responses, where the ductile anchorage contributes to the (i) linear elastic, (ii) yielding transition, and (iii) yielding response (see Figure 15).

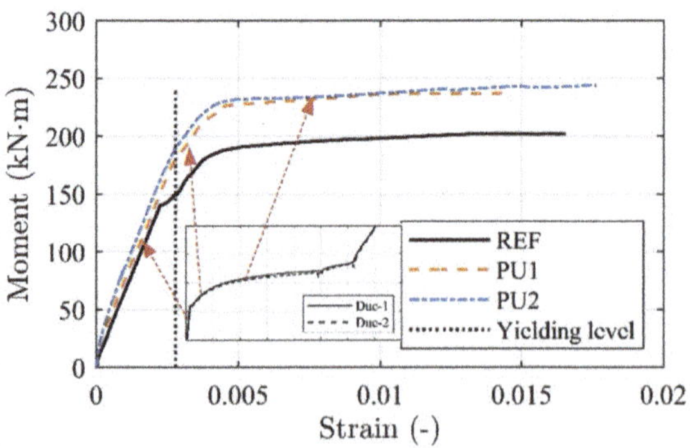


**Figure 15.** Relationship between the moment and measured strain in the steel reinforcement for PU and REF beams.

The high strain magnitude resulted in malfunction of several strain gauges at approximately 0.015. Nevertheless, the data acquisition served well to show, what seems to be, a distinct response control provided by the ductile mechanisms.

The steel reinforcement stress development is slightly different when comparing the PU1 and PU2 configurations. This is deemed to be caused by the adhesive differences, where, in particular, the yielding transition regime is less pronounced for the PU2 configurations.

## 4. Analytical Predictions

The test results show that the ductile strengthening system provides the desired ductile response to the structure and the brittle IC debonding is prevented. It was shown to be possible to activate the anchorages by using the more flexible adhesive, which additionally provided a significantly lower, but more distributed, stress transfer. In addition, the stress development along the CFRP rod was monitored constant until the ultimate capacity level was reached.

Figure 16 depicts the load–deformation development of the tested ductile mechanisms, where specific branches of interest are marked. These are deemed to relate to transition stages of the beam moment–deflection response curves, where (1) depicts the initial activation magnitude, (2) relates to the exposure on the ductile mechanism when yielding in the steel reinforcement initiates, and (3) is a regime from 12.1 to 23.7 mm which correspond to the minimum and maximum ductile mechanism deformation reached in the beam tests at failure (see Table 5). It is seen that the response load is horizontal and linear in the third stage, thus providing a constant load magnitude to include in the theoretical calculations.

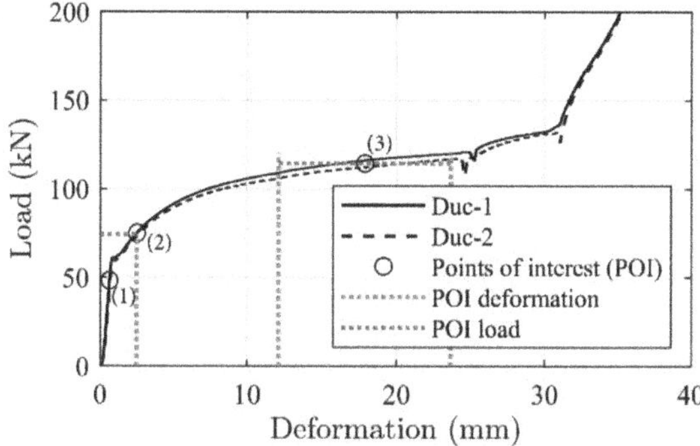


**Figure 16.** Ductile mechanism response magnitudes at the (1), (2), and (3) stages.

Due to the small stress transfer and large load distribution to the anchorage zones, it is hypothesized that the strengthening contribution can be directly related to the ductile mechanism load–deformation response. The highly controlled load transfer opens the opportunity to implement the tailored ductile mechanism load–deformation development in a conventional concrete beam evaluation approach. Figure 17 shows the moment–deflection development with the transition areas of the beam configurations (i) linear–linear cracked, (ii) yielding initiation, and, finally, (iii) the beam failure regime.

In the theoretical evaluation, the transition point between the linear uncracked and linear cracked regimes (1) can be found by evaluating the transformed cross section stresses, where the dead load moment and counteracting strengthening system moment is well known. The applied moment can then be adjusted until the bottom web stress exceeds the concrete tension stress capacity.

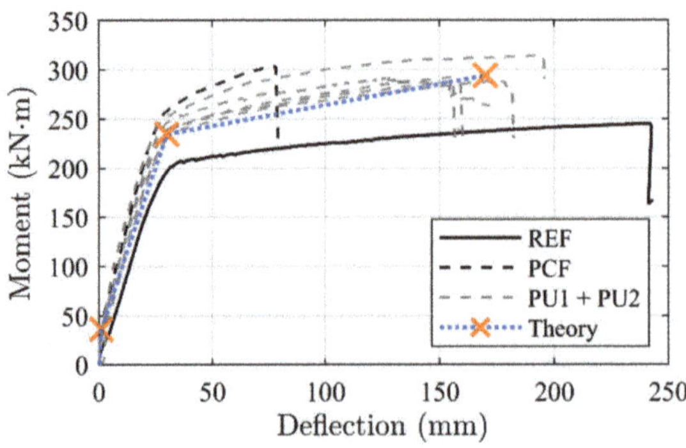


**Figure 17.** Stress variation along the CFRP bar for the different adhesives.

When evaluating the transition point of steel reinforcement yielding (2), a linear elastic cracked transformed cross section can be used for the stress evaluation. A linear relation in the steel reinforcement layers can be used to predict the CFRP strain exposure. The steel reinforcement bar closest to the beam bottom is; thus, presumed to yield.

The force exposure on the ductile mechanism can be found by performing a linear strain extension from the two steel reinforcement locations to the CFRP rod location. Following this analogy, a strain magnitude of 0.0033 was found at the CFRP location based on the estimated internal steel reinforcement strain magnitudes of 0.0029 and 0.0025 in the first and second reinforcement layer, respectively. When calculating the ductile anchorage tension force, the 7.9 mm CFRP rod cross section area and E-modulus of 165 GPa was used in the process to reach 26 kN (540 MPa). Additionally, the initial activation magnitude of 48 kN (980 MPa) was added to this value, giving a final force exposure to the ductile mechanism of 74.5 kN (see Figure 16).

It is seen that this exposure magnitude on the ductile mechanism provides a deformation of approximately 2.5 mm which correlate well with the ductile mechanism response development, depicted on Figure 11, where a deformation of 2.2 mm was measured. Consequently, this force provides a counteracting moment at the anchorage location which can be added to the original moment of the reference beam. The counter acting moment should thus be exceeded by the acting moment from the four-point bending load, before yielding occurs at an identical beam deflection magnitude as seen from the un-strengthened beam.

The ultimate failure can be found by using area (3) of the ductile mechanism (see Figure 16). It is seen that the ductile mechanism yielding regime provided at a deformation from 12.1 to 23.7 mm and thus a corresponding force of approximately 115 kN to the cross section.

The steel reinforcement tension curve (gained by tensile testing) at a deformation magnitude corresponds to the ductile mechanism, showing a stress magnitude of approximately 620 MPa, which is included in the evaluation. The internal steel reinforcement bars thus, yield at this transition point and a cross section force equilibrium can now be utilized to estimate the ultimate capacity.

The estimated capacity is based on a constant stress in the CFRP rod between the anchorages, when using the load magnitude from the ductile mechanism response. In addition, the tests showed a CFRP rod stress increase of 380 MPa (19 kN) at the beam middle, which needs to be added to the moment capacity. Consequently, a calculated ultimate capacity of 294 kNm was found.

A deformation evaluation using the beam rotation at midspan based on the strains was used. Consequently, a summation of the concrete strain of 0.0035, deformations of both ductile mechanisms divided by the anchorage distance (equivalent ductile mechanism strain), and the CFRP strain were used.

In order to provide the rotation magnitude, the summarized strain was divided by the distance from the CFRP rod location to the top concrete surface. The deformation was achieved by applying this value to the estimate equation 1/k × rotation × support distance. k can vary from 8 to 10 depending on the load configuration applied. A k of 9 was used since the applied load contains a mixture of the load configurations. Accordingly, the deformation magnitude varies depending on the deformation changes in ductile mechanisms and adhesive transfer stress.

The theoretical concrete beam transition points (i) cracking, (ii) yielding, and (iii) failure on the moment–deflection curve, related to the PU strengthened beam configurations, are depicted in Figure 17. It is seen that a linear evolution between the transition points provide a good prediction of the response.

Finally, the calculated stress magnitudes in the CFRP rod at each transition point were compared with measured values from strain gauges (5, 7 and 8). Figure 18 shows a good correlation, where a linear development between the thresholds seem to provide a good prediction. Tables 7 and 8 provide an overview of the tested and theoretically predicted values related to the moment–deflection and CFRP stress development, respectively.

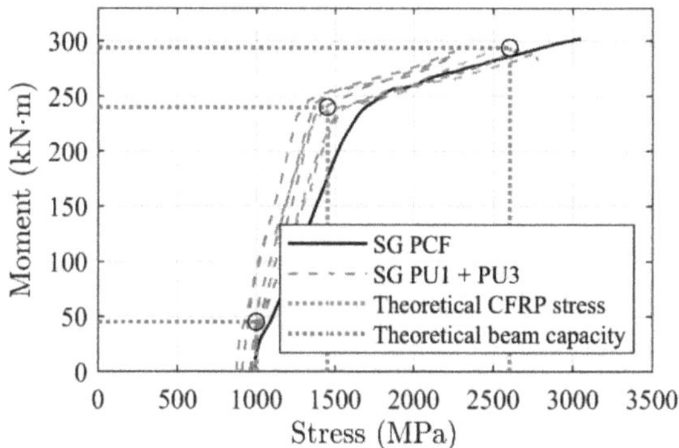


**Figure 18.** Stress variation along CFRP bar for the different adhesives.

**Table 7.** Moment and deflection predictions and test results.

| Beam | Crack Initiation | | | Steel Yielding | | | Ultimate Capacity | | | | | |
|---|---|---|---|---|---|---|---|---|---|---|---|---|
| | $M_c$ (kN·m) | $M_{c,calc}$ (kNm) | Dev (%) | $M_y$ (kN·m) | $M_{y,calc}$ (kN·m) | Dev (%) | $M_u$ (kN·m) | $M_{u,calc}$ (kN·m) | Dev (%) | $\delta_u$ (mm) | $\delta_{u,calc}$ (mm) | Dev (%) |
| PCF | 38.7 | - | - | 241.2 | - | - | 303.3 | - | - | 76.2 | - | - |
| PU1-1 | 37.1 | 35.2 | 5.4 | 233.6 | 234.7 | 0.5 | 291.1 | 293.9 | 0.9 | 155 | 164 | 5.8 |
| PU1-2 | 33.5 | 36.5 | 8.9 | 224.3 | 234.7 | 4.6 | 288.5 | 293.9 | 1.8 | 178 | 172 | 3.5 |
| PU2-1 | 35.9 | 36.5 | 1.6 | 237.1 | 234.7 | 1.0 | 315.1 | 293.9 | 7.2 | 195 | 179 | 8.9 |
| PU2-2 | 36.9 | 36.5 | 1.1 | 223.1 | 234.7 | 5.2 | 292.2 | 293.9 | 0.6 | 156 | 153 | 1.9 |
| PU2-3 | 37.3 | 36.5 | 2.2 | 231.1 | 234.7 | 1.6 | 288.7 | 293.9 | 1.8 | 159 | 156 | 1.9 |

$M_c$: Crack initiation moment, identified in test; $M_{c,calc}$: Calculated crack initiation moment; $M_y$: Steel reinforcement yielding initiation moment, identified in test; $M_{y,calc}$: Calculated steel yielding moment; $M_u$: Ultimate capacity moment, identified in test; $M_{u,calc}$: Calculated ultimate capacity; $\delta_u$: Maximum beam deflection, identified in test; $\delta_{u,calc}$: Calculated maximum beam deflection; Dev: Deviation between test and calculated values.

**Table 8.** CFRP stress predictions and test results.

| Beam | Crack Initiation | | | Steel Yielding | | | Ultimate Capacity | | | |
|---|---|---|---|---|---|---|---|---|---|---|
| | $\sigma_{cf}$ (MPa) | $\sigma_{cf,calc}$ (MPa) | Dev (%) | $\sigma_{cf}$ (MPa) | $\sigma_{cf,calc}$ (MPa) | Dev (%) | $\sigma_{cf}$ (MPa) | $\sigma_{cf,calc}$ (MPa) | Dev (%) | Failure Mode |
| PCF | | 1026 | | 1666 | | | 3049 | | | ICD * |
| PU1-1 | 875 | 867 | 0.9 | 1352 | 1476 | 9.2 | 2361 | 2568 | 8.8 | UDF ** |
| PU1-2 | 998 | 984 | 1.4 | 1369 | 1476 | 7.8 | 2746 | 2568 | 6.9 | MCFO (CRC) *** |
| PU2-1 | 1020 | 984 | 3.6 | 1481 | 1476 | 0.3 | 2594 | 2568 | 1.0 | MCFO (CRC) *** |
| PU2-2 | 912 | 984 | 7.9 | 1386 | 1476 | 6.4 | 2559 | 2568 | 0.4 | MCFO (FO) *** |
| PU2-3 | 996 | 984 | 1.2 | 1501 | 1476 | 1.7 | 2442 | 2568 | 5.1 | MCFO (FO) *** |

* ICD: Intermediate crack debonding; ** UDF: Undefined failure (yielding threaded rods); *** MCFO: Mixed failure mode of CCF: Concrete compression failure and FO: Frontal. $\sigma_{cf}$: Carbon fiber reinforced polymer stress (CFRP) at crack initiation, identified in test; $\sigma_{cf,calc}$: Calculated carbon fiber reinforced polymer (CFRP) stress at crack initiation; $\sigma_{cf}$: Carbon fiber reinforced polymer (CFRP) stress at steel reinforcement yielding initiation, identified in test; $\sigma_{cf,calc}$: Calculated carbon fiber reinforced polymer (CFRP) stress at steel reinforcement yielding initiation; $\sigma_{cf}$: Carbon fiber reinforced polymer (CFRP) stress at ultimate beam capacity, identified in test; $\sigma_{cf,calc}$: Calculated Carbon fiber reinforced polymer (CFRP) stress at ultimate beam capacity; Dev: Deviation between test and calculated values.

## 5. Conclusions

The presented research regards an anchored CFRP NSMR system hypothesized to provide a tailored strengthening and ductile strengthening effect. The novel ductile end anchors were used to activate 8 mm circular CFRP bars to approximately 980 MPa. Five concrete T-beams of 6.4 m (support distance 5 m) were strengthened using the flexible adhesives PU1 (two beams) and PU2 (three beams), and were compared with a conventional epoxy-bonded NSMR strengthened beam and an un-strengthened beam. It was found that the novel system provided the desired strengthening effect to the T-beams, with the following main outcomes:

- The system provided a significant and controllable strengthening effect;
- The ductile anchorages combined with the flexible adhesives ensured the desired response;
- Response curves seemed reproducible, following the tailored ductile anchor behavior well;
- Elastic regime responses of the PU configurations were similar to the one observed in the epoxy-bonded NSNR CFRP beam;
- An average deflection increase of 220% compared to the epoxy-bonded NSMR was reach
- Average yielding regime was extended by 300% compared to the epoxy-bonded NSMR;
- Brittle delamination failure mechanisms were prevented;
- Ductile mechanism response curves could be used to theoretically predict the strengthening magnitude
- The ductile mechanism provides a measure for the applied strengthening effect, since the deformation reflects the stress in the CFRP NSMR. Consequently, it may be possible to predict the strengthening effect (CFRP NSMR stress) throughout the service life;
- A controlled yielding regime is provided by the system when the ultimate capacity is reached. The internal steel reinforcement yields as well, which means that a conventional cross section equilibrium can be performed (due to the constant force provided by the system).

The mounting procedure of the strengthening system worked well and provided a good basis for a robust activation and installation of the stressed CFRP rod. The ductile mechanisms seem to open an opportunity for extreme ductility, since several of these components can be combined and installed in extension. The novel system is currently being tested on several structural levels in order to verify the results further and facilitate a more optimized interaction between the ductile system and flexible adhesives.

**Author Contributions:** Conceptualization, J.W.S.; methodology, J.W.S.; software, J.W.S., C.O.C.; validation, J.W.S., C.O.C., P.G. and J.S.-C.; formal analysis, J.W.S.; investigation, J.W.S.; resources, J.W.S.; data curation, J.W.S.; writing—original draft preparation, J.W.S.; writing—review and editing, J.W.S., C.O.C., P.G. and J.S.-C.; visualization, J.W.S., C.O.C.; project administration, J.W.S.; funding acquisition, J.W.S. All authors have read and agreed to the published version of the manuscript.

**Funding:** The funding source do not wish to be revealed as stated in this section.

**Institutional Review Board Statement:** Not applicable.

**Informed Consent Statement:** Not applicable.

**Data Availability Statement:** All data and information needed for the evaluations discussed, are provided in the paper.

**Acknowledgments:** Sincere gratitude is addressed to S&P Denmark for supporting the ongoing research. Also a great thank you to Jorcks foundation for their acknowledgement and funding. Thank you to Perstrup Beton industri A/S. Furthermore the contributions from former students Frederik Alexander Meinzer Vind, Frederik Jensen, Nathalie Lückstädt Nielsen, Frits Eið, Frederik Munck, Matias Brix Mikkelsen, Klavs Foged Skovby and Nicolaj Jacob Birkebæk Thomsen are greatly appreciated.

**Conflicts of Interest:** The authors declare no conflict of interest.

## References

1. El-Hacha, R.; Rizkalla, S.H. Near-Surface-Mounted Fiber-reinforced polymer reinforcement for flexural strengthening of concrete structures. *ACI Struct. J.* **2004**, *101*, 717–726.
2. Mohamed Ali, M.S.; Oehlersa, D.J.; Griffith, M.C.; Seracino, R. Interfacial stress transfer of near surface-mounted FRP-to-concrete joints. *Eng. Struct.* **2008**, *30*, 1861–1868. [CrossRef]
3. Rashid, R.; Oehlers, D.J.; Seracino, R. IC Debonding of FRP NSM and EB retrofitted concrete: Plate and cover interaction tests. *J. Compos. Constr.* **2008**, *12*, 160–167. [CrossRef]
4. Seracino, R.; Jones, N.M.; Ali, M.S.M.; Page, M.W.; Oehlers, D.J. Bond Strength of Near-Surface Mounted FRP Strip-to-Concrete Joints. *J. Compos. Constr.* **2007**, *11*, 401–409. [CrossRef]
5. Cruz, J.M.S.; Barros, J.A.O.; Coelho, M.R.F.; Silva, F.F.T. Efficiency of different techniques in flexural strengthening of RC beams under monotonic and fatigue loading. *J. Constr. Build. Mater.* **2012**, *29*, 175–182. [CrossRef]
6. Al-Mahmoud, F.; Castel, A.; Francois, R. Failure modes and failure mechanisms of RC members strengthened by NSM CFRP composites—Analysis of pull-out failure mode. *Compos. Part B Eng.* **2012**, *43*, 1893–1901. [CrossRef]
7. Schmidt, J.W.; Krabbe, J.; Sørensen, N.O.; Hertz, K.D.; Goltermann, P.; Sas, G. CFRP strengthening of RC beams using a ductile anchorage system. In Proceedings of the Eighth International Conference on Fibre-Reinforced Polymer (FRP) Composites in Civil Engineering, Hong Kong, China, 14–16 December 2016.
8. Schmidt, J.W.; Hertz, K.D.; Goltermann, P. NSMR strengthening of short rc beams using activated anchorage. In Proceedings of the 9th International Conference on Fibre-Reinforced Polymer (FRP) Composites in Civil Engineering, Paris, France, 17–19 July 2018.
9. Sabau, C.; Popescu, C.; Sas, G.; Schmidt, J.W.; Blanksvärd, T.; Täljsten, B. Strengthening of RC beams using bottom and side NSM reinforcement. *Compos. Part B Eng.* **2018**, *149*, 82–91. [CrossRef]
10. Nordin, H.; Täljsten, B. Concrete Beams Strengthened with Pre-stressed Near Surface Mounted CFRP. *J. Compos. Constr.* **2006**, *10*, 60–68. [CrossRef]
11. Täljsten, B. Strengthening of Beams by Plate Bonding. *J. Mater. Civ. Eng.* **1997**, *9*, 206–212. [CrossRef]
12. Triantafllou, T.C.; Antonopoulos, C.P. Design of concrete flexural members strengthened in shear with FRP. *J. Compos. Constr.* **2000**, *4*, 198–205. [CrossRef]
13. Smith, S.T. Modeling debonding failure in FRP flexurally Strengthened RC Members Using a Local Deformation Model. *J. Compos. Constr.* **2007**, *11*, 184–191. [CrossRef]
14. Teng, J.G. Intermediate crack-induced debonding in RC beams and slabs. *J. Constr. Build. Mater.* **2003**, *17*, 447–462. [CrossRef]
15. Said, H.; Wu, Z. Evaluating and Proposing Models of Predicting IC Debonding Failure. *J. Compos. Constr.* **2008**, *12*, 284–299. [CrossRef]
16. Täljsten, B. *FRP Strengthening of Existing Concrete Structures*; Division of Structural Engineering; Luleå University of Technology: Luleå, Sweden, 2006.
17. Gao, B.; Leung, C.K.; Kim, J.-K. Prediction of concrete cover separation failure for RC beams strengthened with CFRP strips. *Eng. Struct.* **2005**, *27*, 177–189. [CrossRef]
18. Corden, G.; Ibell, T.; Darby, A. Concrete cover separation failure in near-surface mounted CFRP strengthened concrete structures. *Struct. Eng.* **2008**, *86*, 19–21.

19. Smith, S.T.; Hua, S.; Kima, S.J.; Seracino, R. FRP-strengthened RC slabs anchored with FRP anchors. *Eng. Struct.* **2011**, *33*, 1075–1087. [CrossRef]
20. Kalfat, R.; Al-Mahaidi, R.; Smith, S.T. Anchorage Devices Used to Improve the Performance of Reinforced Concrete Beams Retrofitted with FRP Composites: State-of-the-Art Review. *J. Compos. Constr.* **2013**, *17*, 14–33. [CrossRef]
21. Hansen, C.S.; Schmidt, J.W.; Stang, H. Transversely compressed bonded joints. *Compos. Part B Eng.* **2012**, *43*, 691–701. [CrossRef]
22. Brunckhorst, L.; Knudsen, P.J.; Poulson, E.; Thorsen, E.T. *Forstærkning af Betonkonstruktioner Med Bolte-Limede Kulfiberbånd. En Elementær Teknisk Introduktion*; Rosendahls og Trykkeri A/S: Esbjerg, Denmark, 2007. (In Danish)
23. Schmidt, J.W.; Bennitz, A.; Täljsten, B.; Pedersen, H. Development of Mechanical Anchor for CFRP Tendons Using Integrated Sleeve. *J. Compos. Constr.* **2010**, *14*, 397–405. [CrossRef]
24. Schmidt, J.W.; Smith, S.T.; Täljsten, B.; Bennitz, A.; Goltermann, P.; Pedersen, H. Numerical Simulation and Experimental Validation of an Integrated Sleeve-Wedge Anchorage for CFRP Rods. *J. Compos. Constr.* **2011**, *15*, 284–292. [CrossRef]
25. Cuntze, R.G.; Freund, A. The predictive capability of failure mode concept-based strength criteria for multidirectional laminates. *J. Comput. Sci. Technol.* **2004**, *64*, 343–377. [CrossRef]
26. Schmidt, J.W. External Strengthening of Building Structures with Prestressed CFRP. Ph.D. Thesis, Technical University of Denmark (DTU), Lyngby, Denmark, 2011.
27. Bennitz, A.; Schmidt, J.W.; Täljsten, B. Failure modes of prestressed CFRP rods in a wedge anchored set-up. In Proceedings of the Advanced Composites in Construction, Edinburgh, UK, 1–3 September 2009; Volume 4.
28. Schmidt, J.W.; Bennitz, A.; Täljsten, B.; Goltermann, P.; Pedersen, H. Mechanical anchorage of FRP tendons—A literature review. *Constr. Build. Mater.* **2012**, *32*, 110–121. [CrossRef]
29. Schmidt, J.W.; Christensen, C.O.; Goltermann, P.; Hertz, K.D. Shared CFRP activation anchoring method applied to NSMR strengthening of RC beams. *Compos. Struct.* **2019**, *230*, 111487. [CrossRef]
30. Schmidt, J.W.; Christensen, C.O.; Goltermann, P. Ductile response controlled EW CFRP anchor system. *Compos. Part B Eng.* **2020**, *201*, 108371. [CrossRef]
31. ISO 10563:2017. *Buildings and Civil Engineering Works—Sealants—Determination of Change in Mass and Volume*; International Organization for Standardization: Geneva, Switzerland, 2017.
32. ISO 8339:2005. *Building Construction—Sealants—Determination of Tensile Properties (Extension to Break)*; International Organization for Standardization: Geneva, Switzerland, 2005.
33. ISO 868:2003. *Plastics and Ebonite—Determination of Indentation Hardness by Means of a Durometer (Shore Hardness)*; International Organization for Standardization: Geneva, Switzerland, 2003.
34. ISO 37:2017. *Rubber, Vulcanized or Thermoplastic—Determination of Tensile Stress-Strain Properties*; International Organization for Standardization: Geneva, Switzerland, 2017.

MDPI

*Article*

# Effect of TRC and F/TRC Strengthening on the Cracking Behaviour of RC Beams in Bending

**Edoardo Rossi *, Norbert Randl , Tamás Mészöly and Peter Harsányi**

Faculty of Civil Engineering and Architecture, Carinthia University of Applied Science (CUAS), Villacher Straße 1, A-9800 Spittal an der Drau, Carinthia, Austria; n.randl@cuas.at (N.R.); t.meszoely@cuas.at (T.M.); p.harsanyi@cuas.at (P.H.)

* Correspondence: e.rossi@cuas.at

**Abstract:** The increasing demand on the performance of existing structures, together with their degradation, is among the main drivers towards the development of innovative strengthening solutions. While such solutions are generally aimed at increasing the load-bearing capacity of structural elements, serviceability limit states also play an important role in ensuring the performance and durability of the structure. An experimental campaign was performed to assess the cracking behaviour of reinforced concrete beams strengthened with different typologies of Textile-Reinforced Concrete. The specimens were monitored using Digital Image Correlation (DIC) technology in order to obtain a quantitative evaluation of the evolution of the crack pattern throughout the whole test. Results show the beneficial effects of this retrofitting strategy both at ultimate limit states and serviceability limit states, provide detailed insights on the progression of damage in the specimens and highlight how different parameters impact the cracking behaviour of the tested elements.

**Keywords:** textile-reinforced concrete (TRC); fibre/textile-reinforced concrete (F/TRC); strengthening; cracking behaviour; serviceability limit state; digital image correlation (DIC)

**Citation:** Rossi, E.; Randl, N.; Mészöly, T.; Harsányi, P. Effect of TRC and F/TRC Strengthening on the Cracking Behaviour of RC Beams in Bending. *Materials* **2021**, *14*, 4863. https://doi.org/10.3390/ma14174863

Academic Editor: Francesco Fabbrocino

Received: 27 July 2021
Accepted: 23 August 2021
Published: 27 August 2021



## 1. Introduction

Degradation of concrete constructions is an important issue that seriously affects the durability and functionality of structures. The costs related to maintenance, rehabilitation and upgrade of the existing built environment are extremely high. According to a 2013 ASCE report [1], approximately 11% of US bridges were structurally deficient and 25% functionally obsolete. According to the Federal Highway Administration (FHWA), the replacement and repair cost of the deficient bridges alone was almost USD 76 billion, while the investment backlog for the nation's bridges amounted to USD 121 billion. Similar scenarios can be found worldwide as a consequence of structures' aging and increase in performance demand, such as increased traffic levels. According to a 2004 study [2], 16% of European railway bridges were between 50 and 100 years old, while 55% were between 20 and 50 years old, thus resulting in elevated maintenance and upgrade costs. Germany, as an example, planned in 2018 an investment of EUR 3.9 billion, expected to increase to EUR 4.4 billion in 2021, for the maintenance of federal highways, 37% of which was exclusively for bridge maintenance [3].

A promising technique for the rehabilitation and strengthening of concrete structures consists in the use of Textile-Reinforced Concrete (TRC). Such material, also known as Textile-Reinforced Mortar (TRM) or Fabric-Reinforced Cementitious Matrix (FRCM), is composed of a high strength fabric embedded in a cementitious matrix. Several studies have been performed on the application of such materials, both on concrete and masonry elements, showing very promising results [4–11]. More recently, researchers have been studying an enhanced version of these kinds of materials, consisting of the admixture of short dispersed fibres in the cementitious matrix used to bind the textile [12–21]. Such material will be referred to in the present work as Fibre/Textile-Reinforced Concrete.

Most of the studies performed on such strengthening solutions, however, focus mainly on the increase in load-carrying capacity of the retrofitted element. Only a few works reported a detailed description of the evolution of the crack pattern of the strengthened elements. Yin et al. [22] analysed beams strengthened with TRC and subjected to four-point bending focusing on the effect of such strengthening solution in terms of cracking load and maximum crack width. They observed an increase in the cracking load ranging between 11% and 31%, depending on the number of textile layers, presence of short dispersed fibres and mechanical anchorage elements. In terms of maximum crack width at the yielding load, they recorded a reduction ranging between 57% and 78%. Verbruggen et al. [23,24] performed four-point-bending tests on reinforced concrete beams strengthened both with TRC (using randomly oriented glass fibre mats) and CFRP (carbon fibre-reinforced polymer) strips. The crack pattern and its evolution were analysed using a Digital Image Correlation (DIC) system and an Acoustic Emission (AE) system which enabled the characterization of every crack in the pure bending zone in terms of crack width and horizontal displacement of the tensile chord. They observed that the TRC elements had a beneficial effect on the cracking behaviour of the specimens, increasing the number of cracks while reducing their opening. From the comparison of the two publications, it can also be observed that the geometry of the specimens and strengthening elements influences the cracking behaviour. In a subsequent study [25], the effect of the width of the TRC strengthening element was investigated. The results confirmed such dependency upon the geometry, recording an increased number of cracks and reduced crack width for specimens strengthened with wider TRC elements. Park et al. [26] instead focused on the number of cracks that appeared on the specimens, once again tested in a four-point-bending setup. They analysed both cracks on the whole length of the specimens and the ones that appeared in the pure bending zone. They observed that specimens strengthened with TRC tended to concentrate the cracks in the pure bending zone, resulting in an advantageous distribution.

The following work presents a detailed analysis of the evolution of the crack pattern of Reinforced Concrete (RC) beams strengthened with different typologies of TRC and F/TRC, varying in terms of cementitious material and number of textile layers. Different typologies of mechanical anchors were also installed on some of the specimens. The reported results complement a previous work [20] focused on the analysis of such strengthening solutions in terms of Ultimate Limit State (ULS) and increase in load-carrying capacity. In this work, it is shown how the different parameters influence the evolution of the crack pattern. The cracks were analysed in terms of Crack Opening Displacement (COD), monitoring the evolution of both the maximum and average COD with load, and number of cracks. Furthermore, to assess the performance of the members at Serviceability Limit States (SLS), two representative load levels were chosen and the cracking state was analysed and discussed.

## 2. Materials and Methods

### 2.1. Geometry and Test Setup

A total of 11 beams were strengthened with TRC or F/TRC elements differing in the number of textile layers, typology of concrete and mechanical anchorage configuration. Two more beams were tested as a reference. Each beam was 2.5 m long and had a cross-section of 220 × 450 mm. The internal longitudinal reinforcement consisted of 2 Ø20 steel bars located in the tensile zone and 2 Ø10 steel bars in the compression zone. A total of 20 Ø10 stirrups, with a spacing of 125 mm, were used as transversal reinforcement. The strengthening elements were cast, using different methods (lamination, manual pouring and pumping), against the soffit of the beam over a length of 2.1 m. In order to ensure a good bond between the beams and the strengthening elements, the concrete surface was roughened using a waterjet gun until an average roughness of approximately 1.5 mm, measured with the sand patch method, was obtained. A detailed representation of the specimens can be found in Figure 1a.

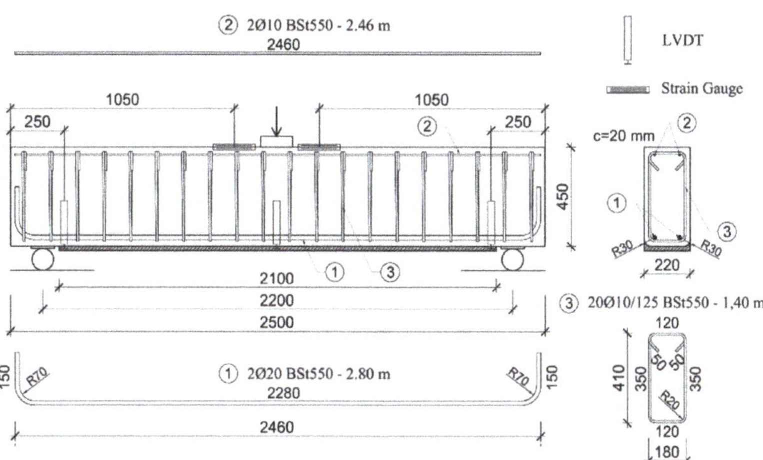


(**a**) Detail of the test set-up

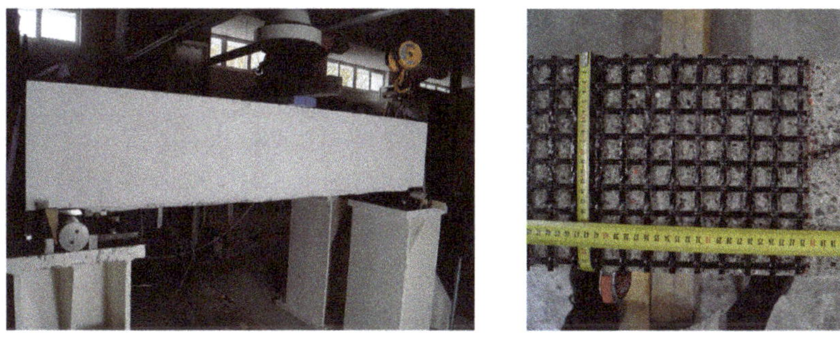

(**b**) Test rig (**c**) Textile reinforcement

**Figure 1.** Geometry and reinforcement of the beam specimens.

All beams were tested using a three-point bending setup, with the supports, consisting of two steel rollers placed at a distance of 2.2 m, thus resulting in a distance of 50 mm from the strengthening layers. A detail of the test setup can be found in Figure 1a, while Figure 1b,c show, respectively, a beam under the test rig, and a detail of the textile. The load introduction system consisted of a hydraulic cylinder, mounted on a steel frame, with a maximum load capacity of 2100 kN and a resolution of 0.01 kN.

Two sets of measuring devices were used to monitor the deformative state of the specimens. The first, more traditional one, consisted of a series of strain gauges and LVDTs (Linear Variable Displacement Transducers). Two strain gauges were placed on top of the compression zone, near the load introduction point, at a distance of 1.05 m from the ends of the beam. One LVDT was placed at the middle of the span to record the deflection. Two more LVDTs were placed at each end of the strengthening layer, fixed on the substrate with the measuring tip resting on aluminium plates which were glued to the strengthening layer. They were used to monitor delamination phenomena. A schematic representation of such measuring instrumentation set can be found in Figure 1a. This set was used to obtain a live evaluation of the behaviour of the specimens during the tests and to assist in a more detailed analysis.

The second set of measuring equipment consisted of a DIC system. Such a solution was chosen due to its ability to accurately monitor the deformative state of the whole surface of the specimens, providing more insights than traditional methods. The reliability of such measuring technology was validated in previous research activities [27–30].

*2.2. Material Characteristics and Strengthening Solutions*

The textile material was the same for every specimen and consisted of one to three layers of a commercially available carbon grid characterised by a 25 mm distance between the rovings' axes, a cross section of the strand of 3.62 $mm^2$ and a textile cross-section of 142 $mm^2/m$ [31]. Two different cementitious materials were used to bind the textile to the specimens. One was a commercially available fine-grain pre-mixed High-Performance Concrete (HPC), while the other was a self-developed UHPFRC (Ultra High Performance Fibre Reinforced Concrete) recipe based on the results obtained in previous research [32] and also successfully used for strengthening flat slabs, however with prefabricated UHPFRC elements [33]. This particular HPC was chosen due to the possibility of applying it by means of lamination, thus avoiding the need for formwork, and because it is explicitly mentioned and mandatory to use in the German technical approval for strengthening with TRC [34]. The UHPFRC material was chosen due to its self-compacting characteristics, higher compressive and tensile strengths, and good bonding performance [35]. The fibres used in such mixture (2.5 vol%), were 5 mm long and had a diameter of 0.15 mm, resulting in an aspect ratio of 33.3. From the basis of these two materials, two additional cementitious matrices were employed: the first one, referred to as HPFRC (High-Performance Fibre-Reinforced Concrete), consists of the admixture of the short dispersed steel fibres with HPC; the second one, referred as UHPC (Ultra High Performance Concrete), has the same components of UHPFRC, without however the steel fibres. The strengthening layer had a length of 2.1 m and a width of 220 mm. Each textile layer consisted of 9 longitudinal yarns distanced 25 mm from each other's axis. The distance between two adjacent textile layers, as well as the distance separating the layer from the concrete substrate or covering the outmost external one, was approximately 5 mm.

Two different kinds of mechanical anchorages were also used. The first ones were short threaded studs, with a diameter of 8 mm and a total length of 62 mm, including 37 mm of embedment length. Such devices were directly shot into the substrate concrete after predrilling a hole of 23 mm depth and 5 mm in diameter. The second ones were resin bonded steel anchors with a diameter of 12 mm and a total length of 220 mm, 160 mm of which were embedded in the substrate. The mechanical anchorages were arranged in three different configurations (Figure 2). Configuration A consisted of four threaded studs positioned 380 mm away from each end of the beams. They had a longitudinal and transversal spacing of 130 mm and 80 mm, respectively. Configuration B consisted of two steel anchors, distanced 110 mm in the longitudinal direction and 30 mm in the transversal one, at each end of the beams. They were placed in a point-symmetrical manner, with the external one at a distance of 270 mm from the end of the beam. Configuration C consisted of the steel anchors positioned as in configuration B with the addition of threaded studs, arranged in a triangular pattern with a distance of approximately 110 mm, covering the remaining span.

The specimens were strengthened using two different application procedures. The first one, involving HPC and HPFRC, was performed with the beams resting on two elevated supports while alternative layers of the cementitious compound and textile fabrics were applied on the bottom side of the beam through lamination. The second one, instead, made use of a formwork holding the textile layers and in which the UHPFRC or UHPC was pumped using a modified snail pump. One specimen strengthened with UHPFRC was also produced without the use of such a pump (manually pouring the fresh concrete in the formwork) in order to assess any influence of the pumping operations. A detailed summary of the strengthening solutions tested in this experimental campaign can be found in Table 1, together with the cube compressive and splitting tensile strengths of the cementitious compounds used to produce the TRC and F/TRC elements (with the exception of the reference beams where the value refers to the substrate concrete). The specimens' ID reflects the different strengthening solutions that were employed. It consists of up to three groups of letters/numbers. The first one (FR—Fibre-Reinforced) indicates F/TRC materials; the second one indicates both the type of cementitious matrix (P—HPC, U—UHPC) and the number of textile layers; the third indicates the configuration of the mechanical anchorage, with the only exception of specimen FR-U1-M, where the last letter refers to the application procedure (M—Manual Pouring). As an example, specimen FR-P2-B refers to a beam strengthened with F/TRC using an HP(FR)C matrix, two layers of textile fabrics and anchorages as in configuration B.

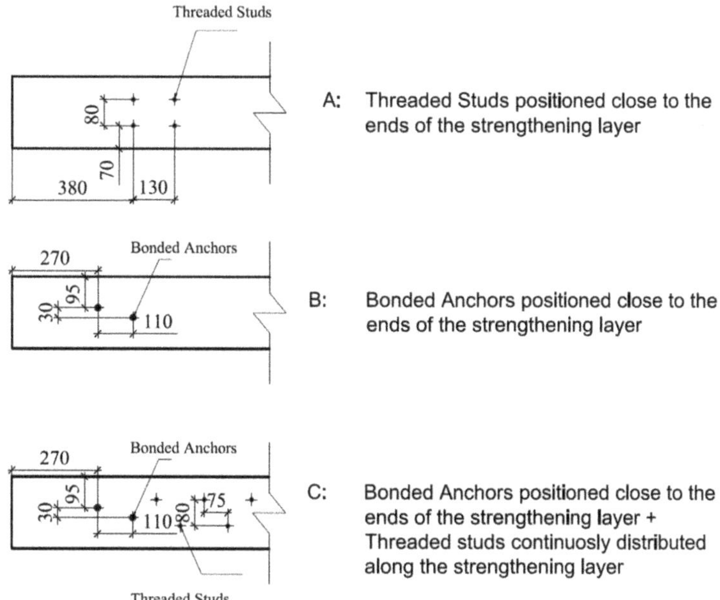


**Figure 2.** Geometrical layout of the mechanical anchorages.

**Table 1.** Strengthening solutions and characteristics of the cementitious materials.

| Specimen ID | Casting Method * | N° of Layers/Anchorage | Cementitious Matrix | Compressive Strength [MPa] | Tensile Strength [MPa] |
|---|---|---|---|---|---|
| REF1 | – | – | – | 55.2 | – |
| P1 | L | 1/– | HPC | 88.7 | 4.8 |
| P2 | L | 2/– | HPC | 81.9 | 4.8 |
| P2-A | L | 2/A | HPC | 80.8 | 5.9 |
| P3-A | L | 3/A | HPC | 92.7 | 6.6 |
| FR-U1-M | M | 1/– | UHPFRC | 169 | 13.0 |
| FR-U1 | P | 1/– | UHPFRC | 159.2 | 11.8 |
| FR-U2-A | P | 2/A | UHPFRC | 146.4 | 13.3 |
| U1 | P | 1/– | UHPC | 184.4 | 5.2 |
| FR-P2-B | L | 2/B | HPFRC | 89.7 | 9.3 |
| FR-U2-B | P | 2/B | UHPFRC | 184.1 | 14.3 |
| P2-C | L | 2/C | HPC | 88.8 | 3.3 |
| REF2 | – | – | – | 59.1 | – |

* L: Lamination; M: Manual Pouring; P: Pumping.

The normal strength concrete (NSC) used to produce the beams was ordered from an external company and was supposed to be a C30/37; however, a higher quality product was delivered, closer to a C40/50. The material of the steel reinforcing bars was a B550 B with a tensile yield strength close to 600 MPa. Both the compressive and tensile tests on the concrete specimens were performed the same day as the beam test. The NSC compressive properties were obtained by testing 150 × 150 × 150 mm cubes. The high compression strength of the compounds used to produce the TRC and F/TRC elements was, instead, obtained using 100 × 100 × 100 mm cubes. The splitting tensile tests were performed on 100 × 200 mm cylinders. Furthermore, in order to ensure that shrinkage of the strengthening elements would not significantly affect the behaviour of the specimens (see, e.g., [36,37]), the beams were carefully inspected and no sign of distress or shrinkage cracks could be detected.

The mechanical and geometrical properties of the textile fabrics, according to its technical sheet, are the following: 3100 MPa and 3300 MPa average tensile strength for the longitudinal and transversal rovings, respectively; and modulus of elasticity greater than 220 GPa and 205 GPa for the longitudinal and transversal rovings, respectively [31].

## 3. Results

### 3.1. Load Carrying Capacity and Failure Modes

As already mentioned, a detailed discussion on how the different parameters that characterize the tested strengthening solutions affect the load-carrying capacity of the beam specimens was presented in a dedicated article [20]. Nevertheless, such results are briefly summarized here to provide a more comprehensive description of the behaviour of the beams and enable a more thorough analysis of the evolution of the crack pattern. All the beams failed in bending, due to the failure of the strengthening layers or, in the case of the reference beams, rupture of one of the longitudinal steel reinforcing bars placed in the tensile zone. A significant degradation of the compression zone was also observed in every specimen. The strengthened specimens exhibited different failure mechanisms that can be summarised as follows:

(a) rupture of the textile fabric;
(b) pull-out of the longitudinal strands from the cementitious matrix;
(c) debonding of the TRC of F/TRC element signalled by the presence of a long horizontal crack at the interface with the NSC;
(d) interlaminar shearing, where a horizontal crack formed along the textile fabric within the strengthening layer;
(e) peeling of the concrete cover, when the debonding of the strengthening element did not happen at the interface between the two materials but inside the NSC, approximately at the height of the longitudinal reinforcing bars.

A schematic representation of the observed failure modes is reported in Figure 3. The results of the test series, in terms of loads associated with the appearance of the first 0.1 mm crack (cracking load), the yielding of the steel reinforcement (yielding load) and maximum loads, together with the observed failure modes, can be found in Table 2.

All the strengthened beams showed an increase both in the yielding and in the maximum loads, ranging between 22.01 kN (P1) and 49.55 kN (P3-A) in the case of yielding loads, and between 19.14 kN (P1) and 153.37 kN (FR-U2-B) in the case of maximum loads, with a more pronounced increase for the specimens strengthened with UHPFRC. Additionally, the introduction of mechanical anchorage systems had, in most cases, a positive effect on the load-carrying capacity of the beams. Such increase is particularly noticeable when the bonded anchors were used, resulting in a gain of the load-carrying capacity between 40.7% and 48%, while, on average, the others solutions experienced an increase of 17.4%.

In terms of failure modes, it can be observed that the beams strengthened with HPC always exhibited interlaminar shearing and, in the case of the specimens without bonded anchors, such distress was the main phenomenon that led to failure. Beams strengthened with UHPC and UHPFRC, on the other hand, exhibited a more diverse pattern of failure modes. When only one textile layer, combined with pumped UHPC or UHPFRC, was used, the main failure mode was rupture of the textile. When, instead, the F/TRC element was applied by manually pouring the fresh concrete (beam FR-U1-M), the failure mode switched to debonding. All the beams with bonded anchors failed by textile pull-out, combined with interlaminar shearing and debonding of the strengthening element, showing that such anchors were able to effectively redistribute the load into the substructure even with a high degree of distress.

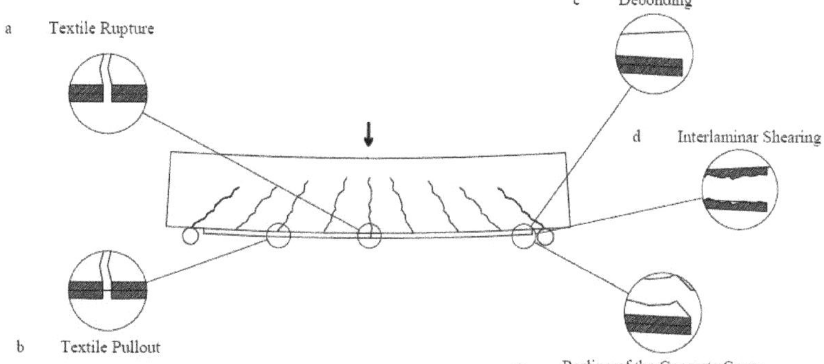


**Figure 3.** Failure modes of retrofitted beams.

**Table 2.** Yielding Loads, Failure Loads and Failure Modes of the tested specimens.

| Specimen ID | Cracking Load (kN) [1] | Yielding Load (kN) | Maximum Load (kN) | Peak Load Increase (%) | Failure Mode |
|---|---|---|---|---|---|
| REF-1 | 73.66 | 275.78 | 316.87 | | – |
| P1 | 92.62 | 297.79 | 337.68 | 6.29 | d |
| P2 | 99.08 | 324.08 | 367.95 | 15.81 | d/e |
| P2-A | 91.76 | 321.90 | 387.71 | 22.03 | d/e |
| P3-A | 90.53 | 357.15 | 416.94 | 31.23 | d/e |
| FR-U1-M | 93.64 | 317.67 | 370.41 | 16.59 | c |
| FR-U1 | 93.27 | 316.56 | 370.46 | 16.61 | a |
| FR-U2-A | 117.66 | 342.36 | 370.35 | 16.57 | e |
| U1 | 84.97 | 304.87 | 362.13 | 13.98 | a/d |
| FR-P2-B | 131.01 | 342.86 | 447.10 | 40.73 | b/d |
| FR-U2-B | 113.65 | 356.74 | 470.24 | 48.01 | b/c |
| P2-C | 95.80 | 334.57 | 447.28 | 40.78 | b/d |
| REF-2 | – | 274.53 | 318.54 | | – |

[1] The cracking load was defined as the load at which a 0.1 mm crack could be detected.

### 3.2. Damage Evolution and Crack Patterns

With the DIC system, it was possible to obtain detailed results on the damage evolution of the beam specimens. Such analysis was performed measuring each crack that appeared on the specimens together with their progressive opening until the maximum load was reached. The cracks were measured by means of virtual sensors just above the height where peeling of the concrete cover would take place, highlighted by a nearly horizontal strain concentration visible from the DIC measurements (Figure 4). Such measurement location was chosen since it is approximately located in the middle of the effective tension zone as defined in Eurocode 2 [38]. Furthermore, several authors suggested that corrosion phenomena might be related to the cracks crossing the reinforcement [39–41]. Several threshold values, representative of the damage state in the beams, were defined. These values, corresponding to crack openings of 0.1 mm, 0.2 mm, 0.3 mm and 0.4 mm, were chosen since they are the limits imposed by several international codes for the protection of concrete structural elements and depend on the aggressiveness of the environment [38,42,43].

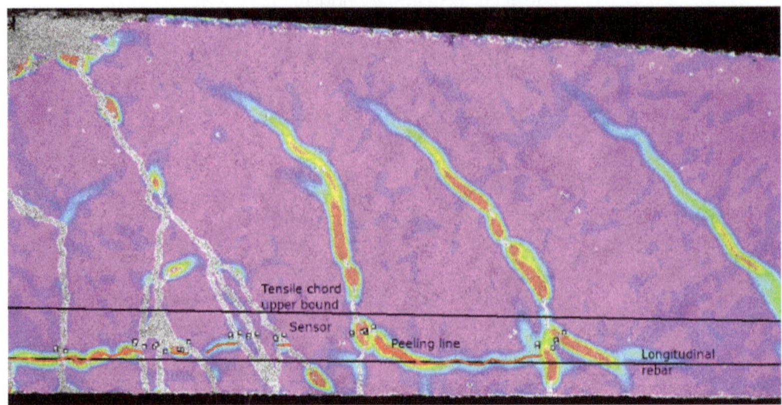


**Figure 4.** Sensor placement.

The crack evolution was monitored both in terms of maximum and average crack opening. Figure 5 reports such data, together with deflection, plotted against load. All the curves showing crack opening can be divided into three stages: the first stage, where the structure can be considered substantially uncracked—in this stage, several micro-cracks are formed but crack opening does not increase with load; the second stage, where cracks start to open with increasing load; and the third stage, where yielding of the steel reinforcing bars occurs and there is a strong increase in the crack opening rate. Each specimen entered the second stage approximately at 75 kN and every strengthening solution resulted in an important reduction in both crack opening at a given load and crack opening rate.

In terms of crack control capability of the strengthening solutions, all beams show the effectiveness of such type of subsequently attached F/TRC layers. In detail, however, the beams strengthened with textile-reinforced UHPC or UHPFRC show a less effective crack control capability when compared to the ones strengthened with HPC, both in terms of maximum and average crack opening, with a more pronounced difference in the former case. Particularly noticeable is the behaviour of the beams where the bonded anchors were installed, which exhibited the overall best behaviour, showing that, even at low loads, these elements enabled a good stress redistribution between the beams and the strengthening layers.

The crack pattern of each specimen can be observed in Figure 6. Each picture was taken from the DIC evaluation at a load level corresponding to the yielding of the longitudinal steel reinforcing bars and the colours represent the major principal strain. It is important to note that the results reported in Figure 6 are only meant to show the crack pattern and, the colour scale was calibrated only to highlight the cracks.

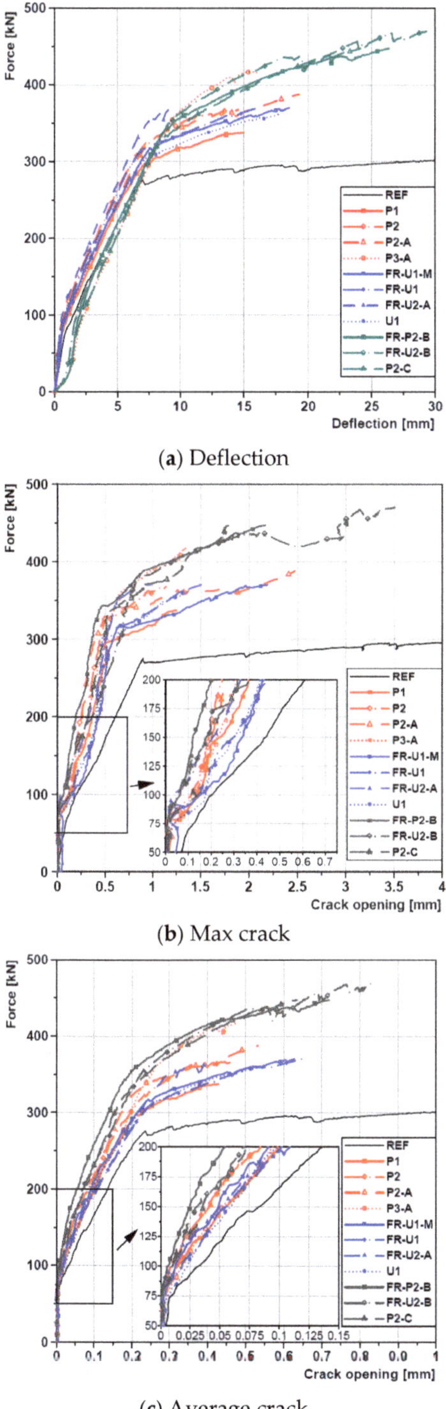


(**a**) Deflection

(**b**) Max crack

(**c**) Average crack

**Figure 5.** Force—Deflection/Crack Opening curves.

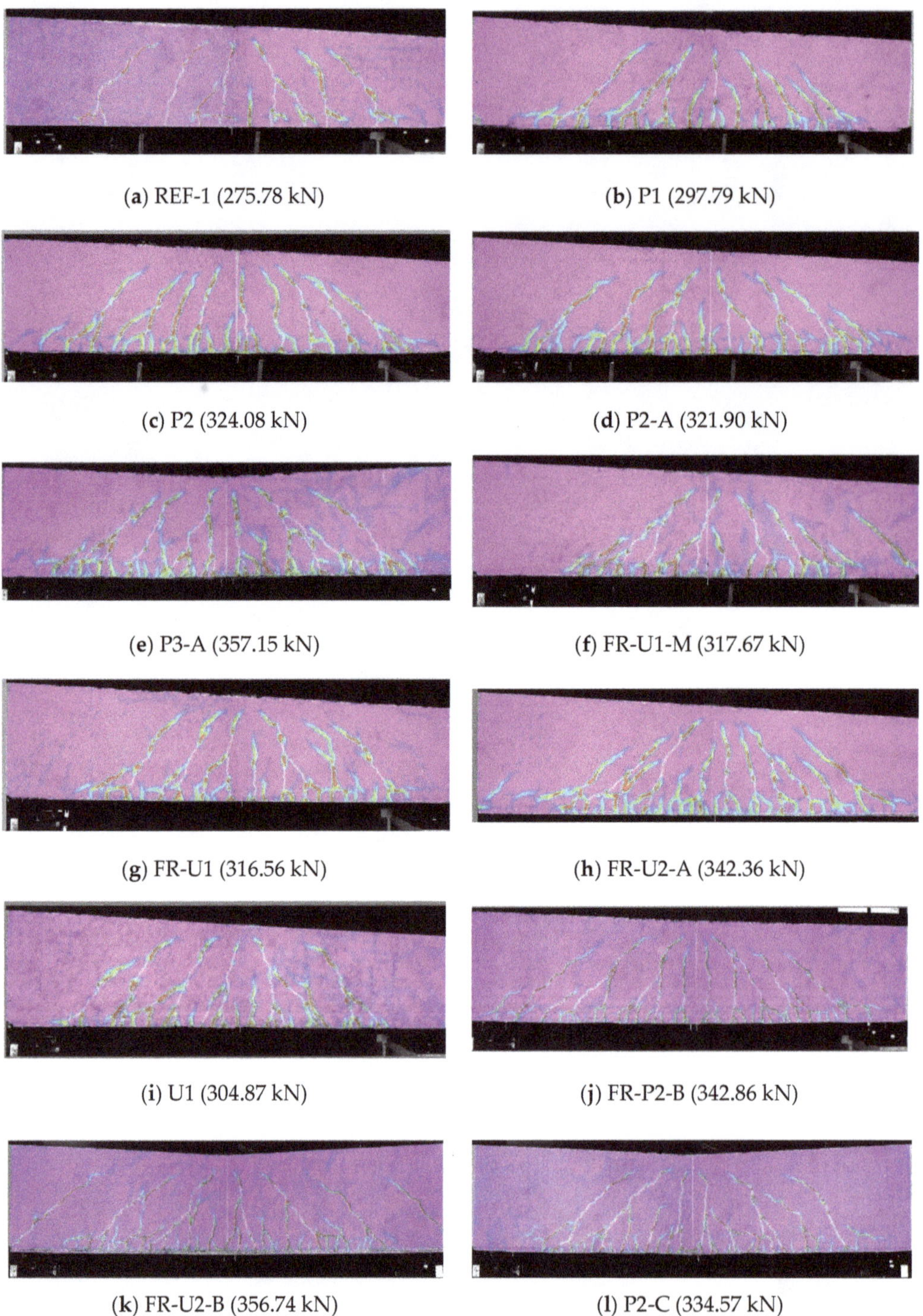

(a) REF-1 (275.78 kN) (b) P1 (297.79 kN)

(c) P2 (324.08 kN) (d) P2-A (321.90 kN)

(e) P3-A (357.15 kN) (f) FR-U1-M (317.67 kN)

(g) FR-U1 (316.56 kN) (h) FR-U2-A (342.36 kN)

(i) U1 (304.87 kN) (j) FR-P2-B (342.86 kN)

(k) FR-U2-B (356.74 kN) (l) P2-C (334.57 kN)

**Figure 6.** Crack patterns of all tested beams at the steel yielding load highlighted by the major principal strain.

Number of Cracks

The results are reported in Figure 7. A crack was counted when its opening overcame the thresholds previously described, furthermore, it is important to mention that crack closure, which was sometimes observed and probably caused by stress redistribution phenomena, was not taken into consideration. The graphs of Figure 7 report the total number of cracks that appeared on the specimens and that were larger than the reference values. The number of cracks that had an opening between two threshold values can be calculated by subtracting the respective curves.

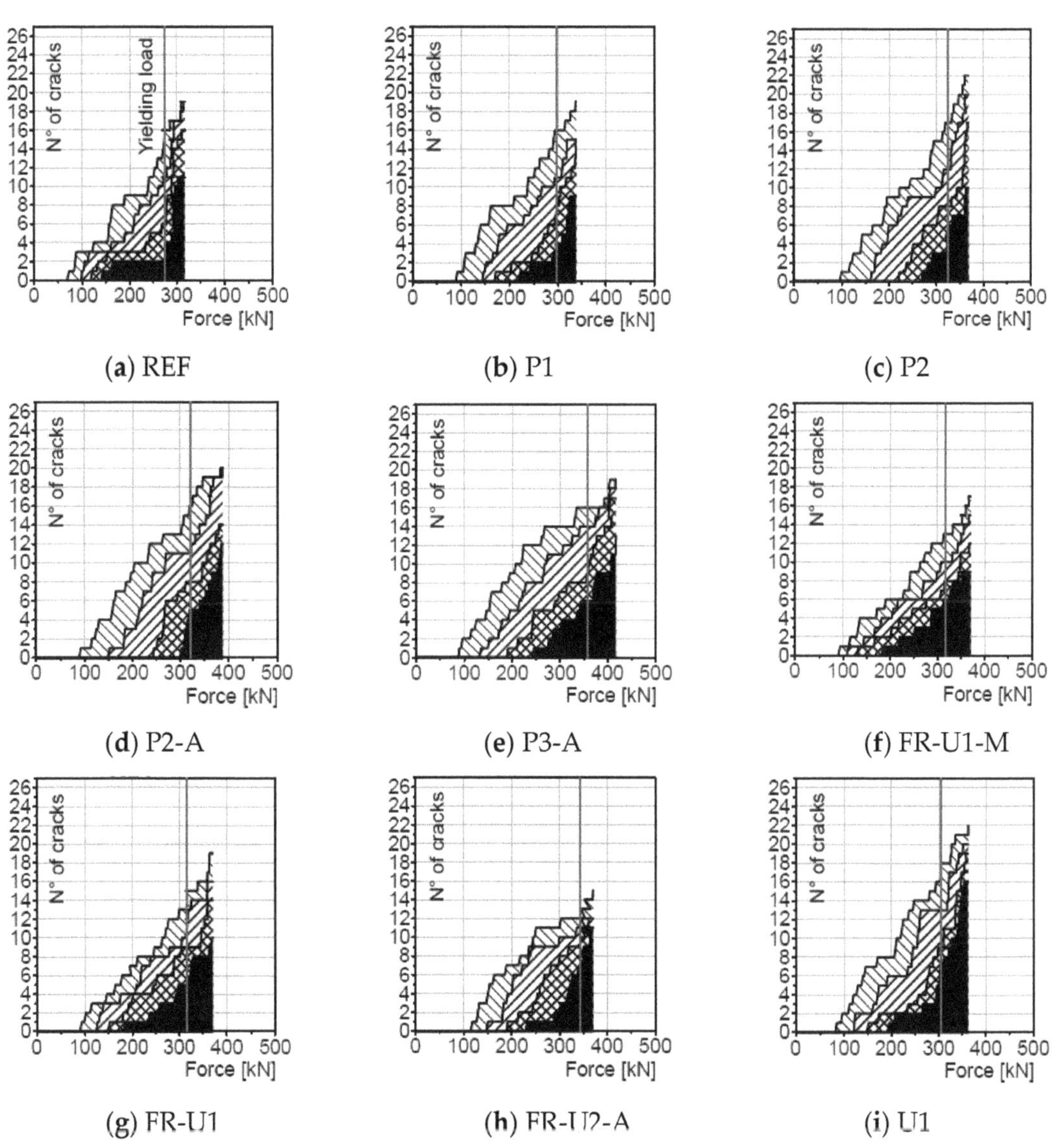


(**a**) REF (**b**) P1 (**c**) P2

(**d**) P2-A (**e**) P3-A (**f**) FR-U1-M

(**g**) FR-U1 (**h**) FR-U2-A (**i**) U1

**Figure 7.** *Cont.*

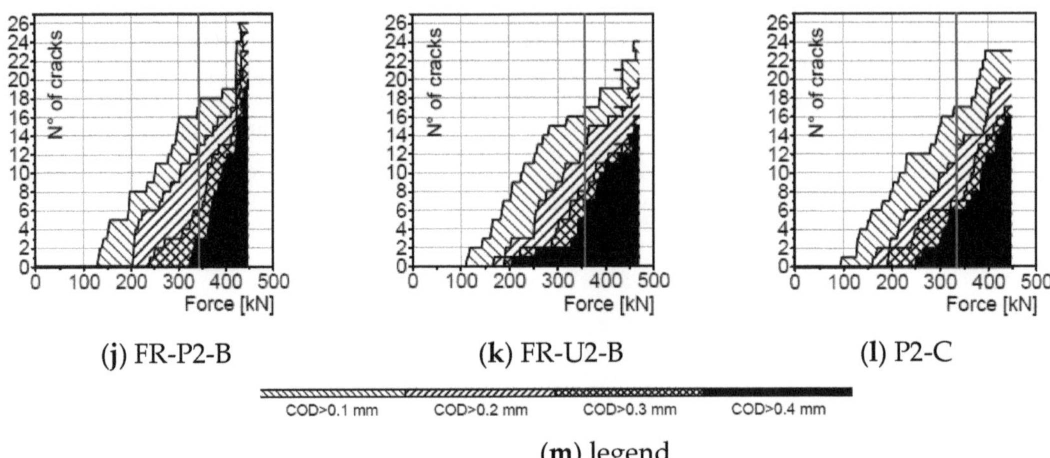


**Figure 7.** N° of cracks—Force diagrams.

The reference beam showed a nearly linear increase in the number of smaller cracks (0.1 mm and 0.2 mm) while the larger cracks (>0.3 mm) tended to form after yielding the steel reinforcement. Upon reaching the maximum load, nearly all the cracks reached an opening bigger than 0.3 mm and more than half overcame the threshold of 0.4 mm Specimen P1 exhibited a more pronounced increase in the larger cracks upon reaching the yielding load and a high percentage of cracks overcoming the 0.3 mm and 0.4 mm thresholds. The formation of the larger cracks happened at significantly higher load levels Referring to specimen P2, the smaller cracks occurred at a load level comparable to beam P1 while the larger ones formed at significantly higher loads. They started forming before the yielding load and their increase rate was nearly linear and not affected by the yielding of the reinforcing bars. Similar consideration can also be taken for beam P2-A, where 0.1 mm and 0.2 mm cracks started to appear more or less at the same load level as the previous specimens. The larger cracks, instead, formed at higher loads; 0.4 mm cracks, in particular, appeared just before the yielding load. Contrarily to what was seen in the previous beams, at the maximum load, a high number of cracks overcame the 0.4 mm threshold. Beam P3-A was characterized by an earlier appearance of both small and large cracks. Upon reaching the yielding load, half of the cracks had a crack opening bigger than 0.3 mm and more than a third overcame the 0.4 mm threshold. It is also possible to see that at the maximum load nearly every crack was wider than 0.3 mm. The lower performance of such a solution could be the result of some early damage in the strengthening layer, possibly caused by the weight of the concrete in its fresh state.

Beams FR-U1-M and FR-U1 had a slightly lower number of small cracks at the yielding and maximum load levels than P1. The 0.4 mm cracks, however, tended to dominate the crack patterns; they appeared at significantly lower loads and increased nearly linearly with the load. This could be the result of the higher tensile strength of the material, which successfully prevented the formation of small cracks while concentrating the elongation of the strengthening element in fewer points leading to a crack localization. With reference to beam FR-U2-A, the number of smaller cracks increased rapidly, shortly after their appearance, and slowed down in the later stages. The largest cracks had a fast appearing rate close to the reaching of the yielding load. Overall, this beam showed the lowest total number of cracks, most of them, however, had already overcome the 0.4 mm threshold at the yielding stage. Beam U1 exhibited a behaviour similar to beam P1. The smaller cracks appeared at relatively small loads with their number increasing nearly linearly, while the larger ones started to increase in number shortly before reaching the yielding load. The total number of cracks, however, is higher than specimen P1, both at the yielding and maximum load. Furthermore, the crack pattern is characterised by the presence of larger cracks, again both at the yielding and maximum load levels.

Beams FR-P2-B, FR-U2-B and P3-C were characterised by the highest number of cracks of the whole test series. In the case of specimen FR-P2-B, before reaching the yielding load, the crack pattern was characterised by the presence of many small cracks (0.1 mm and 0.2 mm) and only a few larger cracks. After reaching such load level the larger cracks increased rapidly, leading to a crack pattern characterised by the presence of mainly 0.4 mm cracks. Referring to beams FR-U2-B and P3-C, an earlier appearance of the larger cracks can be observed. Furthermore, at the maximum load, while 0.4 mm cracks are still predominant, a non-negligible amount of smaller cracks is present.

## 4. Discussion

From the previously described results, several interesting aspects on the ability of TRC and F/TRC materials to control the crack formation and evolution on beams subjected to bending loads arise. The first considerations can be made on the variation of the load values at the reference crack openings. The results are reported in Table 3 and Figure 8 in terms of both load and load increase, calculated as the difference between the load of the strengthened beam ($P_{cr}$) and the reference beam ($P_{cr}^{REF}$) and divided by the cracking load of the reference beam (Equation (1)). The cracking load was defined as the load at which a crack has an opening overcoming one of the previously mentioned thresholds.

$$\Delta P_{cr} = \frac{P_{cr} - P_{cr}^{REF}}{P_{cr}^{REF}} \quad (1)$$

**Table 3.** Cracking loads and relative increase.

| Beam ID | Load at Crack Opening (kN) | | | | Load Increase (%) | | | |
|---|---|---|---|---|---|---|---|---|
| | 0.1 mm | 0.2 mm | 0.3 mm | 0.4 mm | 0.1 mm | 0.2 mm | 0.3 mm | 0.4 mm |
| REF-1 | 73.66 | 100.54 | 121.69 | 143.12 | – | – | – | – |
| P1 | 92.62 | 145.92 | 171.80 | 222.00 | 25.7 | 45.1 | 41.2 | 55.1 |
| P2 | 99.08 | 163.21 | 222.10 | 271.80 | 34.5 | 62.3 | 82.5 | 89.9 |
| P2-A | 91.76 | 151.49 | 245.6 | 302.87 | 24.6 | 50.7 | 101.8 | 111.6 |
| P3-A | 90.53 | 134.95 | 191.94 | 244.88 | 22.9 | 34.2 | 57.7 | 71.1 |
| FR-U1-M | 93.64 | 111.67 | 131.83 | 185.12 | 27.1 | 11.1 | 8.3 | 29.3 |
| FR-U1 | 93.27 | 126.21 | 151.57 | 183.52 | 26.6 | 25.5 | 24.6 | 28.2 |
| FR-U2-A | 117.66 | 150.25 | 191.99 | 232.65 | 59.7 | 49.4 | 57.8 | 62.6 |
| U1 | 84.97 | 123.40 | 151.68 | 195.27 | 15.4 | 22.7 | 24.6 | 36.4 |
| FR-P2-B | 131.01 | 204.07 | 240.89 | 325.05 | 77.9 | 103.0 | 98.0 | 127.1 |
| FR-U2-B | 113.65 | 169.20 | 190.67 | 206.83 | 54.3 | 68.3 | 56.7 | 44.5 |
| P2-C | 95.80 | 160.92 | 192.00 | 252.06 | 30.1 | 60.1 | 57.8 | 76.1 |

All the tested strengthening solutions provided an increase in the cracking load of each crack typology. Higher increases in the cracking load are, in most cases, associated with the larger crack thresholds (0.3 mm and 0.4 mm). Focusing on the beams strengthened with HPC, it is possible to see how increasing the number of layers from one to two resulted in a significant increase in the cracking load, especially for the larger cracks. The installation of the threaded studs resulted in a cracking load increase for 0.3 mm and 0.4 mm cracks, but was less effective for the smaller cracks (0.1 mm and 0.2 mm). The addition of a third layer led to a lower increase in the cracking load. Once again, this phenomenon is attributed to the weight of the strengthening element that, while the concrete was still in its fresh state, might have experienced some damage.

The cracking load increase in the strengthening solutions where UHPC or UHPFRC was used was less effective. Comparing specimens FR-U1-M and FR-U1, it is possible to see that both of them achieved similar increases for 0.1 mm and 0.4 mm cracks. The increase for 0.2 mm and 0.3 mm cracks for beam FR-U1-M, instead was significantly lower. This phenomenon can be attributed to the lower viscosity that fresh UHPFRC exhibited when pumped, which, in turn, enabled better bonding performances. Specimen U1, on the other hand, showed a worse performance concerning 0.1 mm cracks but a higher load increase for 0.4 mm cracks, when compared to specimen FR-U1. The addition of a second textile layer, as in specimen FR-U2-A, resulted in a higher cracking load increase for every threshold.

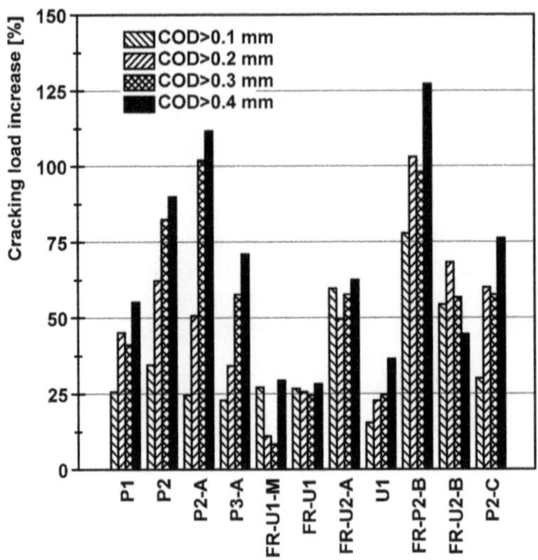


**Figure 8.** Increase in cracking load.

Beam FR-P2-B showed the best results in terms of increase in cracking load, with an increase of more than 125% for 0.4 mm cracks. The results of specimen FR-U2-B are similar to specimen FR-U2-A, with a better performance in the case of 0.2 mm cracks and a worse one for 0.4 mm cracks. The cracking load increase in beam P2-C is similar to P2 and P2-A in terms of smaller cracks. The introduction of the bonded anchors, both in configuration FR-U2-B and P2-C, resulted in lower performance in the case of 0.3 mm and 0.4 mm cracks when compared to specimens FR-U2-A and P2-A, respectively. Analysing the load increase associated with 0.1 mm cracks a particular behaviour arises. Nearly all specimens exhibited very similar values of cracking load increase. The main outliers are the specimens strengthened with a combination of F/TRC and mechanical anchorages (beams FR-U2-A, FR-P2-B and FR-U2-B). This phenomenon suggests that such combination enables an earlier activation of the mechanical anchorage, making such solutions particularly suited for crack control in rather aggressive environments.

To further characterise the behaviour of the specimens, the crack patterns were analysed at two different discrete load levels. Such loads, defined approximately as one-half (150 kN) and two-thirds (200 kN) of the ultimate load of the reference beams, are meant to represent Serviceability Limit State (SLS) loading conditions. The 150 kN load should be representative of the normal usage of the structure, while the 200 kN load should represent a more demanding condition (e.g., increase in traffic loads which were not foreseeable during the design phase) which would justify a strengthening intervention.

### *4.1. Number of Cracks at the Reference Loads*

Referring to Figure 9, it is possible to see that the strengthened beams exhibited, on average, a slightly higher number of 0.1 mm cracks at 150 kN than the reference beam. Cracks with a higher threshold appeared in a significantly lower number and, in the case of 0.4 mm cracks, did not appear at all. The total number of cracks for beams P1 to P3-A is in the same range as the reference beam, however, no cracks larger than 0.3 mm could be observed.

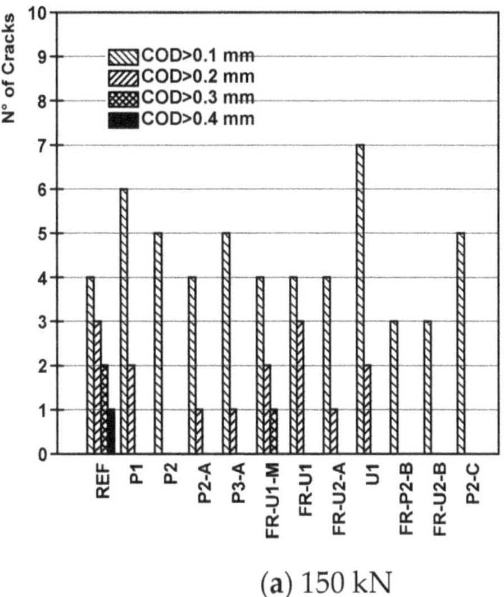


(**a**) 150 kN

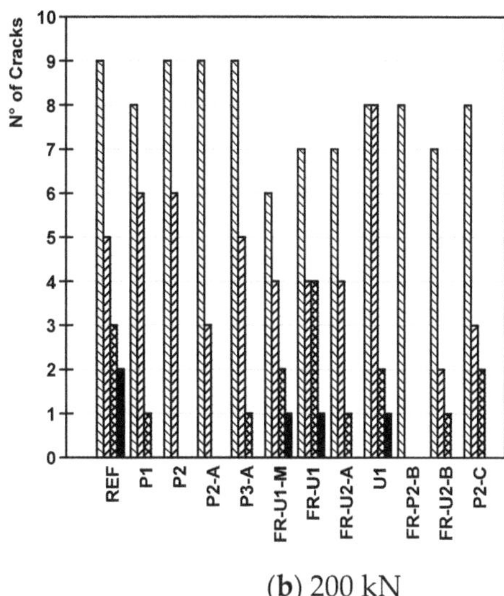


(**b**) 200 kN

**Figure 9.** N° of cracks at reference loads.

Beams FR-U1-M to FR-U2-A had the same number of 0.1 mm cracks of the reference beam at 150 kN. A lower number of 0.2 mm cracks was recorded for beams FR-U1-M and FR-U2-A. No 0.3 mm or 0.4 mm cracks could be observed, with the exception of specimen FR-U1-M, which had a single crack overcoming such threshold. Beam U1 had a significantly higher number of 0.1 mm cracks. Once again, no 0.3 mm crack or larger was recorded. The beams where the bonded anchors were installed exhibited only 0.1 mm cracks at 150 kN. Where short dispersed steel fibres were added to the cementitious matrix (FR-P2-B and FR-U2-B), a maximum of three cracks was recorded. From such results, it is clear that the use of TRC or F/TRC as strengthening solutions is able to increase the performance of the retrofitted elements at the SLS level, independently of the implemented solution. The use of bonded anchors, together with short dispersed fibres and a two-layer configuration, seems the most suited to protect structures against degradation.

At 200 kN, every beam recorded a significant increase in the number of cracks and, contrarily to what was observed at 150 kN, every strengthened specimen exhibited an equal or lower number of 0.1 mm cracks than the reference beam. Focusing on the reference beam, the increase was observed in every crack typology, with a less pronounced effect the bigger the crack. The specimens strengthened with HPC and without bonded anchors (from beam P1 to beam P3-A) had approximately the same number of 0.1 mm and 0.2 mm cracks as the reference beam; however, (P1 and P3-A) 0.3 mm cracks appeared only in two cases. Specimens FR-U1-M, FR-U1 and FR-U2-A showed a less pronounced crack increase for the smallest cracks, recording a total number of cracks lower than the reference beam. Once more, these specimens were characterised by the presence of larger cracks. Beam U1 showed a negligible increase in the number of 0.1 mm cracks with a width of 0.2 mm however significantly increased in numbers from 2 at 150 kN to 8 at 200 kN. Furthermore, every recorded crack showed an opening of at least 0.2 mm. The beams where the bonded anchors were installed also showed an important increase in the number of 0.1 mm cracks. The formation of the larger cracks was strongly reduced, in particular in the case of specimen FR-P2-B, where no crack overcame the 0.2 mm threshold. This phenomenon is probably due to the capability of the bonded anchors to effectively redistribute the stresses from the strengthening layers to the internal reinforcement of the beams.

These results confirm the effectiveness of TRC and F/TRC strengthening solutions, not only for ULS scenarios, but also at SLS levels where crack control can be a governing factor. Within this framework, it is thus possible to attribute to the beams strengthened with fine-grain HPC better results. The installation of the bonded anchors also proved to be an effective way of further increasing the performance. This is particularly evident in the case of beam FR-P2-B, where, at 200 kN, the specimen could still comply with the requirements of most international codes with crack opening smaller than 0.2 mm.

### 4.2. Maximum and Average Crack Opening at the Reference Loads

Lastly, the results were analysed at the reference loads in terms of maximum and average crack opening (Figure 10). All retrofitted beams exhibited a lower maximum crack opening than the reference beam, especially at 200 kN. A reduction in such parameter can be seen between beams P1 and P2 and beams FR-U1 and FR-U2-A, showing that the addition of the second textile layer helps controlling crack opening. Once more, the presence of a third layer resulted in a larger maximum crack opening. The beams strengthened using UHPFRC or UHPC, on average, showed higher values of maximum crack opening both at 150 kN and 200 kN. The installation of the bonded anchors proved to be strongly effective in reducing the maximum crack opening at low load levels (150 kN). The strengthening system used in beam FR-P2-B, in particular, proved to be very effective, recording the smallest maximum crack opening of the whole series, both at 150 kN and 200 kN. The other two specimens where the bonded anchors were placed had a relatively low maximum crack opening at 150 kN, but were subjected to a strong increase when a load of 200 kN was reached, with a more evident effect in the case of beam FR-U2-B. Nevertheless, the maximum crack opening was nearly half of what was observed in the reference beam, showing that such solutions are not only viable options, but can also provide an important performance increase.

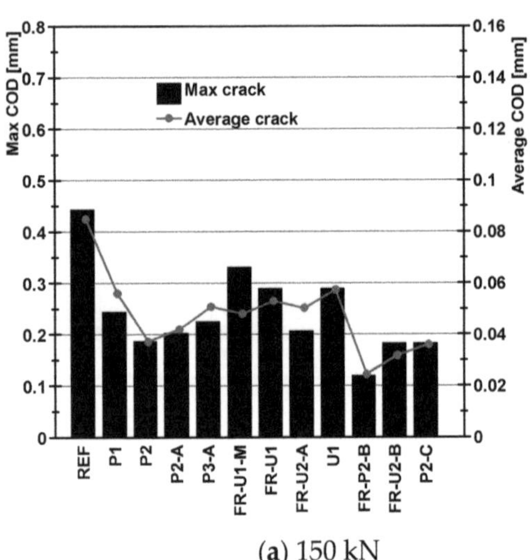


(**a**) 150 kN

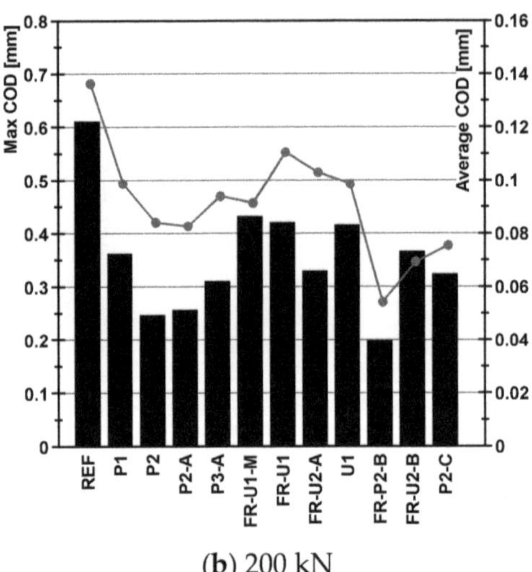


(**b**) 200 kN

**Figure 10.** Maximum and average crack opening at reference loads.

In terms of average crack opening, it can be seen that all the strengthened beams showed reduced values in comparison with the reference specimen. The best results were achieved when the bonded anchors were installed (FR-P2-B, FR-U2-B and P2-C) with the lowest values recorded by specimen FR-P2-B. Comparing them to specimens P1 to P3-A, it can be seen that the performance increase is particularly noticeable at 200 kN, probably due to a more effective stress redistribution. Specimens FR-U1-M to U1, while still showing a reduction in the average crack opening, resulted to be less effective. This can be attributed to the presence of a reduced number of cracks, which, however, resulted in a higher average crack opening.

These results confirm once again the suitability of TRC and F/TRC retrofitting strategies for the load increase and the control of the cracking behaviour in RC beams subjected to bending loads, thus enhancing their performance against environmental actions. Furthermore, it can be observed that fine-grain HPC exhibited generally better performances in reducing the crack width. The introduction of short threaded studs, as in configuration A, seemed to have a counter-productive effect in reducing the maximum crack opening. The longer bonded anchors, instead, proved to be more effective, especially in reducing the average crack opening at higher loads.

## 5. Conclusions

An experimental campaign, aimed at identifying the effect of TRC and F/TRC strengthening elements upon the cracking behaviour of beams subjected to bending loads, was performed at the structural lab of Carinthia University of Applied Sciences. Different strengthening solutions, varying in terms of the number of textile layers, cementitious materials and configuration of mechanical anchorages, were analysed. From these results, the following conclusions can be made:

- Strengthening reinforced concrete beams with TRC or F/TRC materials proved to be an effective and viable solution, not only for retrofitting, but also for a reduction in damage and crack opening control, thus increasing their performance.
- Each strengthening solution resulted in an increase in the cracking load (between 8.3% and 127.1%), depending on the number of textile layers, cementitious matrix, and type and arrangement of mechanical anchorages.
- The combination of short dispersed steel fibres and mechanical anchorage devices has the highest potential in increasing the cracking load (between 59.7% and 77.9%) associated with very small cracks (0.1 mm).
- The use of fine-grain HPC in combination with textile reinforcement proved to outperform UHPFRC in terms of cracking load increase, number of large cracks and ability to reduce both the maximum and average crack opening.
- Bonded anchors, used as end anchorage of the strengthening TRC and F/TRC layer, were very effective in preventing the formation of larger cracks and in reducing both the maximum and average crack opening. They also provided the best performance increase in terms of cracking load and the number of cracks at low load levels only when they were combined with short dispersed steel fibres. Furthermore, all beams strengthened with bonded anchors were able to achieve the highest gain in maximum strength compared to beams without those anchoring elements.

**Author Contributions:** Conceptualization, E.R., N.R., T.M. and P.H.; methodology, E.R., N.R., T.M. and P.H.; validation, E.R., N.R., T.M. and P.H.; formal analysis, E.R.; investigation, E.R., N.R., T.M. and P.H.; resources, N.R.; data curation, E.R.; writing—original draft preparation, E.R.; writing—review and editing, N.R., T.M. and P.H.; visualization, E.R.; supervision, N.R.; project administration, N.R.; funding acquisition, N.R. All authors have read and agreed to the published version of the manuscript.

**Funding:** This research was funded by the Austrian Research Promotion Agency (FFG), grant number 866881.

**Institutional Review Board Statement:** Not applicable.

**Informed Consent Statement:** Not applicable.

**Data Availability Statement:** All data that support the findings of this study are available from the corresponding author upon reasonable request.

**Acknowledgments:** The Authors would like to thank the companies Hilti Corporation, solidian GmbH and Pagel Spezial-Beton GmbH & Co. KG for the supply of material and equipment.

**Conflicts of Interest:** The authors declare no conflict of interest.

## References


1. American Society of Civil Engineers. *Report Card for America's Infrastructure*; American Society of Civil Engineers: Reston, VA, USA, 2013.
2. Bell, B.; Network Rail. European Railway Bridge Demography. 2004. Technical Report Number: WP1-02-T-040531-R-Deliverable D 1.2. Available online: http://bridgeforum.org/files/pub/2006/sustainable-bridges/WP1-02-T-040531-R-Deliverable%20D%201.2.pdf (accessed on 29 July 2021).
3. Bundesministerium Für Verkehr und Digitale Infrastruktur (German Federal Ministry of Transport and Digital Infrastructure), Stand der Modernisierung von Straßenbrücken der Bundesfernstraßen 2018. Available online: https://www.bmvi.de/SharedDocs/DE/Anlage/StB/bericht-stand-der-modernisierung-von-strassenbruecken-2018.pdf?__blob=publicationFile (accessed on 29 July 2021).
4. Awani, O.; El-Maaddawy, T.; Ismail, N. Fabric-Reinforced Cementitious Matrix: A Promising Strengthening Technique for Concrete Structures. *Constr. Build. Mater.* **2017**, *132*, 94–111. [CrossRef]
5. Nobili, A.; Falope, F.O. Impregnated Carbon Fabric-Reinforced Cementitious Matrix Composite for Rehabilitation of the Finale Emilia Hospital Roofs: Case Study. *J. Compos. Constr.* **2017**, *21*, 05017001. [CrossRef]
6. Bencardino, F.; Carloni, C.; Condello, A.; Focacci, F.; Napoli, A.; Realfonzo, R. Flexural Behaviour of RC Members Strengthened with FRCM: State-of-the-Art and Predictive Formulas. *Compos. Part B Eng.* **2018**, *148*, 132–148. [CrossRef]
7. Koutas, L.N.; Tetta, Z.; Bournas, D.A.; Triantafillou, T.C. Strengthening of Concrete Structures with Textile Reinforced Mortars: State-of-the-Art Review. *J. Compos. Constr.* **2019**, *23*, 03118001. [CrossRef]
8. Tarque, N.; Salsavilca, J.; Yacila, J.; Camata, G. Multi-Criteria Analysis of Five Reinforcement Options for Peruvian Confined Masonry Walls. *Earthq. Struct.* **2019**, *17*, 205–219. [CrossRef]
9. Wakjira, T.G.; Ebead, U. Experimental and Analytical Study on Strengthening of Reinforced Concrete T-Beams in Shear Using Steel Reinforced Grout (SRG). *Compos. Part B Eng.* **2019**, *177*, 107368. [CrossRef]
10. Salsavilca, J.; Yacila, J.; Tarque, N.; Camata, G. Experimental and Analytical Bond Behaviour of Masonry Strengthened with Steel Reinforced Grout (SRG). *Constr. Build. Mater.* **2020**, *238*, 117635. [CrossRef]
11. Funari, M.F.; Verre, S. The Effectiveness of the DIC as a Measurement System in SRG Shear Strengthened Reinforced Concrete Beams. *Crystals* **2021**, *11*, 265. [CrossRef]
12. Li, B.; Xiong, H.; Jiang, J.; Dou, X. Tensile Behavior of Basalt Textile Grid Reinforced Engineering Cementitious Composite. *Compos. Part B Eng.* **2019**, *156*, 185–200. [CrossRef]
13. Deng, M.; Dong, Z.; Zhang, C. Experimental Investigation on Tensile Behavior of Carbon Textile Reinforced Mortar (TRM) Added with Short Polyvinyl Alcohol (PVA) Fibers. *Constr. Build. Mater.* **2020**, *235*, 117801. [CrossRef]
14. Dong, Z.; Deng, M.; Zhang, C.; Zhang, Y.; Sun, H. Tensile Behavior of Glass Textile Reinforced Mortar (TRM) Added with Short PVA Fibers. *Constr. Build. Mater.* **2020**, *260*, 119897. [CrossRef]
15. Li, T.; Deng, M.; Dong, Z.; Zhang, Y.; Zhang, C. Masonry Columns Confined with Glass Textile-Reinforced High Ductile Concrete (TRHDC) Jacket. *Eng. Struct.* **2020**, *222*, 111123. [CrossRef]
16. Mészöly, T.; Ofner, S.; Randl, N. Effect of Combining Fiber and Textile Reinforcement on the Flexural Behavior of UHPC Plates. *Adv. Mater. Sci. Eng.* **2020**, *2020*, 9891619. [CrossRef]
17. Yang, X.; Gao, W.-Y.; Dai, J.-G.; Lu, Z.-D. Shear Strengthening of RC Beams with FRP Grid-Reinforced ECC Matrix. *Compos. Struct.* **2020**, *241*, 112120. [CrossRef]
18. Zheng, Y.-Z.; Wang, W.-W.; Mosalam, K.M.; Fang, Q.; Chen, L.; Zhu, Z.-F. Experimental Investigation and Numerical Analysis of RC Beams Shear Strengthened with FRP/ECC Composite Layer. *Compos. Struct.* **2020**, *246*, 112436. [CrossRef]
19. Li, T.; Deng, M.; Jin, M.; Dong, Z.; Zhang, Y. Performance of Axially Loaded Masonry Columns Confined Using Textile Reinforced Concrete (TRC) Added with Short Fibers. *Constr. Build. Mater.* **2021**, *279*, 122413. [CrossRef]
20. Rossi, E.; Randl, N.; Mészöly, T.; Harsányi, P. Flexural Strengthening with Fiber-/Textile-Reinforced Concrete. *ACI Struct. J.* **2021**, *118*, 97–107. [CrossRef]
21. Rossi, E.; Randl, N.; Harsányi, P.; Mészöly, T. Overlapped Joints in Textile Reinforced Concrete with UHPC Matrix: An Experimental Investigation. *Mater. Struct.* **2021**, *54*, 152. [CrossRef]
22. Yin, S.; Lü, H.; Xu, S. Properties and Calculation of Normal Section Bearing Capacity of RC Flexural Beam with Skin Textile Reinforcement. *J. Cent. South Univ.* **2013**, *20*, 1731–1741. [CrossRef]
23. Verbruggen, S.; Aggelis, D.G.; Tysmans, T.; Wastiels, J. Bending of Beams Externally Reinforced with TRC and CFRP Monitored by DIC and AE. *Compos. Struct.* **2014**, *112*, 113–121. [CrossRef]
24. Verbruggen, S.; Tysmans, T.; Wastiels, J. TRC or CFRP Strengthening for Reinforced Concrete Beams: An Experimental Study of the Cracking Behaviour. *Eng. Struct.* **2014**, *77*, 49–56. [CrossRef]
25. Verbruggen, S.; Tysmans, T.; Wastiels, J. Bending Crack Behaviour of Plain Concrete Beams Externally Reinforced with TRC. *Mater. Struct.* **2016**, *49*, 5303–5314. [CrossRef]
26. Park, J.; Park, S.-K.; Hong, S. Experimental Study of Flexural Behavior of Reinforced Concrete Beam Strengthened with Prestressed Textile-Reinforced Mortar. *Materials* **2020**, *13*, 1137. [CrossRef]

27. Mészöly, T.; Randl, N. Derivation of constitutive law for UHPFRC using DIC system. In Proceedings of the AFGC-ACI-fib-RILEM International Symposium on Ultra-High Performance Fibre-Reinforced Concrete, Montpellier, France, 2–4 October 2017; Volume 1, pp. 221–230, ISBN 978-2-35158-166-7.
28. Mészöly, T.; Randl, N. An advanced approach to derive the constitutive law of UHPFRC. *Archit. Civ. Eng. Environ.* **2018**, *11*, 89–96. [CrossRef]
29. Mészöly, T.; Randl, N. Shear Behavior of Fiber-Reinforced Ultra-High Performance Concrete Beams. *Eng. Struct.* **2018**, *168*, 119–127. [CrossRef]
30. Randl, N.; Harsányi, P. Developing Optimized Strengthening Systems for Shear-Deficient Concrete Members. *Struct. Concr.* **2018**, *19*, 116–128. [CrossRef]
31. Solidian GRID Q142/142-CCE-25 Technical Data Sheet, Version: 180703. 2018.
32. Randl, N.; Steiner, T.; Ofner, S.; Baumgartner, E.; Mészöly, T. Development of UHPC Mixtures from an Ecological Point of View. *Constr. Build. Mater.* **2014**, *67*, 373–378. [CrossRef]
33. Ricker, M.; Häusler, F.; Randl, N. Punching Strength of Flat Plates Reinforced with UHPC and Double-Headed Studs. *Eng. Struct.* **2017**, *136*, 345–354. [CrossRef]
34. Deutsches Institut für Bautechnik Verfahren zur Verstärkung von Stahlbeton Mmit TUDALIT (Textilbewehrter Beton) 2016, Technical Approval Number: Z-31.10-182. Available online: https://www.irbnet.de/daten/bzp/2FF929E209/bzp-bfi_3146144.pdf (accessed on 29 July 2021).
35. Graybeal, B.A. *Material Property Characterization of Ultra-High Performance Concrete*; Federal Highway Administration: Georgetown Pike McLean, VA, USA, 2006.
36. Wang, L.; He, T.; Zhou, Y.; Tang, S.; Tan, J.; Liu, Z.; Su, J. The Influence of Fiber Type and Length on the Cracking Resistance, Durability and Pore Structure of Face Slab Concrete. *Constr. Build. Mater.* **2021**, *282*, 122706. [CrossRef]
37. Wang, L.; Jin, M.; Guo, F.; Wang, Y.; Tang, S. Pore structural and fractal analysis of the influence of fly ash and silica fume on the mechanical property and abrasion resistance of concrete. *Fractals* **2021**, *29*, 2140003. [CrossRef]
38. European Committee for Standardization. *EN 1992-1-1, Eurocode 2: Design of Concrete Structures—Part 1-1: General Rules and Rules for Buildings*; CEN: Bruxelles, Belgium, 2004.
39. François, R.; Arliguie, G. Influence of Service Cracking on Reinforcement Steel Corrosion. *J. Mater. Civ. Eng* **1998**, *10*, 14–20. [CrossRef]
40. Castel, A.; Vidal, T.; François, R.; Arliguie, G. Influence of Steel-Concrete Interface Quality on Reinforcement Corrosion Induced by Chlorides. *Mag. Concr. Res* **2003**, *55*, 151–159. [CrossRef]
41. Michel, A.; Solgaard, A.O.S.; Pease, B.J.; Geiker, M.R.; Stang, H.; Olesen, J.F. Experimental Investigation of the Relation between Damage at the Concrete-Steel Interface and Initiation of Reinforcement Corrosion in Plain and Fibre Reinforced Concrete. *Corros. Sci.* **2013**, *77*, 308–321. [CrossRef]
42. ACI Committee 224. *ACI 224R-01 Conctrol of Cracking in Concrete Structures*; American Concrete Institute: Farmington Hills, MI, USA, 2002; ISBN 978-0-87031-056-0.
43. CEB-FIP. *Fib Model Code for Concrete Structures 2010*; Ernst & Sohn: Berlin, Germany, 2013. [CrossRef]

*Article*

# Flexural and Shear Tests on Reinforced Concrete Bridge Deck Slab Segments with a Textile-Reinforced Concrete Strengthening Layer

**Viviane Adam *, Jan Bielak, Christian Dommes, Norbert Will and Josef Hegger**

Institute of Structural Concrete, RWTH Aachen University, 52074 Aachen, Germany; jbielak@imb.rwth-aachen.de (J.B.); cdommes@imb.rwth-aachen.de (C.D.); nwill@imb.rwth-aachen.de (N.W.); jhegger@imb.rwth-aachen.de (J.H.)

* Correspondence: vadam@imb.rwth-aachen.de

Received: 31 August 2020; Accepted: 16 September 2020; Published: 22 September 2020

**Abstract:** Many older bridges feature capacity deficiencies. This is mainly due to changes in code provisions which came along with stricter design rules and increasing traffic, leading to higher loads on the structure. To address capacity deficiencies of bridges, refined structural analyses with more detailed design approaches can be applied. If bridge assessment does not provide sufficient capacity, strengthening can be a pertinent solution to extend the bridge's service lifetime. For numerous cases, applying an extra layer of textile-reinforced concrete (TRC) can be a convenient method to achieve the required resistance. Here, carbon fibre-reinforced polymer reinforcement together with a high-performance mortar was used within the scope of developing a strengthening layer for bridge deck slabs, called SMART-DECK. Due to the high tensile strength of the carbon and its resistance to corrosion, a thin layer with high strength and low additional dead load can be realised. While the strengthening effect of TRC for slabs under flexural loading has already been investigated several times, the presented test programme also covered increase in shear capacity, which is the other crucial failure mode to be considered in design. A total of 14 large-scale tests on TRC-strengthened slab segments were tested under static and cyclic loading. The experimental study revealed high increases in capacity for both bending and shear failure.

**Keywords:** reinforced concrete; bridge deck slabs; strengthening; textile-reinforced concrete; carbon concrete; static loads; fatigue loads; experimental investigations

## 1. Introduction

The age distribution of the existing bridges in combination with several normative changes and significant economic and demographic changes in many industrial countries lead to many structures showing damages. Or else, there are computational deficits due to increased traffic or stricter verification [1–7]. In order to extend the remaining service life of the currently deficient structures, refined design concepts can provide a remedy that allows higher computational load-bearing capacities. In many countries, strategies have been developed for bridge assessment including monitoring, maintenance and design evaluation [8–13]. The aim is to extend the planning horizon in order to postpone the construction of a new replacement bridge for at least a part of the affected bridges. If some of the required verifications cannot be met even after a more detailed structural analysis during bridge assessment, strengthening measures can provide pertinent solutions. For reinforced or prestressed concrete bridges, which represent the majority of the German bridge population [3,14], established strengthening methods are available, such as additional external prestressing [15], insertion of shear connectors, additional concrete in the compression zone or external application of CFRP sheets or lamella [16–18]. Innovative materials, such as UHPC [19–22] or textile-reinforced concrete (TRC) [23–25] enable considerable material savings through increased performance. This results in a more resource-efficient use of materials with a reduced additional dead weight and an extended service life. In consequence, the substructures and foundation of the strengthened bridge are exposed to less extra loading while $CO_2$ emissions from initial construction are spread over a longer period of use. Thus, the ecological footprint of concrete structures might be improved.

The findings presented in the following were part of a project aiming at developing a thin TRC layer which is intended to be added on bridge deck slabs between the RC structure and the road surface. It is called SMART-DECK and offers a three-fold functionality comprising monitoring, corrosion protection and strengthening of the transverse system of concrete T-beam or hollow core bridges. Strengthening is subject to the present paper. In terms of the gain in capacity due to the additional TRC layer, not only regular flexural and but also shear tests were conducted. The results of the experimental investigations are presented later. Prior to the presentation of the conducted investigations, the concept of the strengthening layer is described in Section 2. At the beginning, basic information on TRC is provided to give an understanding of the choice of material and the layout of the TRC layer.

The aim of the study is outlined in detail in Section 2.2, describing the specifics of SMART-DECK and outlining its distinction towards other studies addressing strengthening using TRC. The experimental programme is introduced in Section 3, beginning with small-scale tests. In Section 4, large-scale tests on slab segments under static and cyclic loading are presented. The paper concludes in Section 5 with a brief summary and prospects regarding potential future investigations.

## 2. Concept of the Strengthening Layer

### 2.1. Textile-Reinforced Concrete

Construction with reinforced concrete is rather economical and characterised by its versatile shaping. The two main elements—concrete and steel reinforcement—are combined to exploit their full potential when interacting with each other. One main disadvantage is the predisposition of the steel reinforcement for corrosion. If the concrete surface is damaged or chlorides enter the structure, the reinforcement can be damaged despite sufficient concrete cover. For this reason, research is concerned with the development of corrosion-resistant reinforcement elements. In the last two decades, research on corrosion-resistant textile reinforcements has been intensified. Early applications from Japan on non-metallic carbon mesh reinforcement [26] inspired research in the Collaborative Research Centre (*Sonderforschungsbereich, SFB*) 528 at TU Dresden, Germany, on strengthening of existing structures [27–30], while fundamental research on new construction elements utilising this then new technology was conducted in SFB 532 at RWTH Aachen University, Germany, [31–35].

The reinforcement in TRC is comprised of continuous fibres, which are called filaments (Figure 1). These are characterised by their good mechanical properties such as high tensile strength and, depending on the base material, medium to high modulus of elasticity. Depending on the desired fineness and the application, up to thousands of filaments are assembled to form a roving. The most common production method for non-crimp reinforcements is warp-knitting, where one or multiple rovings are arranged equidistantly and fixed in their position by additional knitting yarns, forming two-dimensional (2D) mesh-like reinforcement textiles. In biaxial textiles, warp rovings run in the longitudinal direction of the mesh, while weft rovings are arranged perpendicular to it.

Alkali-resistant (AR) glass fibres and carbon fibres are preferably used for TRC. Both materials have good mechanical properties, but carbon fibres are more durable than AR glass fibres [36]. Carbon fibres are characterised by their low density, a very high tensile strength and a high modulus of elasticity. They are manufactured in a multi-stage thermal process which determines the strength and stiffness of the fibres. Polyacrylonitrile (PAN) is usually used as the starting material for carbon fibres [37].

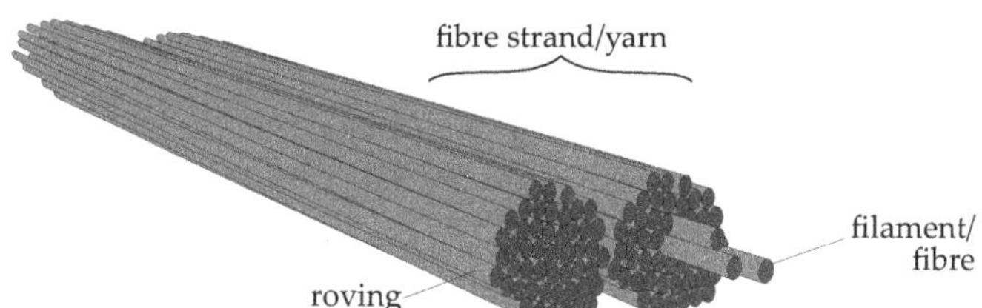


**Figure 1.** Principal sketch of a larger number of single filaments assembled to a roving and then to a yarn.

Bond of non-impregnated textiles is achieved by adhesion and friction rather than mechanical interlock, which is the typical mechanism for conventional deformed steel rebar. Due to the circular shape of the filaments and their small diameter, small voids exist in the roving. Because of their small size, cementitious binder cannot penetrate the roving and only encloses the edge filaments. The core filaments can only be activated for load transfer by friction on the edge filaments. Due to the smooth surface of the fibres, this bonding mechanism is ineffective and results in a telescopic failure with a stress gradient over the cross-section. An impregnation with low-viscosity organic or mineral binders fills the voids and allows for simultaneous activation of the core and sleeve material [34,38]. The most common impregnation materials are epoxy and styrene-butadiene. Also, new concepts with cementitious impregnation exist [39]. While meshes with styrene-butadiene impregnation maintain a certain flexibility but feature a rather soft bond to the concrete or mortar matrix, an epoxy resin provides a relatively rigid material behaviour which forms a strong bond to the cementitious matrix. At the same time, the latter allows for a much more straight-forward production since the reinforcement can be assembled in a formwork similar to reinforcing steel and the concrete can be regularly cast. It only needs to be ascertained that buoyancy of the textiles during compaction is prevented. Also, textile reinforcement must not be walked on due to its sensitivity to lateral compression. To a certain extent, this makes CFRP textiles with epoxy impregnated carbon rovings comparable to carbon FRP reinforcement. After concreting, it can be summarised under the term carbon concrete. The high potential of carbon concrete as an effective construction material has already been given prove of in numerous research projects and applications [40–47].

## 2.2. *Characteristics of SMART-DECK*

### 2.2.1. Intended Functions and Main Questions

The system SMART-DECK presented here was developed in a research project in cooperation with a total of seven project partners from research and building practice as well as administration and material production. It is added to the top of the bridge between the concrete slab and the road surface (Figure 2) and enables bridge strengthening with TRC and offers two other functionalities which were investigated by the project partners: moisture monitoring and preventive cathodic corrosion protection (pCCP) [48,49]. The overall investigations not only foculised the target functionalities but also the feasibility and manufacturing of the system on a bridge [50]. The strengthening effect was investigated at the Institute of Structural Concrete of RWTH Aachen University (IMB) and it is the central content of this paper. In contrast to other investigations with regard to the strengthening potential of bridges with TRC, SMART-DECK not only addresses flexural but also shear capacity of the bridge's transverse system. These represent the main failure cases in ultimate limit state design of the bridge deck slab. So far, the impact of shear strengthening with TRC has been investigated with respect to the longitudinal system of the bridge [25,51–53], i.e., by supplementing the cross-section of the stirrup-reinforced webs. Since shear behaviour of slabs without shear reinforcement differs from that of beams with shear reinforcement [54], insights gained from previous results cannot simply be transferred to the transverse system. Comparable slab strengthening systems [55–57] addressed the increase in flexural capacity. For bending, significantly more pronounced tensile stresses can be assumed in the textile reinforcement compared to shear being the critical load case. Since the very high tensile strength is the essential characteristic of the carbon reinforcement, the question remained to which extent a significant increase in capacity could be achieved by increasing the flexural reinforcement when premature shear failure occurs.

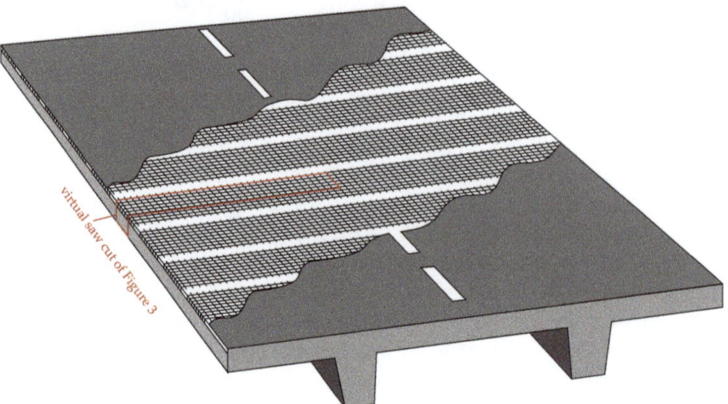


**Figure 2.** Position of the additional textiles on an exemplary bridge deck slab.

Another research question was the manufacturability of SMART-DECK. A common method for TRC production is shotcreting [52,58,59]. Since SMART-DECK is realised in a horizontal plane, shotcreting is not a suitable solution due to ingress of rebound. Therefore, using shotcrete would have conflicted with technical rules for German bridges [60] in this application and was therefore excluded. A production via laminating (layer by layer) is labour-intensive, so regular concreting was preferred from the beginning. Initially, it was not clear whether this method would meet the various technical demands resulting from the implementation of the instrumentation for monitoring and pCCP or a possible device for (partially) automated manufacturing. The manufacturing method was also linked to the development of suitable spacers and fasteners for the reinforcement which do not interfere with the electrical current.

#### 2.2.2. Layout

SMART-DECK is a supplementary TRC layer, which is applied on top of the existing concrete and subsequently covered by the road surface (Figure 3). It is to be processed along the entire width of the bridge deck, but the textiles are to be laid in sections in longitudinal direction of the bridge at a defined distance from each other in order to obtain electrically separated fields. SMART-DECK consists of a high-performance mortar and a textile reinforcement made of carbon fibres impregnated with epoxy resin. The textiles are installed in two layers with a mutual centre distance of 15 mm and an edge distance to the concrete interface and to the upper side of 10 mm, resulting in a total layer depth of approximately 35 mm.

The bond between strengthening layer and existing concrete in this system is not achieved by mechanical connectors but relies on concrete-to-mortar adhesion. This greatly reduces application time and cost but requires proper surface preparation prior to casting as well as a close quality control on site. The basic material carbon of the TRC layer offers the electrical conductivity required for monitoring and pCCP on the one hand. On the other hand, in combination with epoxy resin impregnation, it has good bonding properties to the surrounding concrete and is very efficient due to its high axial tensile strength (5 to 6 times the value of reinforcing steel). The corrosion resistance of the material allows the execution of thin layers with small concrete covers. Especially in combination with a massive existing supporting structure, which provides high stiffness, these advantageous properties of TRC can be used optimally and high increases in load-bearing capacity can be achieved with a minimum use of material.

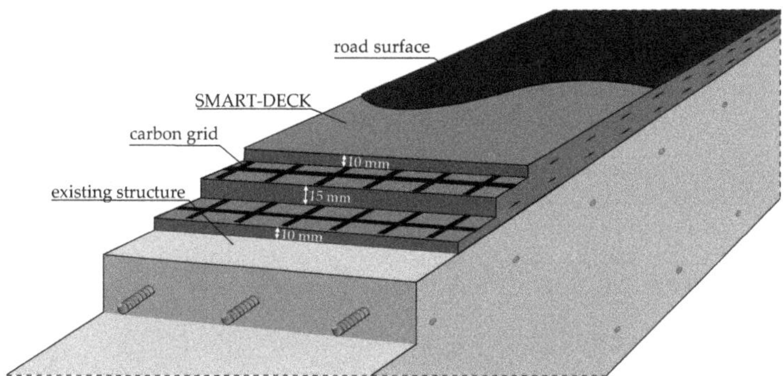


**Figure 3.** Virtual saw cut as schematic sketch of the strengthening layer on a RC slab.

### 2.3. Previous Investigations

As stated before, the flexural strength of regular RC components with small longitudinal reinforcement ratios can be significantly increased by TRC strengthening in the flexural tension zone [57,61–63]. Investigations on the influence of web strengthening with TRC showed that shear capacity of reinforced concrete beams can also be significantly increased by TRC strengthening [23,24,52]. Here, in addition to the high tensile strength of the textile reinforcement, the reduction of the crack width plays an important role, since this can prevent an early confinement of the flexural compression zone and delay the propagation of the shear crack into the compression zone. Furthermore, it could be shown that web strengthening with TRC can also be a useful alternative to existing strengthening methods for cyclically stressed components [23,53,64].

Prior to the start of the project, tests were conducted to estimate whether promising degrees of strengthening for the main failure modes in transverse direction of the bridge can be achieved in the ultimate limit state [65]. It was shown that a supplementary TRC layer can increase shear capacity of the bridge deck. However, different materials and different boundary conditions were applied in those tests compared to the investigation presented in this paper. Therefore, these results only allowed for tendentious statements.

The strengthening effect of SMART-DECK itself was already experimentally examined during realisation of the project demonstrator. For this purpose, a slab of approximately 100 $m^2$ was built which also met the demands resulting from the other two intended functionalities. It featured a height of $h$ = 28 cm and a change of slope was produced to represent realistic conditions for fabrication with respect to an existing bridge slab. The additional TRC layer was applied to 80% of the area, whereby the implementation of the cross-sectional supplement itself and the manufacturing of the measuring device for the monitoring system were tested. Subsequently, the monitoring system (see above) and the achievable increases in capacity were investigated. The latter was done by extracting saw-cuts from the strengthened and non-strengthened areas of the demonstrator slab, which were then tested in a total of eight load tests at IMB until failure. Bending and shear were investigated in the same way as in the series of tests presented here: 24–56% shear strengthening and 90–174% flexural strengthening were achieved [64].

## 3. Experimental Programme and Results of Small-Scale Tests

### 3.1. General

Material and small-scale tests were carried out to characterise the materials intended for SMART-DECK and their interaction with each other. The goal was to select a suitable combination of mortar and textiles for strengthening (I. and II., according to Figure 4). The small-scale tests were divided into three groups: uniaxial yarn tensile tests as well as flexural and uniaxial tensile tests on composite strips. In the uniaxial yarn tensile tests, individual fibre strands extracted from the mesh were investigated, while in the bending and other uniaxial tensile tests, composite components with textile cut-outs embedded in concrete were tested. Also, mortar testing was performed on each batch that was used for SMART-DECK on prisms $W \times H \times L = 40 \times 40 \times 160$ (mm) in accordance with Reference [66].

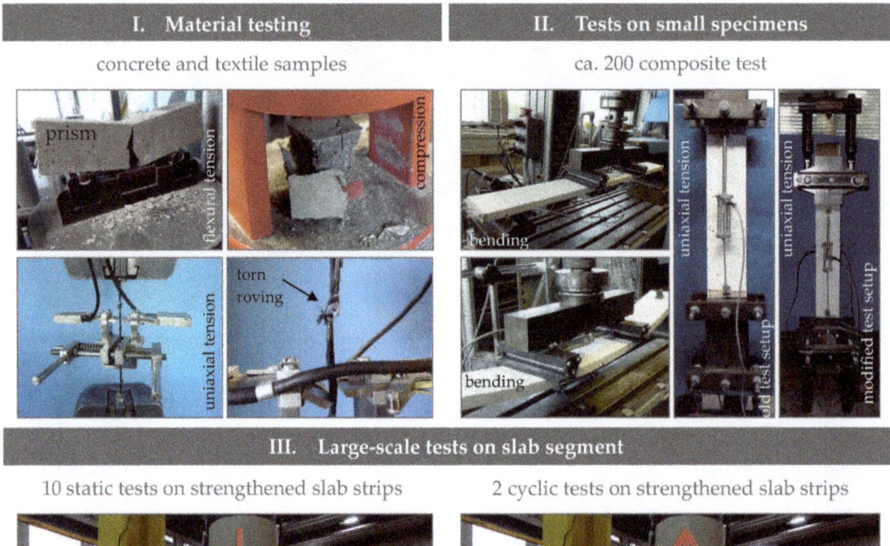

**Figure 4.** Overview of the experimental programme (photographs by IMB, RWTH Aachen University).

The quadratic textile reinforcement meshes were provided by the project partner solidian. The mesh size (centre-to-centre distance of the fibre strands) was either 21 or 38 mm. Depending on the knitting method, different bond properties can occur which not only differ among the various grids but also depend on the direction [38] since weft and warp vary in cross-sectional shape and surface characteristics (Figure 5).

**Figure 5.** Textile grids made of carbon with epoxy impregnation with a mesh size of 38 mm (**a**) and 21 mm ((**b**), photographs by IMB, RWTH Aachen University).

With the findings from the small-body tests, in which various parameter combinations were investigated, the materials for the final SMART-DECK system were pre-selected. This optimised combination was then used as a strengthening layer in the large-scale tests. Small-body tests are particularly suitable for this purpose, as multiple repetitions with the same combination of properties and a large number of different specimens can be tested in a comparatively short time. The tests were labelled X-YZ based on the type of experiment and the materials used as follows:

- X: test type (B: Bending, T: Tensile)
- Y: applied mortar (consecutive number from the material notation)
- Z: applied textile (consecutive number from the material notation)

For some tests, another abbreviation was added to the end in order to mark a more specific property. Table 1 shows the materials used in the course of the investigations on the strengthening effect of SMART-DECK. The notation A-B-C of the materials results from the project consortium's definitions, where A is the material (M: mortar, T: textile), B is a consecutive number for each material and C is a specification (maximum aggregate size, in mm, for mortar and mesh size, in mm, for textiles).

**Table 1.** Details of mortars and textiles.

| Denotation | Specification |
| --- | --- |
| M-1-06 | Base material with a maximum aggregate size of 6 mm |
| M-2-04 | Base material with a maximum aggregate size of 4 mm |
| M-3-04 | Advanced material based on M-2-04 |
| M-4-04 | Advanced material based on M-3-04 |
| T-1-38 | Base material with a mesh size of 38 mm |
| T-2-21 | Base material with a mesh size of 21 mm |
| T-3-38 | Advanced material based on T-1-38 with CNT * |
| T-4-38 | Advanced material based on T-1-38 with modified epoxy resin |

* CNT: Carbo Nano Tubes.

Due to the production process, the weft direction of the textiles is limited to the width of the textile machines. For this reason, the weft direction is positioned in longitudinal direction and the warp direction in transverse direction of the bridge (main load transfer direction of the roadway slab) since separated segments of textile are required in longitudinal direction by means of electrically insulated areas (Figure 2). While the flexural and tensile tests on the composite material were therefore tested in warp direction, the yarn tensile tests were carried out for both fibre strand directions in order to obtain comparative values. Figure 4, Step I, shows the test setups and measuring technology used for the tensile tests on individual fibre strands and an example after failure.

### 3.2. Uniaxial Tensile Tests on Individual Fibre Strands

For test preparation, the fibre strands were extracted from the grid approximately at the centre between two strands (10–15 mm from the stand axis). The textiles have comparatively large yarn cross-sections. At the same time, carbon filaments have high strengths, so high failure loads had to be assumed, which required high lateral compression in the clamps at the ends of the yarns. To prevent them from rupturing in the clamps, the ends were glued into aluminium foil with the aid of epoxy resin mortar. These straps were fixed between the jaws of the clamps after curing.

Only results from tests were considered where the yarn actually ruptured within the free length. Loading was applied displacement-controlled at 1 mm/min. Meanwhile, the applied tensile force and the strain were recorded. Figure 6 shows the results for tests on T-1-38 (left) and T-2-21 (right) as mean value of 5–10 samples.

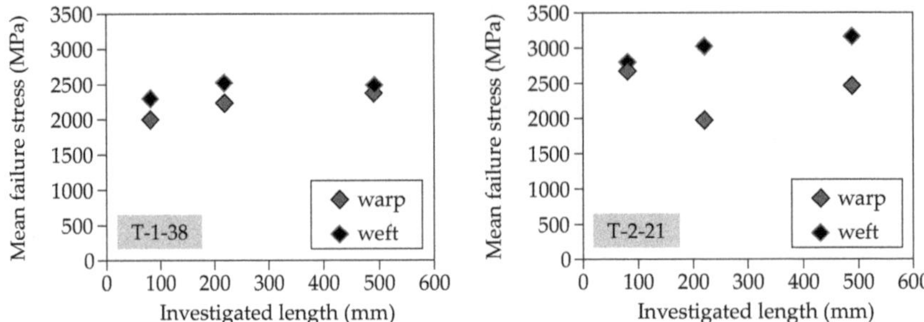


**Figure 6.** Results of tensile tests on yarns for T-1-38 (**a**) and T-2-21 (**b**).

The results differ with regard to the test length (free length between the clamps). Since the probability of imperfections increases with increasing length of investigation, its influence should be checked. For T-1-38, no negative influence due to increasing test length is evident. Only in weft direction, slightly smaller stresses do occur at the largest test length compared to the medium test length. Instead, the tests on the weft yarns of T-2-21 show increasing ultimate stresses with longer test length. Only in warp direction of T-2-21, a decreasing trend is visible, whereby the tests with medium lengths showed comparatively low tensile strengths. Overall, the ultimate stresses of T-1-38 are somewhat lower than those of T-2-21 and the fibre strands in the weft direction have higher tensile strengths than those in the warp direction, which is due to the negative influence of the knitting thread and is commonly known [67,68].

### 3.3. Small-Scale Tests on the Composite

As mentioned before, the composite system was tested in small-body tests under bending and uniaxial tension (Figure 4, step II) to determine the essential properties of the strengthening layer under tension, such as cracking behaviour and stress–strain relationship. The pure tensile stress approximately corresponds to the eventual stress of the TRC layer on the bridge. As considerably more tests and thus more parameter combinations could be tested in a shorter time with bending tests, this method was preferred at the beginning of the project. The initial setup for uniaxial tensile tests ((Figure 4, step II), left test setup for tensile tests) required manual fixing of the specimens between the clamping jaws using several threaded rods, all of which had to be uniformly tightened. For the first materials tested within the scope of the project (M1, M2, T1 and T2 according to Table 2), both methods were therefore used. The dimensions of the specimens were determined according to the geometric properties of the textiles and the layered structure of SMART-DECK. Therefore, one layer of textile was bi-symmetrically positioned. The width of the composite strip was the multiple of the textile's mesh width and the depth was 20 mm to represent a segment taken from a wide strengthening layer.

**Table 2.** Features of the large-scale tests.

| Name | Target Failure Mode | Observed Failure | Material Combination Strengthening Layer | Load Distance | Longitudinal. Reinforcement Ratio | Concrete Compressive Strength ## | Mortar Strength | Strengthening Effect | Failure Load |
|---|---|---|---|---|---|---|---|---|---|
| | | | | $a_i$ (m) | $\rho_{l,s}$ (%) | $f_{cm,cyl}$ (MPa) | $f_{cm,prism}/f_{ct,fl}$ (N/mm²) | η (-) | $F$ (kN) |
| S1-1 | V | V | | 0.7 | 1.0 | 41.0 | 73.6/5.2 | | 155 |
| S1-2 | V | V | | 1.0 | 1.0 | 37.9 | | | 136 |
| S1-2 * | V | V | | 1.0 | 1.0 | 41.0 | | | 145 |
| S2-1 | V | V | M3, T1 | 0.7 | 1.0 | 38.9 | | 1.31 | 203 |
| S2-2 | V | I + V | M3, T1 | 1.0 | 1.0 | 38.9 | | 1.03 | 144 |
| S3-1 | V | V | | 0.7 | 0.5 | 35.6 | 78.7/10.4 | | 118 |
| S3-2 | V | V | | 1.0 | 0.5 | 35.6 | | | 110 |
| S4-1 | V | I + V | M4, T4 | 0.7 | 0.5 | 35.6 | | 1.05 | 124 |
| S4-2 | V | I + V | M4, T4 | 1.0 | 0.5 | 35.6 | | 1.08 | 119 |
| S5-1 | M | V | M4, T4 | 1.0 | 0.2 | 35.6 | | 2.89 | 107 |
| S5-2 | M | V + I | M4, T4 | 1.3 | 0.2 | 35.6 | | 3.63 | 103 |
| S6-1 | M | M + I | M4, T4 | 1.0 | 0.2 | 39.3 | | 2.51 | 93 |
| S6-2 | M | M + I | M4, T4 | 1.3 | 0.2 | 39.3 | | 2.30 | 65 |
| S7-1 | V | V | M4, T4 | 0.7 | 1.0 | 39.3 | | 1.49 | 231 |
| S7-2 | V | V | M4, T4 | 1.0 | 1.0 | 39.3 | | 1.53 | 215 |
| S8-1 # | V | I + V | M4, T4 | 0.7 | 1.0 | 39.3 | | 1.03 | 160 |
| S8-2 # | M | M + I | M4, T4 | 1.3 | 0.2 | 39.3 | | 2.11 | 78 |

# Cyclic tests; ## concrete strength of existing structure. * represents a repetition of S1-2.

The textiles were placed in concrete over a total length of 1000 mm. For the flexural test, the TRC strips were supported over a span of 900 mm, with support overhangs of 50 mm at both ends. The test set-up was a 4-point bending test with centre distances of 300 mm in relation to the position of the supports and load application points. Large deflections occurred (100 to 130 mm in the middle of the span at ultimate load), which correspond to five to six times the depth of the test specimen. Failure always occurred via a propagating crack by spalling of the concrete compression zone without rupture of the textile. As the reinforcement did not fail and due to second-order effects (large deflection of the specimens in relation to the horizontal axis of support), the results of the bending tests were not taken into account in the evaluations and only uniaxial tensile tests were used in the further investigations of the composite load-bearing behaviour.

Further development of the uniaxial tensile test setup [69] minimised execution time. As shown in Figure 4, step II, on the right, the clamping jaws were replaced by a hydraulic device that simplified the installation of the specimens considerably. This setup was used from series 2 onwards. It can be concluded that flexural tests are unfit to determine the mechanical properties of composite specimens that are very thin and simultaneously feature such high flexural slenderness, $l/d$ (here: $l/d = 90$). In this case, the tensile test setup does provide convenient results.

A total of five test series was investigated with their main parameters depending on the progress and open questions regarding the efforts of the project partners to develop advanced materials that meet all, and sometimes contradictory demands resulting from the different targeted functions of SMART-DECK (monitoring, pCCP and strengthening). While all tensile tests aimed at receiving essential insights on mortar to reinforcement combination, each series also aimed at defining another parameter, like, for instance, the following:

- Series 1: manufacturing method → half the specimens were regularly cast (indication C in Figure 7a) with the mortar being fluid while the other half was manually laminated (indication L in Figure 7b) using a stiffer version of the mortar.
- Series 2: number of textile layers → two layers were tested as in the actual strengthening layer (indication 40 mm depth) or one layer as in the description above (indication 20).
- Series 3: water-to-cement ratio of the mortar.

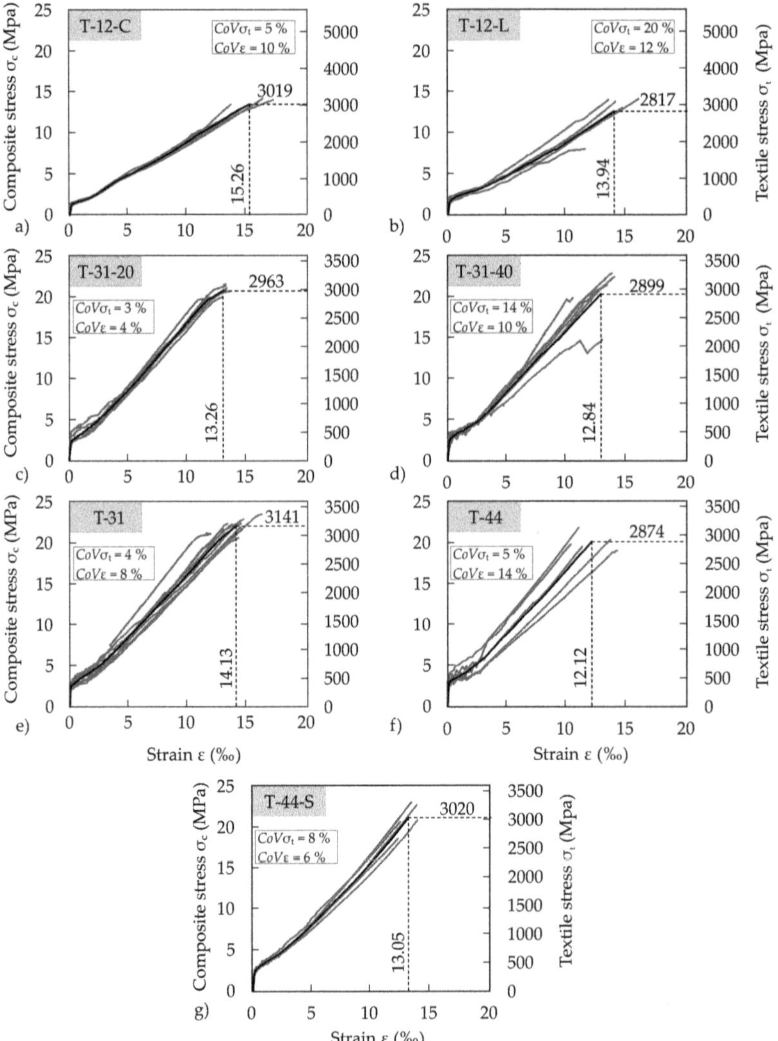


**Figure 7.** Stress–strain relations of seven exemplary groups of uniaxial tensile tests on composite strips with seven different material combinations (**a**–**g**) and indication of the scatter of ultimate stresses, σ, and strains, ε, by means of the coefficient of variation (CoV).

Figure 7 shows the results of some of the uniaxial tensile tests on composite strips by means of their stress–strain relations. While the tensile force is related to both the composite and the textile area (left and right y-axis, respectively), the grey curves show the individual result of one test and the black line represents the mean curve calculated from the total of the individual results correspondent to an established procure [69]. Thus, it was determined by averaging stress values of all test curves over predefined equal strain intervals until the first curves end due to failure of the corresponding specimen. To define the mean increase in length at failure, an additional data point was defined which corresponds to the average strength of the individual specimens.

Those and the other tensile tests that are not shown in detail here, were conducted to characterise the properties in the warp direction, which is decisive for the strengthening effect in the present case. The specimen length was 1000 mm, and the anchor length at each end was 250 mm. Figure 8 shows a uniaxial tensile test until the specimen failed. The cumulative crack openings were measured over a length of 450 mm using displacement transducers attached to the front and back to derive the mean strain of the reinforcement. This is possible because of the large number of cracks in the measurement area (significantly more than five cracks, as indicated in [69]). The tests were carried out displacement-controlled at 2 mm/min.

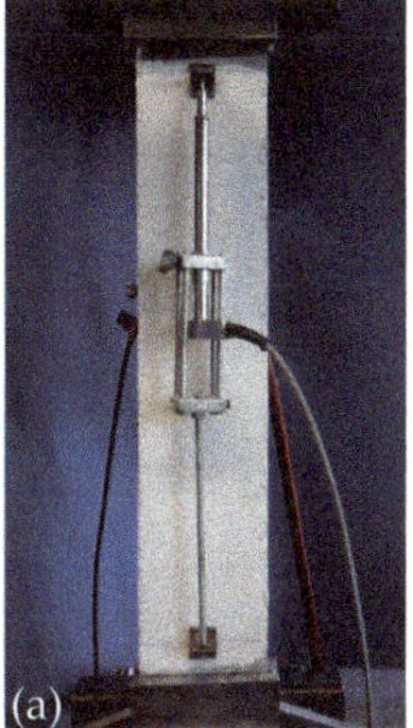

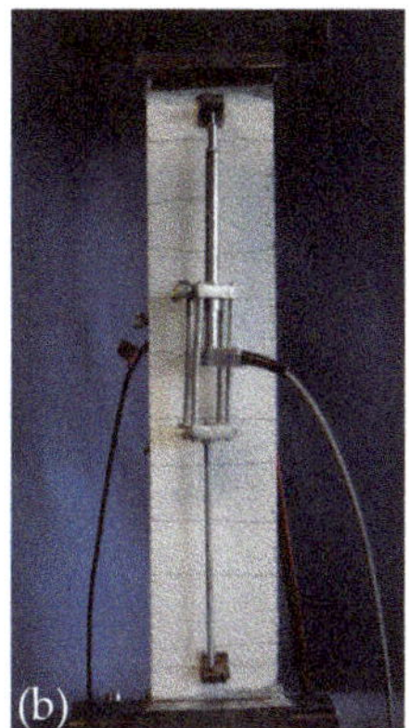

**Figure 8.** Stages of a tensile test: installed test specimen with slight cracking (**a**), specimen featuring complete crack formation (**b**) and specimen after failure (**c**), photographs by IMB, RWTH Aachen University).

Due to the specified layered structure of SMART-DECK with a concrete cover of 10 mm, the concrete cover of the tensile strips was also limited and not optimised regarding maximisation bond or maximum tensile stress development. Therefore, it was expected that a full exploitation of the material's potential might not succeed, which was confirmed by the tests. After completion of the crack formation transverse to the load direction, longitudinal cracks often formed in the textile plane. Failure of the fibre strands occurred abruptly and was accompanied by concrete spalling. In some cases, there was no or only partial spalling. In these cases, however, the matrix clods were no longer bonded to the textile. The fibre strands in tensile direction either ruptured or remained partially intact. Since the specimens that were concreted by means of pouring and compacting (Figure 7a) featured higher ultimate loads and less scattering than comparable specimens that were laminated (Figure 7b), the former manufacturing method was used for the following tests. It was also the preferred procedure in terms of ease-of-use since it is easier and faster to apply to large surfaces like bridge decks. In terms of the specimens' depth, no significant difference was observed (Figure 7c,d). Therefore, specimens that only had one layer of textile reinforcement allow for a representative investigation of the tensile resistance of SMART-DECK. Material combinations 31 and 44 (Figure 7e,f) were used as strengthening layers for the large-scale tests described in Section 4.

In the first two mortars tested, the fibre strands generally failed smoothly. In the more advanced mortars, they were mostly frayed. An improvement in the bond properties was achieved by further development of the textiles (T 4 38). The highest failure stresses, and at the same time the smallest dispersion, were obtained in tests with sanded textiles of T-4-38 in combination with M-4-04 (Figure 7g). However, since sanding results in considerably poorer electrical properties, it was not shortlisted for the project. T4 featured the overall best properties regarding all three target functions of SMART-DECK and was therefore chosen for the large-scale tests.

## 4. Large-Scale Tests

### 4.1. Design and Materials

For investigation of the strengthening effect of SMART-DECK, a total of 17 results from component tests are available, as the overview in Table 2 shows. Twelve of them were carried out on strengthened slab segments. From another project, a double reference test with a longitudinal reinforcement ratio of $\rho_l$ = 1.0% can be referred to [70]. The aim was to investigate the influence of SMART-DECK on the component behaviour for the two decisive failure types, bending (M) and shear (V), as well as failure in the interface (I). In Table 2, a distinction is made between planned and occurred failure types. For mixed forms, the sequence corresponds to the sequence of failure modes in the test. For example, V + I means that primary failure was due to flexural shear and the interface failed secondarily.

To control the bending moment, the load distance $a$ = 0.7/1.0/1.3 m between load and support was modified and to additionally vary the exploitation of the flexural reinforcement, the longitudinal reinforcement ratio was set to either $\rho_{l,s}$ = 1.0/0.5/0.2% in terms of the steel in the existing slab. For the high and medium longitudinal reinforcement ratio, threaded steel bars with a strength class of St900/1000 (Ø15) were used, and for the small longitudinal reinforcement ratio, a ribbed steel of B500 quality (Ø10) was used. The reinforcement layout of the reinforced concrete slab segments is shown in Figure 9. One half of specimen S8 was reinforced like S2 and S7, while the other half was equal to the reinforcement of S5 and S6, respectively. The non-strengthened reference slab segments featured a reinforcement that was identical to the corresponding specimens with TRC-strengthening.

No shear reinforcement was provided in the shear spans, as its installation in slabs is costly and therefore unusual in trunk road bridge slabs for practical reasons. A longitudinal reinforcement of Ø10/20 in transverse direction to the load transfer direction of the test bodies (longitudinal direction of the bridge) was provided in accordance with the normative minimum value for bridge slabs. The height of the RC base specimens was 28 cm, and the concrete cover was 20 mm all around. All slab segments were designed with a width of $b$ = 50 cm, while the load plate's widths were 40 mm to involve the entire cross-section in load transfer. No significant impact was expected by the additional TRC layer on stress redistribution in the slab originating from the concentrated load. In the near past, the influence of the slab width was extensively investigated for reinforced concrete slabs [71–75] and it is assumed that the findings are generally applicable to TRC strengthened slabs.

The reinforced concrete base bodies were concreted indoors with a ready-mixed concrete featuring a target strength corresponding to a C30/37 and a maximum aggregate size $d_g$ = 16 mm. The specimens were compacted by means of an internal vibrator. Material samples (cylinder Ø = 150 mm/$h$ = 300 mm and cubes with an edge length of 150 mm) were produced, which were stored next to the slab segments and tested at the time of the component test in order to be able to draw conclusions about the developed concrete strengths of the slab specimens (Table 2). Prior to the application of the strengthening layer, the upper surfaces of the reinforced concrete components were roughened by means of solid blasting. Using the sand surface method [76], a mean roughness of at least $R_t$ = 1.0 mm was determined (three measurements per slab segment). SMART-DECK was manufactured by the project partner Eurovia under construction site conditions.

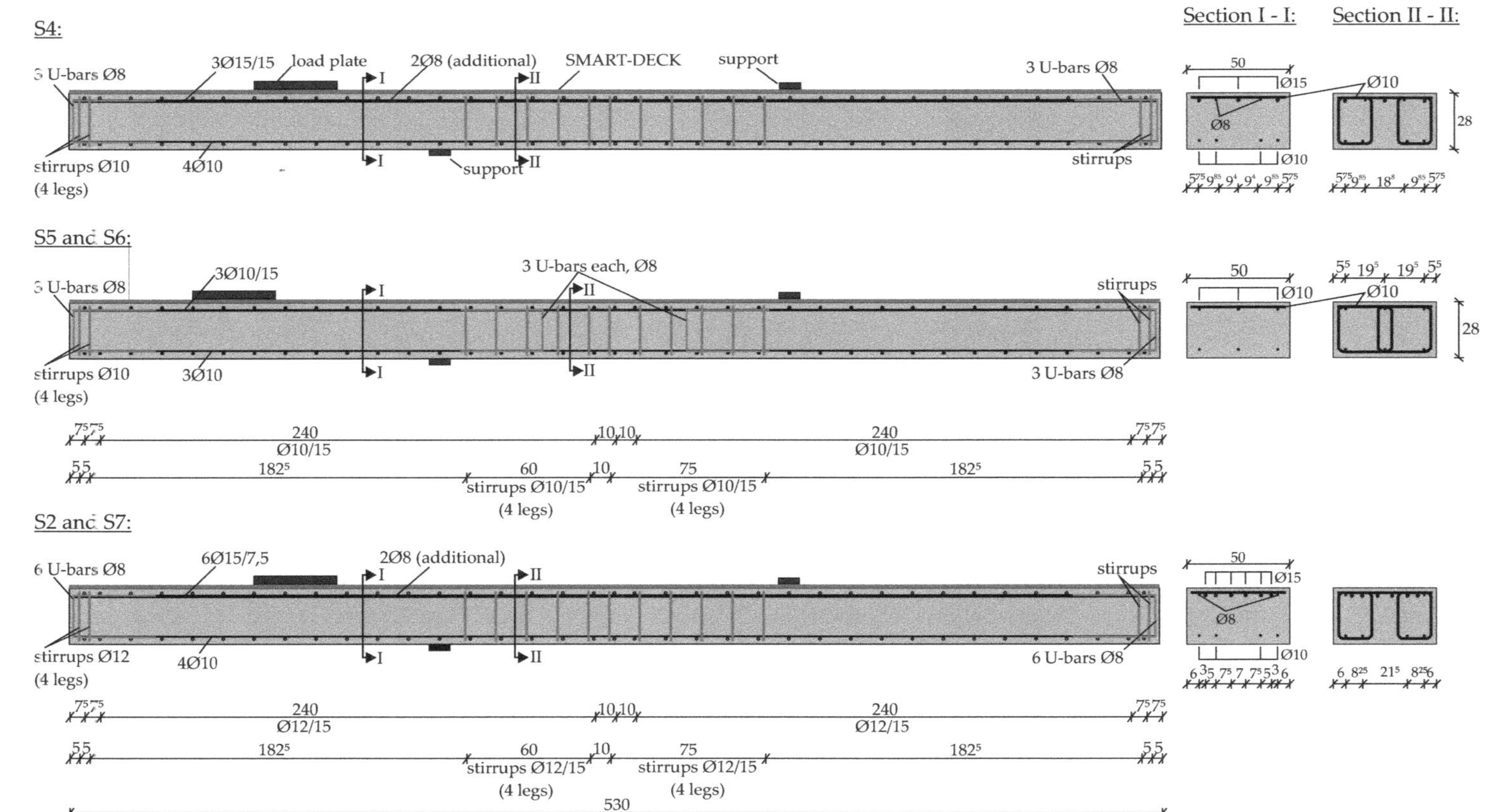


**Figure 9.** Outline of the reinforcement of the strengthened specimens S2, S4, S5, S6 and S7.

Specimens S4, S5, S6, S7 and S8 were produced together and strengthened some weeks later at once. They had the same material combination in the TRC layer. The specimen which was used for tests S2-1 and S2-2 was manufactured at an earlier stage with the material combination then available.

### 4.2. Test Setup

Two separate tests were carried out on each specimen (Figure 10). The load was always applied to represent a load resulting from a truck driving on the outer lane and thus loading the cantilever of the bridge slab. The load distance, $a_i$ (load axis to axis of the support close to the load), was varied according to the specifications in Table 2. The support with a larger distance to the load was designed to take the lifting forces. The load was applied via a hydraulic cylinder and transferred to the specimen via a square load plate of 40 × 40 (cm) corresponding to the wheel contact area for trucks according to European standard [77].

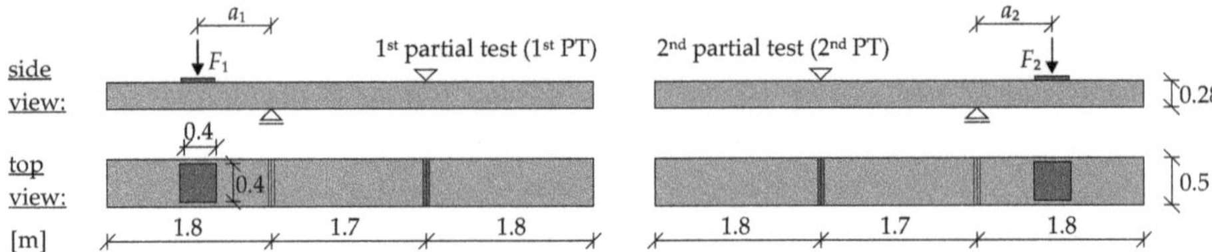


**Figure 10.** Specimen dimensions and test setup for the large-scale tests on slab segments (graphic by IMB, RWTH Aachen University).

All tests were statically loaded with stepwise increments until failure, except S8-1 and S8-2, which were loaded cyclically. The first four load stages were introduced load-controlled. From about half of the expected failure load, the load was applied deformation-controlled.

### 4.3. Results of Tests with Static Loading

#### 4.3.1. Cracking

In some tests, there were production-related imperfections in the interface between old and new concrete. Therefore, the TRC layer was partially detached from the existing concrete during the tests. However, these delaminations only occurred in the concrete interface and always originated from pre-existing imperfections. In case of intact interfaces, the load did not cause the joint to open, which could already be observed in previous tests [78].

Figure 11 shows the crack pattern the specimens exhibited after failure. During loading, a finely distributed crack pattern developed in the TRC layer at the top of the specimens (tension zone). The significantly larger number of cracks with simultaneously reduced crack widths in the strengthened components in comparison to the non-strengthened reference tests is to be regarded positively with respect to the influence of SMART-DECK on the serviceability.

Based on the crack patterns, the tests can essentially be divided into three groups, differentiated by primary cracking that introduced failure, or could be observed at the time when ultimate load was reached (Figure 12). Flexural failure was characterised by a wide vertical accumulated crack in the area of the support (test S6-1 and S6-2 in Figure 11). Relatively few other cracks occurred in the base body. In case of shear failure, an inclined flexural shear crack emerged, which exhibited a large crack opening at failure. This crack pattern was observed among the majority of tests in Figure 11, including the reference specimens. If delamination in the concrete interface occurred prior to failure, the shear crack usually propagated along the flexural reinforcement of the RC structure and the propagation of the crack did not continue into the TRC layer. In case of an intact interface between the existing structure and SMART-DECK, the crack propagated into the interface or the strengthening layer. Then, considerably higher ultimate loads could be achieved than in the case of primary interface failure, which will be discussed in more detail later.

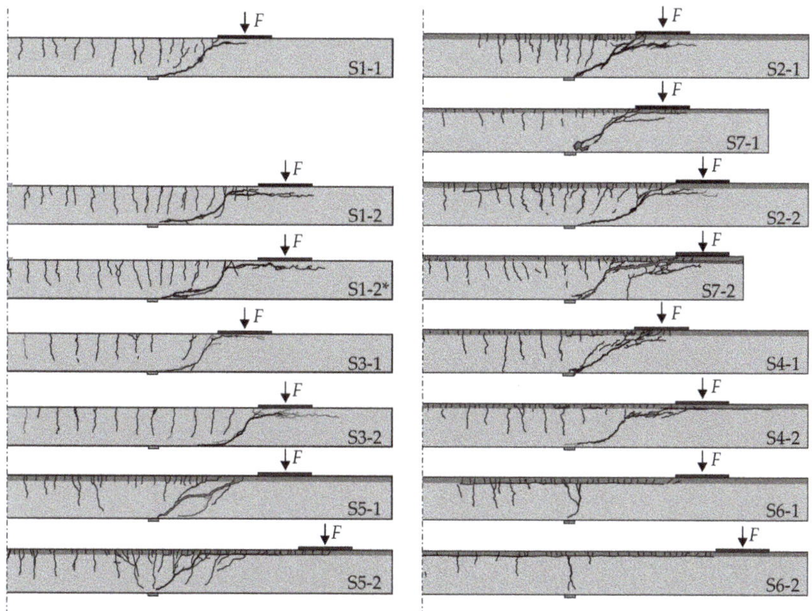


**Figure 11.** Crack patterns of static tests on specimens S1 to S7.

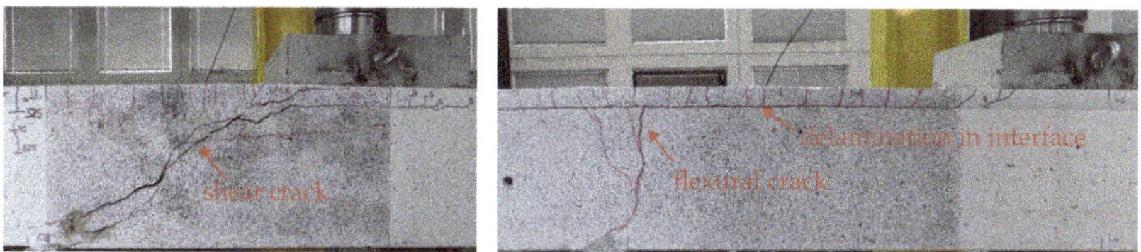


**Figure 12.** Essentially distinguished cracks related to failure (photos and graphic by IMB, RWTH Aachen University).

With a few exceptions, activation of the carbon concrete supplement could be achieved in the tests presented here, as shown in the previous section. However, the findings from manufacturing the strengthening layer for the specimens of this experimental programme provided crucial indications for the further development of the mortar and the production methods. In subsequent applications [50], unimpaired bonded interfaces were achieved so the complete functionality of the system could be realised.

4.3.2. Load-Bearing Behaviour

In addition to crack formation, SMART-DECK influences the load-deflection behaviour of the slab segments (Figure 13 for static tests). Here, a differentiation was made according to the load distance, $a_i$, and the longitudinal reinforcement ratio, $\rho_{l,s}$, referring to the steel reinforcement. Depending on the load distance and the flexural reinforcement, the ranges of abscissa and ordinate were adjusted, which should be kept in mind while comparing the results. The strengthened specimens (solid curves) show less deflection than the plain RC specimens (dashed curves) at the same load level. The deflections, *w*, were measured continuously by means of displacement transducers beneath the specimens in the load axis.

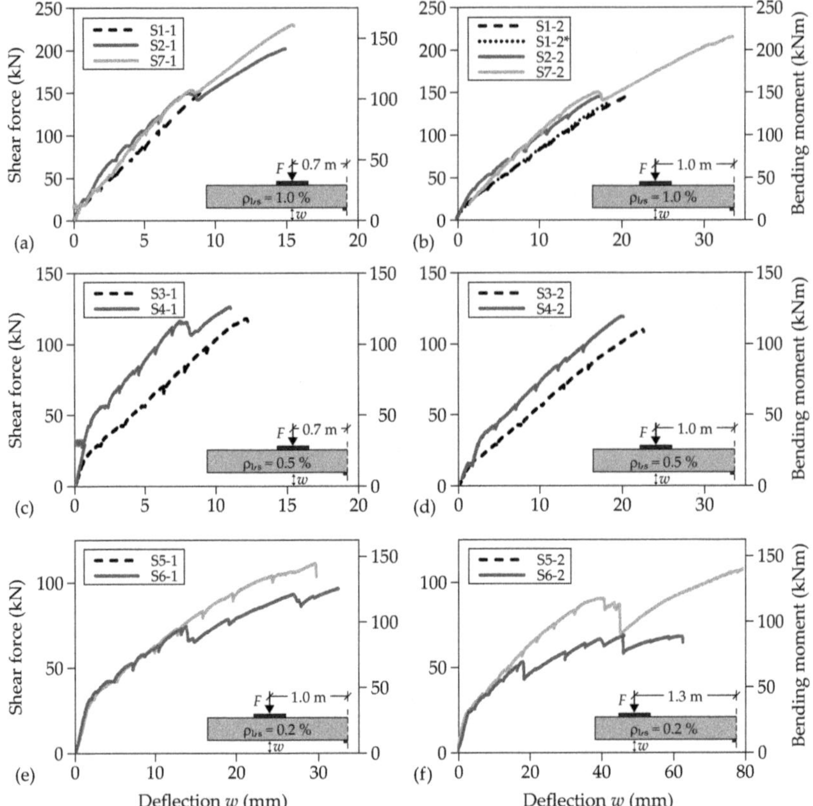


**Figure 13.** Load deflection curves of large-scale tests on slab segments with static loading (**a–f**) separated by combination of flexural reinforcement and load distance.

For quantifying the strengthening effect, increases in ultimate loads were compared to the results of the corresponding non-strengthened reference tests. These quotients are indicated in Table 2. Particularly noteworthy are the shear tests S2-1, S7-1 and S7-2, which illustrate the considerable potential of SMART-DECK for increasing shear capacity. S2 and S7 only differ in the material combination of textile and mortar. The degree of strengthening of specimen S2 could even be increased with optimised materials in the reinforcing layer (specimen S7). The comparability of the results and the observed increases in shear capacity allow the conclusion that a simple top-side supplement of the flexural tensile reinforcement also has a positive influence on shear capacity.

In both partial tests on specimen S5 with a small steel reinforcement ratio, the type of failure shifted due to the strengthening with SMART-DECK. Capacity could be increased to such an extent that shear failure occurred instead of flexural failure.

While testing specimen S6, a loud popping noise occurred several times during the last quarter of the loading process, which was characterised by a slight load drop. This came along with the visible opening of a bending crack that appeared in the area of the cantilever support. Furthermore, the strain gauge attached to the flexural reinforcement in the RC base body failed after very high strains had been measured beforehand. The rebars in the support section (Figure 14) show that the steel reinforcement failed. Despite the partial opening of the interface, the stresses could clearly be redistributed that were released when the steel reinforcement reached its ultimate strength. The subsequent increases in loading provide the evidence that the released stresses could be taken completely by the TRC layer.

**Figure 14.** Plastic deformation of flexural tensile steel reinforcement in the cantilever section of S6 (photo by IMB, RWTH Aachen University).

The results of the tests on S5 and S6 illustrate the enormous strengthening potential of SMART-DECK with regard to flexural capacity of bridge deck slabs in transverse direction. The strengthening degrees determined for the tests on S5 and S6 are beyond the values which could be achieved in bending tests within the previous experimental programme [78]. Since then, advanced materials for the strengthening layer could be provided by the project partners, resulting in better mechanical properties for the mortar and the textile. Delaminations in the concrete interface only occurred in case of pre-existing imperfections. Those represented the origin of cracks along the concrete interface at high-load levels. It was also noticed that such delaminations had a negative influence on the participation of the strengthening layer in shear load transfer (tests on S4), while high degrees of flexural strengthening remained possible (tests on S6). Therefore, it could be assumed that flexural capacity of the increased cross-section is relatively independent of the quality of the concrete interface. Fortunately, imperfections could be prevented later on by enhancing mortar and application method.

### *4.4. Results of Tests with Cyclic Loading*

To investigate the influence of predominantly cyclic loads due to the impact of traffic, two tests were carried out under load collectives (red curves in Figure 15). At least $2 \times 10^6$ load cycles were aimed at. S8 already showed imperfections of the interface in the test area of the first partial test prior to loading. Therefore, an initial load was applied that corresponded to approximately 75% of the failure load of the non-reinforced test. The first partial test featured a high longitudinal reinforcement ratio (shear test). Just as in the static tests, the load was applied stepwise. Subsequently, about 80,000 load cycles with an amplitude of 10 kN were applied at a frequency of $f = 5.243$ Hz. The upper load was 120 kN (maximum peak load) and the lower load 100 kN (minimum peak load). This load range corresponds to about 12.5% of the ultimate load of the RC reference specimen. Hardly any stress changes in the reinforcement were measured and there was no significant change in the crack pattern. Therefore, the amplitude was doubled while maintaining the upper load at 120 kN and approximately $0.5 \times 10^6$ load cycles with a doubled load oscillation width were applied (lower load: 80 kN). In the meantime, the interface between the existing slab and the strengthening layer opened up, starting from the aforementioned imperfection that already existed before the start of the test. Nevertheless, no increase in the strain of the reinforcement and the concrete compression zone could be observed, which is why an increase in the average load was targeted.

www.ingramcontent.com/pod-product-compliance
Lightning Source LLC
LaVergne TN
LVHW070133170726
843515LV00007B/1785

* 9 7 8 3 0 3 6 5 6 0 5 8 8 *